Multi-antenna
Transceiver Techniques for 3G and Beyond

Multi-antenna
Transceiver Techniques
for 3G and Beyond

Ari Hottinen
Nokia Research Center, Finland

Olav Tirkkonen
Nokia Research Center, Finland

Risto Wichman
Helsinki University of Technology, Finland

WILEY

Other Wiley Editorial Offices

John Wiley & Sons, Inc., 111 River Street, Hoboken, NJ 07030, USA

Jossey-Bass, 989 Market Street, San Fransisco, CA 94103-1741, USA

Wiley-VCH Verlag GmbH, Boschstr. 12, D-69469 Weinheim, Germany

John Wiley & Sons Australia Ltd, 33 Park Road, Milton, Queensland 4064, Australia

John Wiley & Sons (Asia) Pte Ltd, 2 Clementi Loop #02-01, Jin Xing Distripark, Singapore 129809

John Wiley & Sons Canada Ltd, 22 Worcester Road, Etobicoke, Ontario, Canada M9W1L1
British Library Cataloguing in Publication Data

Wiley also publishes its books in a variety of electronic formats. Some content that appears in print may not be available in electronic books.

British Library Cataloguing in Publication Data

A catalogue record for this book is available from the British Library

ISBN 0470 84542 2

Produced from files supplied by the author.
Printed and bound in Great Britain by Antony Rowe Limited, Chippenham, Wiltshire.
This book is printed on acid-free paper responsibly manufactured from sustainable forestry in which at least two trees are planted for each one used for paper production.

Contents

Part II   Open-loop Methods

Preface

The target of this book is to present the core ideas behind a very up-to-date research area involving modulation design for multi-input multi-output (MIMO) wireless channels. Our discussion is aimed at presenting the key principles of different mathematical and engineering approaches that have recently emerged in a number of current and upcoming standards. We restrain ourselves from delving into the physical aspects related to the design of practical antenna elements for mobile or fixed wireless communication units. Rather, we choose to explore and develop multi-antenna transceiver techniques from the signal processing perspective. Such an approach is commonly used when proposing and developing new coding or modulation concepts for wireless systems.

Many of the concepts described herein are aimed at improving data rates, signal quality, capacity or system flexibility. To reach this goal, we adopt matrix-valued modulation alphabets, defined over two orthogonal dimensions, usually referred to as *space* and *time*. The space-dimension is realized by using multiple transmit and receive antennas, and involve multi-antenna transceiver structures. Such multi-antenna techniques are generally considered as the most promising avenue for significantly increasing the bandwidth efficiency of wireless data transmission systems. In MIMO systems, multiple antennas are deployed both at the transmitter and the receiver. In ideal situations, this allows signalling over several parallel channels between the transmitter and receiver. These channels can be separated using signal processing means, provided that the channels are sufficiently different. In MISO (multiple-input single-output) systems, the receiver has only one antenna, and the multiple transmit antennas are used for transmit diversity.

This book presents the key aspects of multiple antenna transceiver techniques for evolving 3G systems and beyond. MIMO and MISO (transmit diversity) techniques are explained in a common setting. A special emphasis is put on combining theoretical understanding with engineering applicability.

In particular, the book covers linear processing transmit diversity methods with and without side information at the transmitter, including a description of the current transmit diversity concepts in the WCDMA and cdma2000 standards, as well as promising MIMO concepts, crucial for future high data-rate systems. Furthermore, examples of high throughput, low complexity matrix modulation schemes will be provided, when signalling without side information (open loop concepts). The theory of linear matrix modulations will be developed, and optimal non-orthogonal high throughput schemes will be constructed, both for MIMO and MISO systems.

Performance may be further improved by feedback from receiver to transmitter. The corresponding closed-loop modes in the current 3GPP specifications will be discussed, along with their extensions for more than two transmit antennas. In addition, feedback signalling for MIMO channels will be addressed, as well as optimal quantization methods of the feedback messages. Finally, hybrid schemes are constructed, where the amount of overhead due to feedback is reduced by combining open-loop transmission with closed-loop signalling.

We would like to express our gratitude to a number of colleagues who have helped in preparing this work. We thank Drs Jyri Hämäläinen, Rinat Kashaev and Jussi Vesma and Mr Mikko Kokkonen for fruitful collaboration related to the subject matter of this book. Numerous discussions with colleagues at Nokia Research Center are also acknowledged. Drs Nikolai Nedefov and Kari Kalliojärvi provided a number of constructive comments that enabled us to improve the readability of the text. Financial support from Nokia Foundation is also gratefully acknowledged. A large part of the results documented here have been developed at Nokia Research Center in recent years with support from Dr Jorma Lilleberg at Nokia Mobile Phones. Finally, we appreciate the seemingly unlimited patience of our respective home troops.

Acronyms

AMC	Adaptive Modulation and Coding
AOA	Angle Of Arrival
ARQ	Automatic Repeat reQuest
AS	Azimuth Spread
ASTMA	Adaptive Space–Time Modulation Arrangement
BEP	Bit-Error Probability
BER	Bit-Error Rate
BICM	Bit Interleaved Coded Modulation
BLAST	Bell Laboratories Layered Space-Time architecture
BPSK	Binary PSK
BS	Base Station
CDMA	Code Division Multiple Access
CDTD	Code Division Transmit Diversity
CL	Closed-Loop
CPICH	Common PIlot CHannel
CQI	Channel Quality Indicator
CSI	Channel State Information
DOT	Direction Of Transmission
DSTTD	Double STTD
D-BLAST	Diagonal BLAST

EVD	EigenValue Decomposition
FB	Feedback
FBI	Feedback Back Indicator
FEC	Forward Error Coding
FER	Frame Error Rate
FSM	Feedback Signalling Message
FCS	Fast Cell Selection
FO	Frobenius Orthogonality
FDD	Frequency Division Duplex
HSDPA	High-Speed Downlink Packet Access
HS-DSCH	High-Speed Downlink Shared CHannel
i.i.d.	independent and identically distributed
IC	Interference Cancellation
ID	IDentification
ISI	Inter-Symbol Interference
LAN	Local Area Network
LLR	Log-Likelihood Ratio
LMMSE	Linear MMSE
LOS	Line Of Sight
MRC	Maximal Ratio Combining
MF	Matched Filter
MIMO	Multiple-Input Multiple-Output
MISO	Multiple-Input Single-Output
ML	Maximum Likelihood
MMSE	Minimal Mean-Square Estimate
MS	Mobile Station
MSD	Maximal Symbolwise Diversity
Node-B	Base station
OFDM	Orthogonal Frequency Division Multiplexing
OD	Orthogonal Design
OL	Open Loop
OSIC	Orderered SIC
OTD	Orthogonal Transmit Diversity
PAM	Pulse Amplitude Modulation
PAS	Power Azimuth Spread
PHOP	Phase Hopping

PIC	Parallel Interference Cancellation
PSK	Phase-Shift Keying
PSTD	Phase Sweep Transmit Diversity
QAM	Quadrature Amplitude Modulation
QPSK	Quadrature Phase Shift Keying
QOML	Quasi-Orthogonality assisted ML
QoS	Quality of Service
RF	Radio Frequency
RH	Radon–Hurwitz
RHO	RH Orthogonal
Rx	Receiver
SIC	Successive Interference Cancellation
SIMO	Single-Input Multiple-Output
SISO	Single-Input Single-Output
SNR	Signal-to-Noise Ratio
SH	Symbol Homogeneity
SSDT	Site Selection Diversity Transmission
STBC	Space–Time Block Code
STC	Space–Time Code
STTC	Space–Time Trellis Code
STD	Selective Transmit Diversity
STTD	Space–Time Transmit Diversity
SVD	Singular Value Decomposition
TDD	Time Division Duplexing
TDTD	Time Division Transmit Diversity
TrSTBC-OTD	Transformed STBC-OTD
TSTD	Time Switched Transmit Diversity
TTI	Transport Time Interval
Tx	Transmitter
UB	Upper Bound or Union Bound
UE	User Equipment
ULA	Uniform Linear Array
UTRA	Universal Terrestrial Radio Access
V–BLAST	Vertical BLAST
WCDMA	Wideband CDMA
XTRM	eXTRemuM

2G	Second Generation
3G	Third Generation
3GPP	3G Partnership Project
4G	Fourth Generation

Part I

Introduction

1

Background

The research and standardization of 3rd generation (3G) wireless systems has been ongoing for about a decade. Initially standardization work was carried out in national standardization bodies. Different national bodies, major corporations and several research institutes initiated the 3rd Generation Partnership Project (3GPP), which has been active since the end of 1998 (see http://www.3gpp.org). The first WCDMA based 3G solutions developed by 3GPP are Release '99 and Release 4 [1,2]. These constitute the basis for the first commercial WCDMA systems. The most recent update, Release 5, is optimized for high-speed downlink packet access [3].

3GPP is continuously enhancing the WCDMA specifications to further enhance the performance of 3G systems. It is anticipated that some future physical physical layer standard release will contain further enhancements related to multi-antenna transmission techniques. In particular, future releases are likely to include support to novel transmit diversity concepts and high-rate transmission schemes that explicitly support multiple transmit antennas. A similar project, 3GPP2, is currently ongoing in the USA in an attempt to develop and define a cdma2000-based solution for 3G [4–6] (see http://www.3gpp2.org). Many of the recent technical solutions in the cdma2000 family of systems are similar to those in the WCDMA system, and certain solutions have been developed hand-in-hand. For example, both systems include support for multi-antenna transmission.

Compared to 2G systems such as GSM or IS-95 [4], 3G systems provide enhanced services and significantly higher data rates. In addition, the overall capacity of 3G systems has significantly increased beyond 2G systems through the adoption of latest technological achievements. Many of these technological achievements were discovered in the 1990s after 2G systems were already operational. Examples of such techniques include major leaps in coding theory via the invention of turbo codes [7,8]. Turbo codes have been demonstrated to approach the theoretical channel capacity limit, as derived by Shannon in the 1940s [9]. Eventually they found a way to both WCDMA and cdma2000 systems. Another important development has been the introduction of novel multi–antenna transmission techniques [10].

Multi-antenna transmission techniques provide transmit diversity to 3G systems and enable a significant increase in downlink capacity. Transmit diversity is not an entirely new concept, although significant breakthroughs have been made in recent years. Simple space-diversity techniques are already applied in 2G systems, such as GSM or IS-95. As an example, diversity reception using multiple receive antennas is a mature technology which is often applied in 2G base stations to improve uplink coverage. However, due to implementation costs and space constraints, receive diversity methods are not as applicable for mobile handsets. For this reason, the first release of the 3G wideband CDMA standard [11] applies transmit diversity schemes at base stations to improve downlink reliability. These schemes are specifically designed for two co-channel antennas [10, 12, 13]. Recently, such schemes have been suggested for more than two Tx antennas [14]. These transmit diversity solutions mitigate the need to deploy multiple antenna in mobile handsets solely for the purpose to increase diversity. They are in many respects even simpler than well-known downlink beam-forming concepts, in which directional beam patterns are formed towards the desired user. The availability of such simple approaches for downlink capacity enhancement is pleasing, since many of the proposed services, like wireless web browsing, are likely to be downlink-intensive.

With multiple antennas at the base station, and one antenna at the mobile, the uplink is a SIMO (single-input multiple-output) radio channel, whereas the downlink is MISO (multiple-input single-output). In a MIMO (multiple-input multiple-output) system, one has multiple antennas both at transmitter and receiver. In fading channels, these create respectively Tx- and Rx-diversity. The attractive characteristic of a MIMO channel is that it may be used to increase the data rate by transmitting multiple streams simultaneously using different spatial channels. Loosely speaking, Rx-diversity is used to separate these multiple streams from each other, while Tx-diversity may be used to improve performance. Thus, in a sense, a MIMO multi-antenna channel increases the effective bandwidth of a wireless channel. This challenges the conventional thinking which suggests that extremely high data rates either require extremely wide frequency bandwidth and/or extremely high transmit power. On the other hand, even if multi-antenna modems may avoid some of these problems, they tend to require rather heavy signal processing in an attempt to gain access to this projected spatial bandwidth. Hence, the complexity of the terminals and network elements will increase, when compared to traditional modems offering similar data rates with a larger bandwidth. Despite the implementation cost, Moore's Law suggests that at some point in the future multi-antenna techniques may also enable high data rate services for those that do not have bandwidth in abundance. On the other hand, if extremely high data rates are not required, the increased fading resistance (diversity) inherent in many multi-antenna transceiver concepts can be used to increase system capacity or coverage.

Although several transmit diversity and general multiple-input multiple-output (MIMO) transmission techniques have been known for some twenty years [15], the theoretical capacity results developed in [16, 17] revitalized the research area. Essentially, it was shown in [16, 17] that under certain conditions the capacity increases linearly with $\min(N_t, N_r)$, where N_t is the number of deployed transmit antennas and N_r is the number of receive antennas. Communication theorists and engineers have since developed coding and modulation concepts that realize a large portion of this gain. In light of these recent results, multi-antenna transmission and reception techniques are currently seen as the most promising avenue for significantly increasing the capacity and spectral efficiency of wireless systems. Novel bandwidth efficient multi-antenna modulation concepts can be used to enable power efficient signalling at data rates beyond 10 Mbps using the WCDMA 3.84 MHz Chip rate occupying 5 MHz bandwidth. Recently, MIMO concepts have entered the standardization discussion of future releases of 3G standards.

This book deals with the impact of these multi-antenna techniques on the design of 3G and future mobile communication systems. Applications to WCDMA downlink have been our primary motivation, but the considered methods are not restricted to downlink direction nor WCDMA system evolution. Most of the developed solutions are general and applicable to generic 3G and 4G systems, as well as for example wireless LANs as long as multiple transceivers are deployed. The approach will mainly be one of baseband signal processing. The number of transmit and receive antennas is assumed to be equal to the number of Tx and Rx radio frequency (RF) front ends. The problem of multiple physical antennas coupled to a smaller number of RF front ends, interesting as it is, is not considered in this book.

Transmission Resources: The methods discussed in this book divide transmission resources into two complementary categories. Often these will be called "space" and "time". In the spatial dimension, the discrete unit is referred to as an "antenna", whereas in the temporal dimension, the discrete unit is referred to as "symbol period" or "symbol epoch". The essential difference between these two dimensions is that the "time" dimension is substantially orthogonal, whereas the "space" direction is typically non-orthogonal—symbols transmitted simultaneously from two co-channel antennas typically interfere.

Substantially Orthogonal: Time, Frequency, Code: Instead of (or in parallel to) time division multiplexing, the substantially orthogonal "time" dimension may indicate frequency division multiplexing or code division multiplexing. To keep the near orthogonality of the "time" dimension, inter-symbol interference in multipath channels should be mitigated. This means proper equalization when using time-division multiplexing, or proper guard intervals when using, for example, orthogonal frequency division multiplexing (OFDM).

Spatially Separate, or Polarization: The "space" dimension may indicate antennas operating in spatially separate locations and/or in different polarizations. Due to different local scattering environments, sufficiently separated antenna elements provide almost independent fading channels. What sufficient means, depends on the environment. In rural macro cells, separations of many wavelengths may be required to de-correlate antennas, whereas in indoor environments a half-wavelength separation may be sufficient. For polarization, the cross-polarization coupling ratio determines whether polarizations provide diversity, or whether they provide near-orthogonal parallel channels.

1.1 MODULAR SYSTEM DESIGN

It has gradually become clear that one of the most efficient ways to answer the demand of ever increasing spectrum efficiency is to deploy multiple antennas at the transmitter and receiver end of a radio link. In this book we advocate a modular design solution in which the number of antennas is increased so that the impact to other system parts is minimized. The (single user) MIMO/MISO transmission chain considered in this book is as follows.

1. The source emits information bits **b** at a source rate R_{source}. From the source, the information bits go to an *encoding and interleaving circuit*. Typically, it includes a

binary encoder, with code rate R_c and an interleaver with interleaver depth N_i. The output is a stream (or a vector) of coded bits **c**.

2. The coded bits go to a *modulator*, which maps M bits into complex modulation vector **x**.

3. The stream of symbols goes into a *space–time modulator*, which takes sequences of $R_s T$ symbols and maps them to a $T \times N_b$ matrix **X**, preparing the symbols for transmission over T substantially orthogonal transmission resources (time, subcarrier, spreading codes) and N_b beams. The symbol rate of the space–time modulator is R_s symbols per transmission resource. The simplest form of a space–time modulator is a serial-to-parallel multiplexer, which constructs a vector modulation.

4. The output of the space–time modulator is conveyed to the *beam-forming circuit*. This constructs N_b beams out of N_t spatial transmission resources (antennas, polarizations). The action of the beam-former can be described by a $N_b \times N_t$ matrix **W**.

5. Finally, the signals to be transmitted on the beams are upconverted to the radio frequency, and transmitted.

A diagram of the transmission chain can be found in Figure 1.1. This transmission chain is general enough to cover most MISO/MIMO transmission schemes discussed in the literature, with or without channel information at the transmitter:

- For *feedback modes* [2, 12], the beam-forming matrix **W** constructs a single beam, which is used to weight the transmission using N_t antennas. The beam-forming matrix is determined using feedback information conveyed by the receiver to the transmitter, or estimated at the transmitter from reverse direction signalling. The space–time modulator is trivial, **X** collapses to a 1×1 matrix for each modulation symbol.

- For *space–time block codes* [10, 18], and more generally for linear MISO/MIMO modulation, the space–time modulator maps the modulation symbols x to a space–time code matrix **X** in a linear fashion. The space–time block length is T. The beam-forming matrix is trivial, $\mathbf{W} = \mathbf{I}_{N_t}$, when used as a pure open-loop concept.

- For *space–time trellis codes* [19], $T = 1$, the space–time modulator is a vector modulator, and the beam-forming matrix is trivial. Thus **XW** outputs sequences of $R_s = N_t$ symbols from the N_t transmit antennas during each symbol period. In the coding

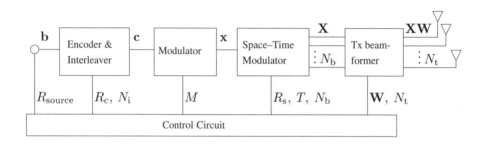

Fig. 1.1: Multi-antenna transmission chain.

circuit, the interleaver is trivial, and the encoder and modulator may be thought of as a joint trellis-coded modulator.

- *Multistream MIMO systems,* or spatial multiplexing, in the spirit of [15, 16, 20, 21], differ from space–time trellis codes in that the coding and interleaving circuit includes de-multiplexing and multiplexing units with the effect that the information stream is split into multiple parallel streams (or layers), which are independently encoded and transmitted simultaneously from the multiple antennas. The number of these streams is R_s, the symbol rate. Depending on the multiplexing units, one has diagonally (DBLAST) [20] or vertically (VBLAST) [15, 21] encoded vector modulation, the latter either with horizontal or vertical coding.

- For *unitary space–time modulations* [22], the modulator is trivial, and it outputs just a sequence of bits. The space–time modulator takes this sequence and maps it in a non-linear fashion to the matrix $\mathbf{X}$.

- *Randomization techniques* apply a (pseudo)randomly chosen beam-forming matrix $\mathbf{W}$. Some have a trivial space–time modulator $\mathbf{X}$, and apply phase or antenna hopping [23, 24], which can be described by choosing different $1 \times N_t$ beam-forming vectors $\mathbf{W}$. Some have a non-trivial space–time modulator $\mathbf{X}$, and the beam-forming matrix performs multibeam-forming by applying antenna permutations or multi-antenna hopping.

- *Hybrid open/closed-loop schemes* [25–29] resemble randomizing techniques with non-trivial $\mathbf{X}$, except that $\mathbf{W}$ is chosen based on feedback.

As indicated above, various parts of the transmission chain can be merged and jointly optimized. When merging the encoder and the modulator, one may discuss trellis-coded modulation. Similarly, merging the symbol modulator and the space–time modulator, one may discuss unitary (non-linear) space–time modulations. The main paradigmatic split is between jointly or separate design of channel coding and space–time modulation. This is the difference between space–time trellis coding and the modular approach in 3G standards [1], where space–time block codes are applied.

In this book, the modular approach of the 3G standards is adopted. In this approach, the parts responsible for exploiting the spatial structure of the channel are separate modules, $\mathbf{X}$ and $\mathbf{W}$, that are designed separately from binary channel codes. These two modules are joined with the coding module by an interleaver, the design of which may take into account possible periodicities produced by $\mathbf{X}$, $\mathbf{W}$ and the modulator, when systematic bits of the channel coders are mapped.

The reasons for favouring a modular approach are

- *Flexibility.* Several rates can be treated with the same space–time modulator. The principle of digital convergence suggests that a multitude of different services converge to be operated by the same device. In the case of mobile communication, speech and different kinds of streaming/packet data services operate at widely differing data rates and QoS requirements. In a modular system, different services with different rates may apply the same transmit diversity scheme without losing too much in performance. The rate-matching algorithms adopted in WCDMA and cdma2000 systems enable power-efficient transmission regardless of the packet size delivered to the encoding chain. The space–time modulation methods should not impact on the performance of the rate matching procedure.

- *Robustness against change of channel conditions.* This is the underlying reason why bit-interleaved coded modulation is favourable in mobile communication. A mobile communication link may operate under a wide variety of channel conditions, from completely specular line-of-sight to rich scattering environments with severe multi-path and deep fades, with corresponding fluctuations in SNR. Bit-interleaved coded modulation (BICM) schemes have a tendency to be more robust against changes in channel conditions [30].

- *Implementation simplicity.* Hardware implementation complexity should not be underestimated. In a modular system, the circuitry is divided into smaller pieces that can be designed separately.

For these reasons, the bulk of this book is dedicated to the modules **X** and **W**, which form the core of exploiting the spatial structure provided by a MISO/MIMO channel. Clearly, joint optimization of beam-forming matrix **W** and the symbol matrix **X** for a given multiple-access or modulation method is a challenging problem. In practice, when multiple users or services share the same channel, the corresponding matrices should be jointly defined to mitigate interference between different users and to maximize system capacity. Recent multiuser scheduling and power allocation solutions involving one or multiple transmit antennas [31–33] lead to orthogonal multiple-access signaling concepts. This allows to mitigate the signal processing aspects of a multiuser communication system, while increasing the complexity of control and access protocols. In decentralized systems, e.g. in ad-hoc networks, the control problems are even more challenging and new paradigms may be needed [34–36]. For ease of exposition, we only address in this book the signal processing aspects, and consider a signal model incorporating one point-to-point wireless link.

Signal Model: The baseband signal model corresponding to these modules is formulated concisely as follows. The coding, interleaving and multiplexing unit is left out from the signal model. The space–time modulator matrix **X** transmits $R_s T$ complex modulation symbols over N_b beams during a block of T symbol epochs. The number of parallel streams R_s is defined as the (average) number of complex symbols transmitted per symbol epoch, i.e. the symbol rate. In a space–time modulator with block of length T, altogether $R_s T$ complex symbols are thus transmitted. The beam-forming unit prepares the N_b beams for transmission from N_t antennas.

Much of this book will suppress delay spread considerations in favour of concentrating on the spatial structure of the channel. In a one-path channel the MIMO/MISO signal model covers one transmission block:

$$\underset{T \times N_r}{\mathbf{Y}} = \underset{T \times N_b}{\mathbf{X}} \ \underset{N_b \times N_t}{\mathbf{W}} \ \underset{N_t \times N_r}{\mathbf{H}} + \underset{T \times N_r}{\text{noise}} \qquad (1.1)$$

Here N_r is the number of receive antennas, **Y** is the $T \times N_r$ matrix of received signals, **X** is the $T \times N_b$ transmission matrix (the space–time modulation) and **W** is the $N_b \times N_t$ beam-forming matrix. The channel **H** is a matrix where each column is a channel vector from the multiple transmit antennas to one receive antenna,

$$
\mathbf{H} =
\begin{bmatrix}
h_{11} & h_{12} & \cdots & h_{1N_r} \\
h_{21} & h_{22} & \cdots & h_{2N_r} \\
\vdots & \vdots & \ddots & \vdots \\
h_{N_t 1} & h_{N_t 2} & \cdots & h_{N_t N_r}
\end{bmatrix},
\tag{1.2}
$$

as depicted in Figure 1.2. In a MISO system, $\mathbf{H}$ is a column vector $\mathbf{h}$. When a multi-path channel model is considered, the channel matrix $\mathbf{H}$ is extended to cover the multipath components, and $\mathbf{X}$ is extended to cover multiple transmission blocks.

BICM in WCDMA: The WCDMA [1, 3, 11] system supports three transmit diversity concepts on dedicated traffic channels. The open loop scheme is the 2×2 space–time block code proposed by [10], known as STTD (Space–Time Transmit Diversity) in WCDMA. The closed loop schemes, known as Mode 1 and 2, apply a 2- or 4-bit quantization for the feedback weight, respectively, to parameterize the matrix $\mathbf{W}$. There are four channel coding options; rate 1/3 and 1/2 convolutional and turbo codes (for speech and data services, respectively). Between the channel code and interleaver, there is a rate-matching circuit, which performs additional puncturing/padding so that transmitted data packets are made to fit into a radio frame.

From the point of view of coding theory, the space–time coding scheme in WCDMA can thus be interpreted as bit-interleaved coded space–time modulation, in the spirit of [30]. This is a simple and efficient way of exploiting two transmit antennas. In [37], it was proven that STTD reaches channel capacity when $N_t = 2$, $N_r = 1$, if the channels are independent and identically distributed (i.i.d.) Rayleigh block fading, with a channel coherence time longer than the delay of $T = 2$ symbol periods required for transmitting STTD. Inherent capacity is turned into reliable signalling by efficient concatenated binary codes. Similarly, the 2- and 3-bit quantization concepts of Modes 1 and 2 provide the optimal increase in SNR [38] for the number of feedback bits used.

1.2   DIVERSITY TECHNIQUES IN 3G SYSTEMS

To increase downlink and uplink capacity, the Universal Terrestrial Radio Access (UTRA) WCDMA system, cdma2000 and GSM evolutions incorporate various diversity techniques. These mitigate the effects of disadvantageous channel conditions, by enabling access to multiple (independent) channel realizations.

1.2.1   WCDMA Rel99 and Rel4

Diversity benefit can be obtained by a number of different technical solutions in 3G systems. WCDMA Release '99 and Release 4 support the following diversity techniques:

- multipath diversity

- time diversity (Automatic Repeat ReQuest)

- Rx diversity, using multiple receive antennas

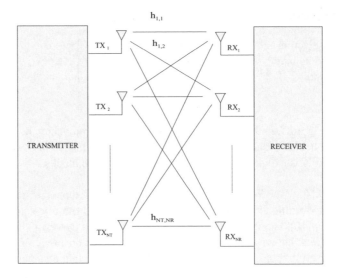

Fig. 1.2: MIMO model with N_t transmit and N_r receive antennas.

- Tx diversity, with one open and two closed loop solutions

- macro diversity (soft handover)

- Site Selection Diversity Transmission (SSDT)

1.2.1.1 Multipath Diversity   A wideband channel enables the receiver to resolve a large number of multipath components. This increases diversity in channels with sufficient delay spread and boosts performance when compared to narrowband CDMA or TDMA systems. With multipath diversity the independent signal copies are sampled in time domain, with the assumption that signals arriving at different delays do not fully correlate. Diversity can be captured using the RAKE or some other linear detector or channel equalizer. Multipath diversity is not available in all environments. For example, a single tap channel is typical in indoor environments.

1.2.1.2 Macro Diversity   Macro diversity can be used both in uplink and downlink to combine signals transmitted to or received from multiple base stations (Node Bs). In uplink signal copies are sampled from spatially separate sensors or antennas with independent fading. In downlink, multiple copies of the same signal are transmitted from spatially separate source locations, again to result in independent fading.

1.2.1.3 SSDT   The specification also includes an additional downlink macro diversity option, known as Site Selection Diversity Transmission (SSDT). In SSDT the UE maintains a list of active set cells, and determines the "primary" cell. All other cells are labelled as "non-primary". Each cell is assigned a temporary identification (ID) and UE periodically informs a primary cell ID to the base stations using an uplink signalling field. The dedicated channel in non-primary cells turn off the transmission power. The primary cell ID can be signalled 1-5 times per 10 ms frame, with different signalling formats. SSDT is activated by higher-layer signalling. In addition, the cell ID assignment is all carried out by higher layer signalling. Site selection can thus be carried out without network intervention, differentiating it from conventional hard handover.

The main objective of SSDT is to reduce interference due to multiple transmissions in a soft handover mode. It is essentially an antenna selection concept combined with efficient power allocation, and together these increase both diversity and power efficiency.

1.2.1.4 *Time Diversity*

UTRA Release '99 supports Type I Automatic Repeat Request (ARQ) protocol, where erroneous frames are discarded and the frame is repeated later on. If the frame is repeated after a sufficiently long time interval (beyond channel coherence time), ARQ provides time diversity. Time diversity can also be exploited via the combined use of interleaving and forward error correction (FEC) codes.

1.2.1.5 *Receive Antenna Diversity*

Multiple receive antennas can be implemented in both Node Bs and UE to capture spatial receive diversity (Rx diversity). Rx diversity can easily be utilized in the base station to improve uplink capacity or coverage. Often, however, due to cost and space considerations manufacturers tend to avoid implementing multiple antennas in the smallest handheld terminals. Nevertheless, Rx diversity is one of the most efficient diversity techniques. Moreover, in addition to the diversity benefit, the received aggregated signal power is theoretically N_r fold, when compared to single-antenna reception.

1.2.1.6 *Transmit Diversity*

A significant effort has been devoted in 3GPP to develop efficient transmit diversity solutions to enhance downlink capacity. Transmit diversity methods also provide space diversity for terminals with only one receive antenna, and in that sense retain the complexity at the base station. Typically, in a 3G base station, the transmitting antenna elements are relatively close to each other. In this case the delay profile is essentially the same for each transmitting element. The closed loop Tx diversity solutions developed for the FDD mode support two transmit antennas. Both open-loop and closed-loop Tx diversity solutions are specified for UTRA FDD and TDD modes.

Open-loop Mode: The first open-loop concepts proposed in 3G standardization were based on Code Division Transmit Diversity (Orthogonal Transmit Diversity [39]) and Time Switched Transmit Diversity [40]. Time Switched Transmit Diversity (TSTD) is applied in the WCDMA standard for certain common channels. In TSTD the transmitted signal hops across two transmit antennas, according to [23]. TSTD can be considered as a special case of the Time Division Transmit Diversity (TDTD) concept described in Chapter 3. Eventually a more efficient Space–Time Transmit Diversity (STTD) solution, based on the space–time block code developed by Alamouti [10], was adopted for Release '99 [41].

The Alamouti code used in STTD is

$$\mathbf{X}_{Ala}(x_1, x_2) = \begin{bmatrix} x_1 & -x_2^* \\ x_2 & x_1^* \end{bmatrix}, \tag{1.3}$$

where column 1 is transmitted from antenna 1 and column 2 from antenna 2. The symbols are QPSK modulated in Rel. 99 and Rel. 4. The transmitter structure (omitting spreading and scrambling) is shown in Figure 1.3. In the TDD mode of the WCDMA system a variant called Block STTD (B-STTD) is used. The principle of B-STTD is the same as that of STTD

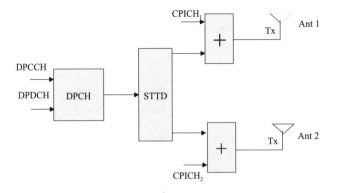

Fig. 1.3: WCDMA open-loop transmit diversity.

but the encoding operation is performed over symbol sequences. The motivation for B-STTD is to simplify receiver processing when applying multiuser or multichannel detection. In the TDD mode the spreading factor is reduced and therefore such advanced baseband receiver algorithms are often required.

Closed-loop Modes: It was noticed early in the 3G WCDMA standardization that even crude feedback signalling can be extremely useful in improving the downlink performance. The first feedback mode introduced to 3G systems was based on selective transmit diversity (STD), where only one additional feedback bit is used to select the desired transmit antenna [14, 42].

Closed-loop methods that provide beam-forming gains, as opposed to antenna selection gains, are more efficient. These gains are obtained through coherent signal combining or co-phasing in different transmit antennas, provided that the antennas share a common delay profile. In analogy with STD, co-phasing coefficients, matched to the instantaneous downlink channel, can be signalled from the UE to the BS using a fast feedback channel.

The WCDMA Release '99 and Release 4 specifications include two closed-loop transmit diversity concepts. In both closed-loop transmit diversity modes co-phasing information, contained in a fast feedback signal (of rate 1500 bps), is used to select one of 4 or 16 possible beam weights. These two modes approximate coherent transmission or channel-matched beam-forming. In both modes the terminal selects a transmission antenna route or a beam. However, they use different channel quantization and feedback signalling strategies. The related transmitter architecture is depicted in Figure 1.4.

The transmit weight is selected using the following approach. First, a given terminal obtains channel estimates for the l-path channels $\mathbf{h}_1 \in \mathbb{C}^l$ and $\mathbf{h}_2 \in \mathbb{C}^l$ for antenna 1 and antenna 2, respectively. These are estimated using antenna-specific orthogonal common channel pilot signals (CPICH). Then each terminal determines how the desired transmit weights (beam coefficients) for the dedicated channel should be modified in order to maximize the signal-to-noise ratio (or to minimize the transmit power). Thus, the problem

$$\mathbf{w} = \arg \max_{\mathbf{w}} (w_1 \mathbf{h}_1 + w_2 \mathbf{h}_2)^\dagger (w_1 \mathbf{h}_1 + w_2 \mathbf{h}_2)$$

is solved by the UE. In an attempt to reduce feedback signalling we make the *a priori* assumption that only one complex weight is signalled.

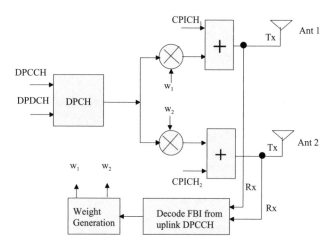

Fig. 1.4: WCDMA closed-loop transmit diversity.

The transmit beam is given by $\mathbf{w} = (w_1, w_2)$. Without loss of generality it can be assumed that w_1 is real. Thus the problem is converted to

$$w_2 = ze^{j\phi} \tag{1.4}$$

$$(z, \phi) = \arg \max_{z \in \mathbb{A}, \phi \in \mathbb{B}} ||(\sqrt{1 - z^2}\mathbf{h}_1 + ze^{j\phi}\mathbf{h}_2)||^2$$

where $\mathbb{A} = [0, 1]$ and $\mathbb{B} = [0, 2\pi)$. In feedback modes w_2 is quantized in certain ways and the quantized values are transmitted to the network (base station) using the $FSMph$ field of the uplink signal. The Feedback Signalling Message (FSM) is part of the FBI field of the uplink dedicated physical control channel (DPCCH). Each message is of length $N_{ph} + N_{po}$ bits, and one bit is transmitted in each slot resulting in a 1500 Hz signalling overhead, as will be explained in detail in Chapter 10.

1.2.1.7 Beam-forming   Beam-forming is supported for both fixed and steerable beam concepts by enabling different channel estimation solutions. Either primary common pilot, secondary common pilot or dedicated pilot can be used for channel estimation in a down-link. Dedicated pilots support dedicated beam-forming, and common pilots are generally applicable with fixed beams.

1.2.2 WCDMA: Recent Releases

WCDMA Rel. 5 specification contains a highly optimized downlink data transmission concept. This has recently been developed within the High Speed Downlink Packet Access (HSDPA) work item in 3GPP [3].

The HSDPA concept applies advanced physical layer solutions such as

- adaptive modulation and coding (link adaptation),

- optimized hybrid ARQ solution,

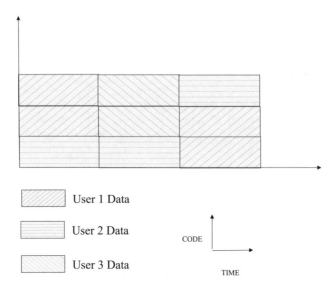

Fig. 1.5: Code and time multiplexing in HS-DSCH.

- reduced length (2 ms) Transport Time Interval (TTI),

- high order modulation and

- Fast Cell Selection (FCS)

to enable high throughput and reduced service delays. It utilizes a new type of transport channel, the High Speed Downlink Shared Channel (HS-DSCH) with a fixed spreading factor of length 16. For FDD the HS-DSCH Transport Time Interval is fixed and equal to 2 ms. This reduces service delays, when compared to 10 ms TTI in Release '99 and Release 4. A terminal may be assigned multiple channelization codes in the same Transport Time Interval (TTI), depending on terminal capability. TTI defines the frame length. In addition, it is possible to multiplex multiple UEs in the code domain within a HS-DSCH TTI, as depicted in Figure 1.5.

The control of HS-DSCH is terminated in the base station to enable a rapid reaction to changing channel conditions via link adaptation. Link adaptation is implemented with a large set of possible transport format configurations, each associated with a unique combination of modulation, coding and block size parameters. Release '99 and Rel. 4 specifications support only QPSK modulation, but an HSDPA terminal (Rel. 5) also supports 16-QAM. This increases the peak rates achievable in good channel conditions.

Data rates are assigned to a UE based on channel state information. Channel state information is measured at the UE and signalled to the base station in the form of a Channel Quality Indicator (CQI). The base station can use the CQI information at its discretion and, for example, allocate transport formats, channelization codes and time slots to maximize system capacity. Thus, CQI feedback provides the necessary information for efficient real-time link adaptation. Higher instantaneous data rates can be provisioned to users with better channel conditions. This enables a form of multiuser diversity, when applied together with downlink scheduling. Fast Cell Selection is used in downlink to select the desired base station.

The Hybrid ARQ (HARQ) combining scheme is based on Incremental Redundancy while Chase Combining is considered as a particular case of Incremental Redundancy. The adopted Hybrid ARQ solutions include Type I, Type II and Type III (using Chase Combining) with an N-channel stop and wait principle. HARQ provides implicit rate matching, while AMC is used to maximize instantaneous throughput given side information on channel state.

A number of proposals have emerged in which multiple-input multiple-output (MIMO) and multiple-input single-output (MISO) channels are considered as ways to further increase the instantaneous data rates or spectral efficiency and the power efficiency of future high capacity wireless systems. The maximum bit rates currently under study for High Speed Downlink Packet Access (HSDPA) are around 10-20 Mbps. Upgrading the spectral efficiency tenfold from what was originally envisioned for 3G is a challenging task. It is likely that such data rates cannot be implemented efficiently with just one single technology, but will require a combination of technologies. As an example, efficient exploitation of a MIMO channel, feedback concepts (ARQ, Closed-loop Tx diversity, AMC) and multiuser diversity (scheduling) might pave the way for power and bandwidth-efficient communications with the desired data rates.

1.2.3   cdma2000

The cdma2000 standard [43,44] applies space–time spreading (STS) [45] in place of STTD. STS separates Alamouti-coded symbols using two orthogonal codes rather than two (orthogonal) time slots. In addition, cdma2000 applies Orthogonal Transmit Diversity (OTD), which is in a sense analogous to TSTD or antenna hopping. In cdma2000 both STS and OTD are optional for both terminals and the network. This is different from the UTRA solution, where all terminals are required to support all three transmit diversity concepts. However, in UTRA only STTD is mandatory for the network.

The cdma2000 system further includes a high data-rate data-only solution known as 1xEV-DO (single carrier cdma2000 EVolution-Data Only) that provides a peak data rate of 2.457 Mbps in downlink using 1.25 MHz spectrum [46]. A separate carrier is needed for the service. The high data rates are achieved using link adaptation, and the highest possible data rate is provisioned to a given user by always operating the BS at full power. A more flexible solution, in which data and voice can be multiplexed in a single carrier, is called 1xEV-DV [44] (cdma2000 EVolution- Data & Voice). The technical physical layer solutions in 1xEV-DV are similar to those in the WCDMA/HSDPA concept.

1.3   GSM/EDGE

The GSM/EDGE standard itself does not suggest the use of multi-antenna transmit diversity concepts. However, in practical implementations implicit transmit diversity solutions, such as frequency sweep, frequency hopping, antenna hopping and delay diversity can easily be used. These need to be applied so that signal processing for a given received burst need not be changed. As an example, the transmitter (base station) can send different bursts using different transmit frequencies or different transmit antennas to gain time-diversity via channel interleaving.

Possible improvements for TDMA systems may apply some variant of space–time coding. However, a complex receiver would be needed in a typical TDMA channel with significant intersymbol interference, if the Alamouti code were applied to individual symbols. A modification using time-reversal codes [47, 48] may be more readily applicable for selected scenarios, provided that sufficient energy for channel estimation exists. The word "Time-reversal" is apparent from the code structure, which can be written as

$$
\begin{bmatrix}
x_1 & x_3 & \cdots & x_{2N-1} & \text{pilot}_1 & -x_{2N}^* & \cdots & -x_4^* & -x_2^* \\
x_2 & x_4 & \cdots & x_{2N} & \text{pilot}_2 & x_{2N-1}^* & \cdots & x_3^* & x_1^*
\end{bmatrix}^T ,
\tag{1.5}
$$

for a frame comprising $2N$ symbols. Antenna-specific pilot/training sequences, pilot_1 and pilot_2, are used to estimate channel coefficients, and eventually the coefficients of the space–time decoding matrix. With space–time block codes, the number of channel estimates increases along with the number of transmit antennas, and tradeoffs between channel estimation performance and diversity order need to be properly balanced.

1.4 MULTI-ANTENNA MODEMS FOR 3G AND BEYOND

1.4.1 Motivation

Some transmit diversity and MIMO transmission techniques are applicable for both second- and third-generation systems. Receive diversity using multiple receive antennas is already a mature technology, and applicable in any wireless system. However, receive diversity alone does not provide full access to the promised capacity gains. The transmitter structure and modulation and coding concept must match and exploit the underlying MIMO channel properties. This must be taken into account early on when designing a new wireless system or standard.

Generally, multi-antenna transmission and reception techniques provide

- improved fading resistance, or deliberate exploitation of fading,

- interference mitigation (e.g. using beam-forming and null steering at both transmitter and receiver),

- reduced transmitter power levels per transmit antenna path, which simplifies power amplifier design problems,

- a new dimension for rate and power allocation problems,

- theoretically higher system capacity.

Transmit diversity techniques provide fading resistance without requiring complex signal processing. In particular, open-loop transmit diversity solutions reduce signal power fluctuations in the receiver, by increasing the number of spatial diversity components. In this way, multipath-induced fading can be turned into a benefit. However, very sophisticated signal processing (e.g. multiuser detection) is needed in order to be able to capitalize on theoretical gains.

Multi-antenna transceivers that support more than two transmit antennas are currently studied for possible inclusion to future 3GPP physical layer specifications. One motivation for the increasing interest is that a MIMO channel can be used to create a virtual (spatial) spectrum, and thus pave the way for future services or increased system capacity. MIMO systems achieve high spectral efficiency in rich scattering environments with uncorrelated fading across each transmit and receive antenna.

The channel correlation depends on both the environment and the spacing of the antenna elements. A terminal, surrounded by a large number of local scatterers, can achieve relatively low correlation values even if the antennas are separated by half the wavelength. This, combined with polarization diversity, suggests that significant diversity benefits can be reaped even if the elements are separated by 7.5 cm, assuming 2 GHz carrier frequency. In outdoor base stations the antennas are significantly higher than the scatterers, and sufficiently low correlation is likely to require 10 wavelengths between neighbouring dual-polarized antenna elements. In indoor base stations the required antenna separation is likely to be in between these two extremes. These requirements are not impossible, and it is likely that certain high-end terminals could benefit from MIMO techniques, when properly taken into account in system definition. Of course, deploying multiple receive antennas necessarily increases the cost of the handset, as it increases both the RF and signal processing complexity significantly.

1.4.2 Examples of Recent Multi-antenna Transmission Methods

The 3GPP work (MIMO study item) currently focuses on improving the downlink performance beyond that achievable by STTD. A number of proposals assume that four transmitting elements are used, either with terminal to base feedback (enhanced closed-loop mode) or without feedback (enhanced open-loop mode). Current transmitter proposals that support four transmit antennas include

- combined STTD and Orthogonal Transmit Diversity (OTD), and

- combined STTD and Phase Hopping.

The relevance of symbol rate one multi-antenna concepts is the following: the symbol rate remains unchanged when compared to STTD or single-antenna transmission. Hence, the effect to transport block sizes and to the overall transmission chain is minimized. However, by upgrading the two-antenna transmission concept to support four elements, power efficiency and eventually system throughput is significantly improved.

Both of the aforementioned open-loop solutions use STTD as an integral part of the multi-antenna transmitter. When STTD is concatenated with Orthogonal Transmit Diversity (OTD), we obtain an STTD-OTD transmitter [49]. STTD-OTD achieves only a fraction of maximal diversity gain. On the other hand, the transmitted signals are orthogonal and therefore a relatively simple receiver suffices. When STTD is concatenated with phase hopping, the corresponding method STTD-PHOP, also called "Trombi", arises. This was proposed first in [24]. The receiver for Trombi is essentially identical to that in STTD, making it a strong candidate for evolved 3G systems. Both STTD-OTD and STTD-PHOP are transmit diversity solutions, designed for a MISO channel. It is assumed that they are used as an inner code in a concatenated encoding chain. Without a suitable outer code the performance of both concepts is identical to that of STTD with only two transmit antennas. We will return to both of these concepts later.

Multiple-input multiple-output (MIMO) and multiple-input single-output (MISO) transmission techniques are under consideration for future releases, where the target is for significantly higher data rates and improved power efficiency. MIMO transmission methods explicitly assume that both the transmitter and the receiver have more than one antenna and that the channel is sufficiently uncorrelated between different antennas. Modern code design for MIMO systems takes into account the antenna domain, although a portion of performance gains can also be achieved with a canonical BLAST transmission scheme [17] applying vector modulation.

Some of the proposed or possible open-loop MIMO modems include

- unconstrained signalling (BLAST),

- double STTD (DSTTD),

- non-orthogonal space–time codes (or linear dispersion codes).

In principle, all of the above are instances of non-orthogonal space–time block codes. BLAST typically presumes that the number of receive antennas at least equals the number of transmit antennas. Typically more receive antennas are needed to ensure reliable signal reception. Double STTD (DSTTD) involves two parallel STTD streams, where individual STTD streams are transmitted from different antennas. DSTTD assumes that at least two (uncorrelated) receive antennas are deployed.

In contrast to BLAST and DSTTD, non-orthogonal space–time block codes deliberately break the rate limitations associated with orthogonal space–time block codes, and attain high symbol rates with minimal or reduced self-interference. The design process leading to high-performance non-orthogonal block codes takes into account self-interference. This leads to better performance than is achievable with BLAST or DSTTD. Also, in some cases fewer receive antennas are needed to achieve a given performance level or diversity order. As an example, the code proposed in [50] outperforms DSTTD by about 1-2 dB in a flat Rayleigh fading channel, using four transmit and two receive antennas. An efficient solution deploying two transmit and two receive antennas was proposed in [51]. The solutions in [51] outperform related Linear Dispersion codes, which are optimized with respect to mutual information, rather than performance. In addition to the codes described above, an increasing number of different high-rate MIMO transmission techniques are proposed in the open literature, see e.g. [50, 52, 53], and subsequent chapters in this book.

Clearly, MIMO transmission methods are by no means restricted to 3G air interface developments. For example, the data rate and bandwidth efficiency targets suggested for 4G systems [54] are even more demanding. These may warrant the use of MIMO modulation techniques, possibly in conjunction with Orthogonal Frequency Division Multiplexing (OFDM). Also other applications, such as Wireless Local Area Networks (WLAN) and Digital Audio/Video Broadcasting (DAB/DVB-T) systems, provide attractive case studies for novel multi-antenna systems [55, 56].

1.5 SUMMARY

Wireless systems can exploit diversity by various different means, many of which are applicable without changing the standard. Receive diversity, while highly effective, increases the complexity and cost of a terminal. Transmit diversity can be used to reduce terminal complexity and to enable significant increase in downlink capacity in environments where additional diversity resources are not available. 3G system standards (WCDMA and cdma2000) were developed at a time when novel and efficient transmit diversity methods had already matured.

Therefore, both 3G systems include explicit support for two transmit antennas using space–time coding. In addition, both 2G and 3G systems may utilize more primitive (yet efficient) forms of transmit diversity. Many of the diversity methods can be used in conjunction with each other, and the relevant design issue is to make sure that at least one effective solution is supported in all relevant environments.

Modular design principles allow incremental improvements to prevailing systems without the need to redesign large portions of the system when changing one single part. As an example, space–time block codes can be incorporated into the WCDMA transmission chain without the need to re-optimize channel coding, the bit interleaver or symbol mapping. In the same vein, feedback mode transmit diversity touches only the block $\mathbf{W}$ in the system model, and this can be separately optimized for each user and transport block.

Future evolutions of wireless systems are likely to provide explicit support for more than two transmit antennas. Increased number of receive antennas improves both power and spectral efficiency, especially when the modulation/coding design is optimized for the MIMO channel. Realizable capacity gains obtained by modulation designed for a MIMO channel are yet unknown in practice. However, the projected gains are significant enough to warrant detailed analysis and code design for possible inclusion for future wireless standards.

Novel multi-antenna techniques tend to explicitly utilize the random characteristic of the wireless medium. Fading need not always have a detrimental effect on system or link capacity. Rather, fading can be considered as an additional multiplexing resource for wireless systems. A multi-antenna channel, with multiple transmit and receive antennas, can address the new multiplexing dimension in a rigorous and efficient manner, provided that the coding and modulation solutions are properly developed. In this book, we shall review the state-of-the-art in this field, and develop new solutions from a signal processing perspective. In particular, several transmission, modulation and coding techniques capable of power-efficient information transfer shall be described in detail.

The MIMO and MISO transmission methods may not yet have reached full maturity. New techniques pop up at regular intervals in different forums, for example in this book. However, when the number of transmit antennas is limited to, say, four and the number of receive antennas to, say, two, we believe that several strong practical and robust candidates are already available, as will be shown in detail in the following chapters. A desirable solution is one that is flexible enough that it does not depend on any particular channel model, transmitter-receiver architecture or feedback channel configurations.

Based on the lessons learned, multiple-antenna transmission and reception methods are widely considered as one of the most promising techniques for significantly improving the spectral efficiency of wireless systems. The solutions included in the current 3G wireless systems, however, represent only the tip of the iceberg. Similar solutions are currently being considered for other wireless standards, including WLAN, GSM/EDGE and 4G. For example, projected 4G data rates of 100 Mbps/1 Gbps using 100 MHz bandwidth in wide-area high-mobility/local-area low-mobility environments are achievable with the MIMO transmission methods presented in the book. In this example, 1 Gbps data rate is achieved using at least four transmit and receive antennas, combined with a symbol rate four code, where each symbol is 16-QAM modulated.

2

Diversity Gain, SNR Gain and Rate Increase

Diversity gain occurs when the receiver obtains multiple copies of the same information bearing signal via uncorrelated channels. In a MIMO wireless system, with multiple transmit and receive antennas, the number of propagation paths is at least the product of N_t and N_r, where N_r and N_t denote the number of transmit and receive antennas. Additionally, in a multipath channel the receiver receives delayed copies of the same signal. Delayed copies may be combined using a channel equalization algorithm, or for example, a RAKE receiver. Open-loop multi-antenna modulation techniques, considered in the following chapters, show practical approaches on how to exploit space-diversity by proper modulation design using multiple transmit antennas, and how to exploit MIMO channel structure to increase the data rate.

Beam-forming is another well-known capacity enhancing concept in which partial or complete channel state information is used in the transmitter to increase the signal power through coherent signal combining. Performance limits for beam-forming solutions, as considered here, reflect the closed–loop transmit diversity solutions studied later in this book.

In this chapter we abstract from the engineering design principles, and do not look at how the system is constructed. Instead, the tradeoffs and relations between diversity, beam-forming and multistream concepts are addressed. The purpose of this chapter is to provide insight into relations between open-loop and closed-loop solutions. First, we introduce the channel models used in the sequel.

2.1 CHANNEL MODELS

Wireless channels generate time-varying attenuations and delays, which degrade the performance of a communication system. In this book we assume that large-scale fading and shadowing resulting from motion over large geographic areas is compensated for by outer-loop power control, and we are only concerned about small-scale fading.

Suppose that the duration of a symbol or chip is much larger than the delays caused by multipath propagation. Thus, the receiver is able to distinguish only a single propagation path, which results in frequency non-selective or frequency flat fading, where all the spectral components of the transmitted signal are affected in a similar manner. Different propagation paths with different phase shifts and attenuations sum up in the receiver antenna, and assume a dense scattering model, where the number of the paths becomes large, and the resulting sum approaches a complex Gaussian random variable, whose real and imaginary components are independent with zero-mean and equal variance. Thus, the envelope of the signal becomes Rayleigh-distributed (power is then exponentially distributed), and the corresponding probability density function is given by

$$f(r) = \left(\frac{1}{2\sigma^2}\right) r e^{-r^2/2\sigma^2}, \ r \geq 0 \tag{2.1}$$

whereas the phase of the signal is uniformly distributed in $[-\pi, \pi)$. In addition to Rayleigh probability density function, commonly used models for the envelope of the signal are Nakagami and Rice.

With Nakagami fading, we denote $h = re^{j\varphi}$. The angle φ is uniformly distributed on $[-\pi, \pi)$, r and φ are mutually independent, and the probability density function of r is given by

$$f(r) = \frac{2}{\Gamma(\kappa)} \left(\frac{\kappa}{2\sigma^2}\right)^\kappa r^{2\kappa-1} e^{-\kappa r^2/2\sigma^2}, \ r \geq 0, \tag{2.2}$$

where $2\sigma^2 = \mathrm{E}\langle r^2 \rangle$, $\Gamma(\cdot)$ denotes the Gamma function, and $\kappa \geq \frac{1}{2}$ is the fading figure, which can be intuitively seen as a number of degrees of freedom, corresponding to a number of added Gaussian random variables. The instantaneous received power is distributed according to a Gamma distribution. However, in Nakagami distribution κ is continuous, and lies in the interval $[1/2, \infty)$ because it was originally developed empirically based on field measurements [57]. In a Gamma distribution κ is a positive integer. Analogously to the concept of degrees of freedom, when κ increases, the number of Gaussian random variables added up increases and the probability of deep fades or tails in the corresponding probability density function decreases In case of L independent Rayleigh fading paths with equal average powers, setting $\kappa = L$, the corresponding Gamma distribution gives the distribution of the output power of the RAKE receiver (maximal ratio combining).

Comparing Equations (2.1) and (2.2) shows that the probability density function for the Rayleigh distribution is given by (2.2) when $\kappa = 1$. Equation (2.2) is often referred to as the Nakagami-m density function. Figure 2.1 shows the behaviour of the Nakagami density function with different values of κ.

In Rayleigh fading models, the worst case scenario is, when there is no line of sight (LOS) between transmitter and receiver. In case of Ricean fading we denote $h = \alpha e^{j\phi} + \nu e^{j\psi}$, where α follows the Rayleigh distribution, and $\nu > 0$ is a constant such that ν^2 represents the power of the line of sight signal component. Angles ϕ and ψ are assumed to be mutually independent and uniformly distributed on $[-\pi, \pi)$. The probability density function of $r = |h|$ is of the form

$$f(r) = \frac{r}{\sigma^2} e^{-(r^2+\nu^2)/2\sigma^2} I_0\left(\frac{r\nu}{\sigma^2}\right), \ r \geq 0, \tag{2.3}$$

where $2\sigma^2 = \mathrm{E}\langle \alpha^2 \rangle$ and $I_0(\cdot)$ is the modified Bessel function of order zero. Note that angle $\varphi = \arg(h)$ is uniformly distributed on $[-\pi, \pi)$, and r and φ are mutually independent.

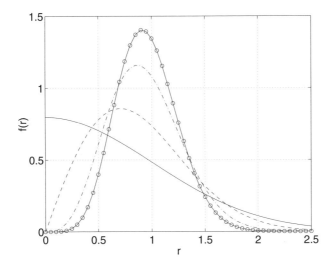

Fig. 2.1: Nakagami probability density function, (—) $\kappa = \frac{1}{2}$, (--) $\kappa = 1$ (Rayleigh), (- · -) $\kappa = 2$, (o) $\kappa = 3$

These properties follow from the assumption that ψ is uniformly distributed on $[-\pi, \pi)$. This assumption is valid when uncalibrated antenna branches are used, which will be the case in practice in the systems exploiting diversity techniques instead of downlink beamforming techniques. The notation $K = \nu^2/(2\sigma^2)$ is used for the so-called Rice factor, which is the ratio between the power of the fixed-path component and the power of the Rayleigh component. Thus, when $K \to -\infty$ in logarithmic scale there is no LOS component, and (2.3) becomes the same as (2.1). Figure 2.2 shows the behaviour of the Ricean density function for different values of K.

Weibull distribution represents another generalization of Rayleigh distribution. When X and Y are i.i.d.zero-mean Gaussian variables, the envelope $R = (X^2 + Y^2)^{\frac{1}{2}}$ is Rayleigh distributed, whereas for Weibull distribution the envelope is defined as $R = (X^2 + Y^2)^{\frac{1}{\kappa}}$, and the corresponding probability density function is given by

$$f(r) = \frac{\kappa r^{\kappa-1}}{2\sigma^2} e^{\frac{-r^\kappa}{2\sigma^2}}, \tag{2.4}$$

where $\mathrm{E}\langle r^2 \rangle = 2\sigma^2$. Weibull distribution has also been suggested for modeling the distribution of amplitudes in fading channels [58].

Probability density functions are sufficient to describe and simulate memoryless fading channels. However, in practice we are also interested in cases where the relative motion between transmitter and receiver is small, which introduces memory between adjacent received symbols. A Doppler shift occurs when the transmitter and receiver are in relative motion. The power spectrum of the signal consists of a sum of impulses which have a different frequency shift. This causes frequency dispersion, and the largest deviation of the nominal frequency is referred to as Doppler spread of the signal.

Let us assume the model in [59], where the amplitudes of the impinging signals to the antenna are all equal, and the phases are random and uniformly distributed on $[-\pi, \pi)$.

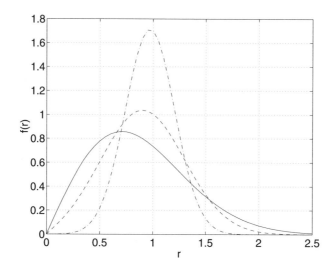

Fig. 2.2: Ricean probability density function, (—) $K = -\infty$, (--) $K = 3$ dB, (-··-) $K = 9$ dB.

Suppose that the mobile is moving to some direction with a constant velocity v during time τ, and denote the phase of the ith signal by ϕ_i at the beginning of the movement. After time τ the phase becomes $\phi_i + \frac{2\pi}{\lambda}v\tau \cos \psi_i$, where ψ_i is the direction of the impinging signal component with respect to the direction of the movement of the mobile. The normalized autocorrelation of the signal becomes

$$R(\tau) = \mathrm{E}\left\langle e^{-j\phi}e^{j(\phi+2\pi/\lambda v\tau \cos \psi)}\right\rangle \tag{2.5}$$

where the expectation is over uniformly distributed ϕ and ψ on $[-\pi, \pi)$. (The direction of the movement of the mobile is random as well.) Thus,

$$R(\tau) = \mathrm{E}\left\langle \cos(2\frac{\pi}{\lambda}v\tau \cos \psi) + j\sin(2\frac{\pi}{\lambda}v\tau \cos \psi)\right\rangle , \tag{2.6}$$

and since $\sin(\cdot)$ is an odd function,

$$R(\tau) = \frac{1}{2\pi}\int_{-\pi}^{\pi}\cos(\frac{2\pi}{\lambda}v\tau \cos \psi)d\psi = J_0\left(\frac{2\pi}{\lambda}v\tau\right) \tag{2.7}$$

where J_0 is the zero-order Bessel function of first kind. In frequency domain the corresponding power spectrum is represented by a bowl-shaped function with a bandwidth $(-f_{\mathrm{D}}, f_{\mathrm{D}})$, where $f_{\mathrm{D}} = v/\lambda$ refers to the maximum Doppler shift. The evolution of the time-varying channel with some velocity can then be simulated using, e.g., the model in [60] or shaping i.i.d.complex Gaussian random variables with a Doppler filter. The fading is said to be slow if the channel coherence time is longer than the symbol duration, whereas the opposite is true in case of fast fading, where the channel decorrelates between successive time intervals.

2.1.1 Multipath Channels

Various propagation paths of the signal cause different delays in the receiver. This delay spread then causes time dispersion and frequency selective fading. The simplest approach is to model each channel tap as an independent process with the corresponding power spectrum.

Performance requirements and the specification of test cases, e.g. channel models, have been outlined in [61]. Different multipath propagation environments in [61] are presented in Table 2.1. Power profiles refer to mean power, and all taps have a classical Doppler spectrum. With a chip rate 3.84, the delays in cases 3 and 6 approximately correspond to multiples of the chip length. While it is possible to map the second tap in cases 1, 4 and 5 to the two adjacent chips we simply associate the tap to the nearest chip, because practically the cases correspond to single path channels.

Table 2.1: Multipath fading environments in WCDMA specifications

Case	Speed (km/h)	Delay profile (ns)	Power profile (dB)
1	3	0, 976	0, -10
2	3	0, 976, 20000	0, 0, 0
3	120	0, 260, 521, 781	0, -3, -6, -9
4	3	0, 976	0, 0
5	50	0, 976	0, -10
6	250	0, 260, 521, 781	0, -3, -6, -9

2.1.2 Spatial Channels

In case of two transmit antennas, it has been assumed in link level simulations that the two channels are uncorrelated. However, when the number of transmit antennas increases the assumption is no longer justified. 3GPP contibution [62] proposed three spatial channel models for simulation studies of closed-loop transmit diversity for four transmit antennas and one receive antenna. Independent time-variant multipath channels with a certain Doppler spread are generated in the same vein as in SISO channel models. To impose the desired spatial correlations between the signals transmitted from different antennas, the signals are multiplied with a Cholesky or matrix square root decomposition [63] of the spatial correlation matrix. This assumes a dense scattering model where the angles of arrival (AOA) of the mobile station are uniformly distributed on $[-\pi, \pi)$ so that fading in time domain can be generated independently from the spatial correlations. A general form of Doppler spectrum depending on the probability density function of AOA has been presented in [64]. Stochastic channel models [62] are easy to generate and use, and the stochastic approach results in relatively fast link level simulations. Other spatial channel modelling approaches include geometric modelling and ray tracing.

In general, the spatial correlations in the base station (in downlink operation) are given by

$$
\mathbf{R}_{\mathrm{Tx}} =
\begin{bmatrix}
\rho_{11}^{\mathrm{Tx}} & \rho_{12}^{\mathrm{Tx}} & \cdots & \rho_{1N_t}^{\mathrm{Tx}} \\
\rho_{21}^{\mathrm{Tx}} & \rho_{22}^{\mathrm{Tx}} & \cdots & \rho_{2N_t}^{\mathrm{Tx}} \\
\vdots & \vdots & \ddots & \vdots \\
\rho_{N_t 1}^{\mathrm{Tx}} & \rho_{N_t 2}^{\mathrm{Tx}} & \cdots & \rho_{N_t N_t}^{\mathrm{Tx}}
\end{bmatrix},
\tag{2.8}
$$

where ρ_{ij}^{Tx} refer to the correlation between the antenna elements i and j and $\rho_{ij}^{\text{Tx}} = (\rho_{ji}^{\text{Tx}})^*$. In rural areas the impinging signals on the base station arrive from a narrow azimuth sector, and the scatterers are mainly located close to the mobile station. In micro and pico cells there are also scatterers near the base station, leading to a wider angular spread of the impinging signals. The results gathered from literature in [65] for median angular spread give less than $10°$ spread in rural macro cells, less than $20°$ spread in micro cells and $20°$-$60°$ spread in indoor cells.

When evaluating correlation coefficients according to [62] it is assumed that the power of planar waves is uniformly distributed on an angular region centred around a fixed angle of arrival/transmission measured with respect to the perpendicular direction of a uniform linear antenna array (ULA). An macro cell example with $10°$ azimuth spread (AS) and AOA $75°$ is in Figure 2.3. Correlations between the antenna elements can then be calculated according to [66]. Other distributions for power azimuth spread (PAS) presented in the literature are, e.g., truncated Gaussian [67], Laplacian [68] and n^{th} power of the cosine function [69].

The base station macro cell correlation matrix in s [62] corresponds to a $10°$ angular region around $75°$ AOA. In an (indoor) micro cell spatial channel model [62] the power of planar waves may be assumed to be uniformly distributed on an $45°$ angular region and centred around $60°$. In a pico cell model transmit antennas are assumed to be uncorrelated, and the correlation matrix is simply the identity matrix $\mathbf{I}_{N_t}$.

Although the mobile station is moving, spatial correlation matrices are fixed in the model according to [62]. This decreases simulation times, because modelling slowly varying effects would result in long simulations in order to obtain sufficient statistics. However, when such effects are ignored simulations are not able to answer the interesting question, which combination of short-term and long-term feedback transmit diversity is the most suitable one for the given propagation environment.

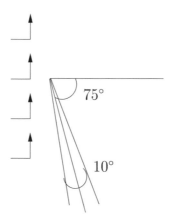

Fig. 2.3: Planar waves with uniformly distributed power within $10°$ and centred around $75°$.

2.1.3 MIMO Channel Models

In the case of a MIMO system the number of parameters needed to describe the channel further increases when compared with a MISO case.

3GPP contributions [70, 71] proposed standardized MIMO radio channel models for WCDMA link-level simulations. The principle is the same as with the MISO models, but here the type of antenna array in the mobile station effects the model parameters. Furthermore, the models also include the direction of travel of the mobile stations so that the spatial properties of the channel change during a simulation run. One motivation behind the introduction of the MIMO models [65, 70, 71] is that they can be used and parameters can be modified without a thorough knowledge of channel modelling issues. A Matlab implementation of the MIMO models can be obtained from http://www.ist-imetra.org.

The parameters of the MIMO model [65, 70, 71] are summarized in Table 2.2. Case 1 is a simple uncorrelated flat Rayleigh fading case used as a reference in link level simulations. An example of channel amplitudes across each transmitter–receiver pair is depicted in Figure 2.8. Cases 2 and 3 correspond to typical urban macro cellular environments with different degrees of time dispersion, where the angle of arrival (AOA) is assumed to be identical for all delays. Case 4 models micro-cellular and bad urban environments, assuming different angle of arrivals for different delays. Power delay profiles refer to ITU models, which are presented in Table 2.3.

Table 2.2: Summary of MIMO channel models

case	1	2	3	4
#paths	1	4	6	6
delay profile	N/A	Pedestrian A	Vehicular A	Pedestrian B
speed [km/h]	3, 40, 120	3, 40, (120)	3, 40, 120	3, 40, (120)
MS topology	N/A	$\frac{1}{2}\lambda$ element spacing		
MS PAS	N/A	Ricean component ($K = 6$) for the first path, uniform $360°$ for the remaining power	1) Laplacian distribution with $35°$ standard deviation, 2) Uniform $360°$	
MS DOT	N/A	0	22.5	-22.5
MS AOA [deg]	N/A	0	0	0
BS topology	N/A	uniform linear array with $\frac{1}{2}\lambda$ or 4λ element spacing		
BS PAS	N/A	Laplacian, AS $5°$	Laplacian, AS $10°$	Laplacian, AS $15°$
BS AOA [deg]	N/A	20, 50	20, 50	2, -20, 10, -8, -33, 31

Table 2.3: ITU multipath channel profiles

Model	delay profile [ns]	power profile [dB]
Pedestrian A	0 110 190 410	0 -9.7 -19.2 -22.8
Pedestrian B	0 200 800 1200 2300 3700	0 -0.9 -4.9 -8.0 -7.8 -23.9
Vehicular A	0 310 710 1090 1730 2510	0 -1 -9 -10 -15 -20

For example, in macro cell (case 2) with $\frac{1}{2}\lambda$ element spacing, $20°$ direction of transmission (DOT) and $5°$ AS, the spatial correlations for a 4-element linear antenna array are given by

$$
\mathbf{R}_{\text{Tx},5°} =
\begin{bmatrix}
1 & 0.97\,e^{0.34\pi j} & 0.89\,e^{0.68\pi j} & 0.77\,e^{0.99\pi j} \\
0.97\,e^{-0.34\pi j} & 1 & 0.97\,e^{0.34\pi j} & 0.89\,e^{0.68\pi j} \\
0.89\,e^{-0.68\pi j} & 0.97\,e^{-0.34\pi j} & 1 & 0.97\,e^{0.34\pi j} \\
0.77\,e^{-0.99\pi j} & 0.89\,e^{-0.68\pi j} & 0.97\,e^{-0.34\pi j} & 1
\end{bmatrix}
\tag{2.9}
$$

It is assumed that DOT and AOA are equal. In a micro cell model (case 4) with $10°$ DOT, $15°$ AS and $\lambda/2$ antenna element spacing, the spatial correlation matrix in the base station becomes

$$
\mathbf{R}_{\text{Tx},15°} =
\begin{bmatrix}
1 & 0.76\,e^{0.17\pi j} & 0.43\,e^{0.35\pi j} & 0.25\,e^{0.53\pi j} \\
0.25\,e^{-0.53\pi j} & 1 & 0.76\,e^{0.17\pi j} & 0.43\,e^{0.35\pi j} \\
0.43\,e^{-0.35\pi j} & 0.25\,e^{-0.53\pi j} & 1 & 0.76\,e^{0.17\pi j} \\
0.76\,e^{-0.17\pi j} & 0.43\,e^{-0.35\pi j} & 0.25\,e^{-0.53\pi j} & 1
\end{bmatrix}
\tag{2.10}
$$

A typical assumption in a mobile station is that there is no dominant direction of the impinging signals. Assumption of uniformly distributed angles of arrival in $[-\pi, \pi)$ and $\frac{1}{2}$ wavelength antenna spacing produces spatial correlations $J_0(\pi k)$, $k = 0, 1, 2, 3$

$$
\mathbf{R}_{\text{Rx}} = \mathbf{R}_{\text{Tx},360°} =
\begin{bmatrix}
1 & -0.3043 & 0.2203 & -0.1812 \\
-0.3043 & 1 & -0.3043 & 0.2203 \\
0.2203 & -0.3043 & 1 & -0.3043 \\
-0.1812 & 0.2203 & -0.3043 & 1
\end{bmatrix},
\tag{2.11}
$$

which can also be used to model the angles of departure in a pico-cell base station.

A $N_r N_t \times N_r N_t$ spatial MIMO correlation matrix can be written as

$$
\text{vec}(\mathbf{H})\text{vec}(\mathbf{H})^{\dagger}
\tag{2.12}
$$

where $\text{vec}(\mathbf{H}) = [\mathbf{h}_1, \cdots, \mathbf{h}_{N_r}]$ stacks the vectors $\mathbf{h}_m = [h_{1m}, \cdots, h_{N_t m}]^{\text{T}}$, and the element h_{ij} models the channel between the ith transmit and the jth receive antenna element.

A simplified, physically motivated model for MIMO correlations may be based on the assumption that all antenna elements have the same radiation pattern and that spatial correlation is independent from the position of the antenna element in the array so that all elements illuminate the same scatterers. With these assumptions, it has been reported that the $N_t N_r \times N_t N_r$ channel correlation matrix can often be well approximated by a Kronecker product of the transmit and receive correlation matrices as [72–75]

$$
\mathbf{R}_{\text{MIMO}} = \mathbf{R}_{\text{Tx}} \otimes \mathbf{R}_{\text{Rx}},
\tag{2.13}
$$

where the spatial correlation matrix in the mobile station is given by

$$
\mathbf{R}_{\text{Rx}} =
\begin{bmatrix}
\rho_{11}^{\text{Rx}} & \rho_{12}^{\text{Rx}} & \cdots & \rho_{1N_r}^{\text{Rx}} \\
\rho_{21}^{\text{Rx}} & \rho_{22}^{\text{Rx}} & \cdots & \rho_{2N_r}^{\text{Rx}} \\
\vdots & \vdots & \ddots & \vdots \\
\rho_{N_r 1}^{\text{Rx}} & \rho_{N_r 2}^{\text{Rx}} & \cdots & \rho_{N_r N_r}^{\text{Rx}}
\end{bmatrix}
\tag{2.14}
$$

and $\rho_{ij}^{\text{Rx}} = (\rho_{ji}^{\text{Rx}})^*$. It can be shown [72] that from (2.13), the channel matrix becomes

$$
\mathbf{H} = \mathbf{R}_{\text{Tx}}^{1/2}\mathbf{N}(\mathbf{R}_{\text{Rx}}^{1/2})^{\dagger},
\tag{2.15}
$$

where $\mathbf{N}$ is a stochastic $N_t \times N_r$ matrix with i.i.d. complex Gaussian elements and $(\cdot)^{1/2}$ denotes a matrix square root such that $\mathbf{R}^{1/2}(\mathbf{R}^{1/2})^\dagger = \mathbf{R}$. In uplink transmission $\mathbf{R}_{\text{MIMO}} = \mathbf{R}_{\text{Rx}} \otimes \mathbf{R}_{\text{Tx}}$.

This was just a quick brush on the surface of the large literature on MIMO channel modeling. Other important contributions are [76, 77], which examine resolvable multipaths in broadband MIMO systems, and [78, 79] where first a multipath-inspired random matrix model was constructed for MIMO channels [78], and then it was shown by measurements that the properties of measured eigenvalue spectra were accurately predicted by the number of dominant scatterers creating non-resolvable multipaths.

2.1.4 Polarization Diversity

Analogously to sufficiently remotely placed spatial antennas, polarized antennas with orthogonal polarizations typically provide independently fading channels in urban environments. Suppose that the receiver and the transmitter utilize vertically polarized antennas. Then the ratio

$$XPR_{VH} = \frac{\sigma_{VV}^2}{\sigma_{VH}^2} \tag{2.16}$$

between received vertical power σ_{VV}^2 and horizontal power σ_{VH}^2 is referred to as a cross-polarization coupling ratio. The usage of the subscript emphasizes that XPR_{VH} presents a leakage of power from vertical to horizontal polarization. The leakage from horizontal to vertical polarization is not necessarily equal to XPR_{VH}. The cross-polarization coupling XPR_{VH} depends on the environment and values of 0-12 dB have been given in the literature [80]. When XPR_{VH} is low, the polarizations mix completely in the channel, giving rise to polarization diversity. When XPR_{VH} is high, increasing the data rate by sending two independent symbol streams using orthogonal polarizations becomes attractive. In this case, the "space" direction becomes orthogonal.

2.2 PERFORMANCE LIMITS OF TRANSMIT DIVERSITY

Figure 2.2 motivates the study of diversity enhancements for wireless systems. In a Rayleigh fading channel, performance improves dramatically as the number of independent signal copies is increased. On the other hand, the figure shows that increasing the diversity order results in diminishing gains, especially in the case of coherent communications when the number of parameters to be estimated in the receiver increases and consequently the variance of estimates increases. Therefore, it would be interesting to know when an improvement in performance does not justify the additional complexity when increasing the diversity order.

Let us consider a SIMO system corresponding to a single transmit antenna and N_r receive antennas with equal received SNR $\eta = \frac{P}{\sigma^2}$ in each branch, where P and σ^2 refer to transmit and noise powers, respectively. Here we have absorbed the channel amplitude into P for simplicity. Assuming maximal ratio combining (MRC) at the receiver, the exact bit-error probability for BPSK modulation in a flat Rayleigh fading channel with independent diversity branches is [81]

$$g(N_r, \eta) = \left(\frac{1-\mu}{2}\right)^{N_r} \sum_{l=0}^{N_r-1} \binom{N_r-1+l}{l} \left(\frac{1+\mu}{2}\right)^l, \quad \mu = \sqrt{\frac{\eta}{1+\eta}} \tag{2.17}$$

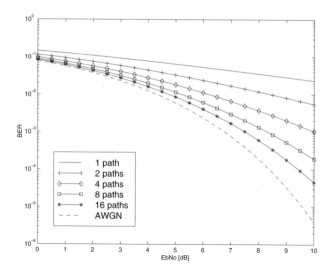

Fig. 2.4: Bit Error Rate (BER) with different number of diversity paths in a Rayleigh fading channel.

The asymptotic BEP, for large SNR ($\geq$ 10 dB/diversity branch) is given by [81]

$$f(N_{\mathrm{r}}, \eta) = \left(\frac{1}{4\eta}\right)^{N_{\mathrm{r}}} \binom{2N_{\mathrm{r}} - 1}{N_{\mathrm{r}}} \tag{2.18}$$

When the diversity order is incremented by one, the ratio of bit-error probabilities becomes

$$\frac{f(N_{\mathrm{r}}, \eta)}{f(N_{\mathrm{r}} + 1, \eta)} = \eta \left(1 + \frac{1}{2N_{\mathrm{r}} + 1}\right),$$

showing that the advantage of diversity becomes larger when SNR increases and the larger N_{r} is, the less impact the additional diversity branch has on BEP.

The same error functions can be used for a number of cases.

- For transmit diversity, the diversity degree is N_{t}, and the SNR per (transmitted) diversity branch is η/N_{t}. The BEP is given by $g(N_{\mathrm{t}}, \eta/N_{\mathrm{t}})$.

- For ideal receive beam-forming, the diversity degree is one (full correlation between the branches), but the beam-forming gain from coherent combining of signals gives a received SNR N_{r} times larger than that of the diversity system. This is the well-known 3dB beam-forming gain per doubling of the number of antennas. The BEP is given by $g(1, N_{\mathrm{r}}\eta)$.

- For transmit beam-forming, the beam-forming gain is the same; the BEP is given by $g(1, N_{\mathrm{t}}\eta)$.

With these functions, performance of diversity and beam-forming methods may be analysed. Thus, the ratio of asymptotic BEPs for receive diversity and receive beam-forming is

$$\frac{f(N_{\mathrm{r}}, \eta)}{f(1, N_{\mathrm{r}}\eta)} = \left(\frac{1}{4\eta}\right)^{N_{\mathrm{r}} - 1} \binom{2N_{\mathrm{r}} - 1}{N_{\mathrm{r}}}$$

With high SNR and large N_r the expression shows that receive diversity is clearly better than receive beam-forming.

In practical wireless systems we are interested in the performance when the received SNR is relatively small. The principle of diminishing returns when adding transmit diversity is seen in Figure 2.5, displaying the ratio $g(N_t, \frac{\eta}{N_t})/\ g(N_t + 1, \frac{\eta}{N_t+1})$ for $\eta = 0, 3, 5$ dB as a function of transmit antennas N_t. For example, when $\eta = 5$ dB, adding a second transmit antenna almost reduces BEP by a factor of two, while BEP with $N_t = 4$ is less than 1.2 times larger than that with $N_t = 5$.

Figure 2.6 compares transmit diversity and beam-forming, depicting the ratio $g(N_t, \frac{\eta}{N_t})/\ g(1, N_t\eta)$. Beam-forming is always better when $\eta = 0, 3$ dB while with $\eta = 5$ dB, transmit diversity is better than beam-forming when $N_t < 5$. Thus, Figures 2.5 and 2.6 confirm that transmit diversity is useful when the number of transmit antennas is of the order of 2–4 and when the system is operating in a relatively high SNR region. It should be noted that this comparison is based on the assumption of perfect channel state information at the receiver. At low SNR, this is a rather unrealistic assumption. The first-order effect of channel estimation noise is simply to decrease η, which further favours beam-forming solutions.

Closed-loop transmit diversity provides both diversity gain and SNR gain, and in principle it outperforms beam-forming and transmit diversity with any N_t and η. However, when uplink and downlink operate on different frequencies closed-loop transmit diversity requires a separate feedback channel. The limited capacity of the feedback channel, feedback latency and errors in the feedback channel deteriorate performance, as further discussed in Chapter 11.

Figure 2.7 depicts the received SNR of a single antenna transmission, STTD, closed-loop Mode 1 and an ideal beam-forming scheme with complete channel-state information as a function of time in a flat Rayleigh fading environment. Diversity schemes avoid the deep fades present in a flat Rayleigh fading channel. However, sometimes STTD may also decrease the received SNR when compared to single antenna transmission, which is only natural, because the received SNR is the same with both schemes. WCDMA closed-loop transmit diversity Mode 1, without feedback errors and zero feedback latency, always results in better received SNR than STTD or single antenna transmission, but is still not as good as the ideal case, because of the quantization of the feedback message.

2.3 THEORETICAL MIMO CHANNEL CAPACITY

Spatial multiplexing gain is a feature specific to MIMO channels. The fundamental observation [16,17,20] is that when the number of Tx and Rx antennas is increased, the link capacity grows, at least in theory, as $\min(N_t, N_r)$. This means that when increasing the number of Tx and Rx antennas in parallel, the bandwidth efficiency grows linearly; with the same total power, $\min(N_t, N_r)$ times the information may be transmitted . This property is referred to as "rate increase" in this chapter.

MIMO signalling has been investigated during the last couple of decades, starting with [15, 82]. The observations in [16,17,20,76] revitalized the field, and spurred an enormous amount of recent research. It is an interesting task for communication engineers to investigate to what extent and by which methods this rate increase may be realized in realistic environments.

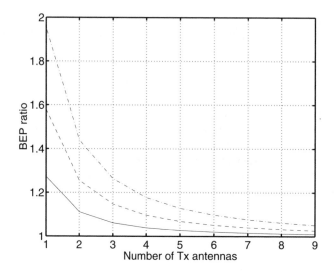

Fig. 2.5: The improvement in bit-error probability when incrementing the transmit diversity order by one, (—) 0 dB SNR, (– –) 3 dB SNR, (- · -) 5 dB SNR.

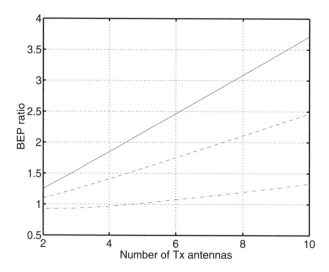

Fig. 2.6: The improvement in bit-error probability of beam-forming vs. transmit diversity, (—) 0 dB SNR, (– –) 3 dB SNR, (- · -) 5 dB SNR.

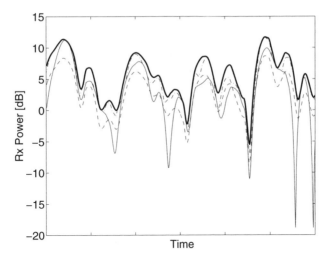

Fig. 2.7: Received power as a function time in flat Rayleigh fading environment: (—) single antenna, (– –) STTD, (- · -) closed–loop Mode 1, (▬▬) full channel state information in the transmitter.

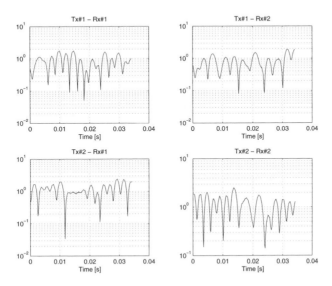

Fig. 2.8: Received amplitudes for signals between each transmitter–receiver antenna pair, with $N_t = N_r = 2$ in a flat Rayleigh fading channel.

2.3.1   No Channel State Information at Transmitter

To understand the fundamentals of the field, consider a ideal system with N_t Tx antennas and N_r Rx antennas, with i.i.d. frequency flat fading channels between each pair of Tx and Rx antennas. Thus in the $N_t \times N_r$ channel matrix, there are altogether $N_t N_r$ independently fading channels (see Figure 1.2). Essentially, this means that we have a radio environment with lots of scatterers but no macroscopic multipath components, e.g. a typical indoor/pico cell.

For simplicity, we first assume no channel state information (CSI) at the transmitter. The MIMO transmission applies vector modulation $\mathbf{x} = [x_1 \ x_2 \ \dots \ , x_{N_t}]$, i.e. transmitting an independent symbol from each channel. From the signal model

$$\mathbf{y} = \mathbf{x} \, \mathbf{H} + \text{noise} \tag{2.19}$$

the N_t transmitted signals may be solved, as long as $\mathbf{H}$ is full-rank.

Using this signal model, the (link) capacity of the channel may be computed. For a scalar channel, capacity is achieved with continuous Gaussian input, and it is

$$\mathcal{C} = \text{E} \left\langle \log(1 + \eta |h|^2) \right\rangle_h . \tag{2.20}$$

The expectation is over the channel realizations h, and $\eta = P/\sigma^2$ is the average SNR, expressed in terms of the total transmit power P and the average noise power σ^2. If the logarithm is taken in base 2, $\mathcal{C}$ is expressed as bits/s/Hz, if it is in base e, $\mathcal{C}$ is nats/s/Hz. Thus $\mathcal{C}$ actually measures bandwidth efficiency. Capacity in bit/s for a bandwidth B is $\mathcal{C}B$.

For a matrix channel with continuous Gaussian input (2.20) generalizes to [1]

$$\mathcal{C} = \max_{\text{Tr } \mathbf{Q} \leq P} \text{E} \left\langle \log \det \left(\mathbf{I}_{N_r} + 1/\sigma^2 \, \mathbf{H}^\dagger \mathbf{Q} \, \mathbf{H} \right) \right\rangle_\mathbf{H} , \tag{2.21}$$

which should be maximized over the $N_t \times N_t$ input covariance matrix $\mathbf{Q} = \text{E} \left\langle \mathbf{x}^\dagger \mathbf{x} \right\rangle$ and averaged over the channel. Note that here $\mathbf{x}$ is a row-vector.

For Rayleigh fading, capacity with no CSI at the transmitter is achieved if the transmitted vector symbol consists of i.i.d. symbols. Thus $\mathbf{Q} = P/N_t \, \mathbf{I}_{N_t}$, and the transmit power is divided evenly among the N_t transmit antennas. The N_t-dimensional identity matrix is $\mathbf{I}_{N_t}$.

The channel matrix may be singular value decomposed (SVD) as

$$\mathbf{H} = \mathbf{W}^\dagger \, \Sigma \, \mathbf{V} . \tag{2.22}$$

Here $\mathbf{W}$ is a $N_t \times N_t$ unitary matrix, Σ is a $N_t \times N_r$ matrix with $\min(N_t, N_r)$ singular values on the main diagonal, and $\mathbf{V}$ is a $N_r \times N_r$ unitary matrix. Corresponding to the singular value decomposition, the channel correlation matrix may be eigenvalue decomposed as

$$\mathbf{H}^\dagger \mathbf{H} = \mathbf{V}^\dagger \, \Lambda \, \mathbf{V} , \tag{2.23}$$

with $\Lambda = \Sigma^2$ a diagonal matrix with the N_t eigenvalues λ_i of the channel correlation matrix on the diagonal. As det is cyclical, the matrix $\mathbf{V}$ vanishes from (2.21). The capacity becomes

---

[1]Note that the capacity expression can equally well be written in the Tx antenna dimensions as $\mathcal{C} = \max_\mathbf{Q} \text{E} \left\langle \log \det \left(\mathbf{I}_{N_t} + 1/\sigma^2 \, \mathbf{Q} \, \mathbf{H} \, \mathbf{H}^\dagger \right) \right\rangle_\mathbf{H}$. These expressions are equivalent due to the cyclicity of det.

a sum over $\min(N_t, N_r)$ *parallel channels*, corresponding to the $\min(N_t, N_r)$ eigenvalues of $\mathbf{H}^\dagger\mathbf{H}$:

$$\mathcal{C} = \sum_{i=1}^{\min(N_t,N_r)} \mathrm{E} \left\langle \log\left(1 + \eta/N_t \, \lambda_i\right)\right\rangle_\mathbf{H} . \tag{2.24}$$

Note that this calculation does not depend on whether $N_r \geq N_t$ or $N_t \geq N_r$, the number of possibly non-zero eigenvalues is always $\min(N_t, N_r)$. Using the Jensen inequality, we see that capacity is upper-bounded by

$$\mathcal{C} \leq \sum_{i=1}^{\min(N_t,N_r)} \log\left(1 + \eta/N_t \, \mathrm{E}\left\langle\lambda_i\right\rangle_\mathbf{H}\right) . \tag{2.25}$$

To asses the order of magnitude of this upper bound, consider the following equality valid for i.i.d. fading,

$$\mathrm{E}\left\langle \sum_{i=1}^{\min(N_t,N_r)} \lambda_i \right\rangle = \mathrm{E}\left\langle \mathrm{Tr}\,\mathbf{H}^\dagger\mathbf{H} \right\rangle = \sum_{k=1}^{N_t}\sum_{l=1}^{N_r} \mathrm{E}\left\langle|h_{kl}|^2\right\rangle = N_t N_r \, \mathrm{E}\left\langle|h|^2\right\rangle , \tag{2.26}$$

meaning that the expectation value of an eigenvalue is $\frac{N_t N_r}{\min(N_t,N_r)}$ times the expectation value of an individual channel power. This is a consequence of the proliferation of independent channels when the number of Tx and Rx antennas is increased. Using Jensen upper bounding once more, this time to the expectation value (sum) over eigenvalues, we get

$$\mathcal{C} \leq \min(N_t, N_r) \, \log\left(1 + \eta \, \frac{N_r}{\min(N_t,N_r)} \mathrm{E}\left\langle|h|^2\right\rangle\right) . \tag{2.27}$$

With $N_r = N_t$ this is exactly N_t times the Jensen upper bound of the scalar channel capacity (2.20). To prove that this behaviour of the upper bounds captures the essence of the behaviour of the true capacities is a very non-trivial calculation performed in [16]. This simple exercise, however, manages to capture the salient features of the MIMO capacity problem.

- With perfect channel state information at the receiver, MIMO capacity grows linearly in $\min(N_r, N_t)$.

- The matrix channel decomposes to $\min(N_r, N_t)$ independent parallel channels.

- The increase in capacity comes about from adding *similar* antenna elements. With the assumption that the elements are sufficiently uncorrelated, the antenna array takes up ever more space. Characterizing the capacity when the area of the transceiver arrays are constrained was considered in [73]. There it was observed that an area constraint removes the linear growth of capacity in $\min(N_t, N_r)$. Instead, capacity growth reduces to an expression of the Tx beam-forming type, with SNR multiplied by N_t. Another way to look at this aspect of MIMO systems was discussed in [83]. It was argued that conventional assessment of MIMO capacity does not take into account that part of the capacity increase is just due to the additional antenna gains of the multiple antennas.

- Realizing the rate increase is based on assuming perfect CSI at the receiver. With increasing numbers of Tx antennas, the channel estimation overhead, however, increases progressively. This aspect was addressed in [84]. It was observed that it is fruitless to make N_t bigger than the coherence length of the channel. This restricts MIMO (and Tx-diversity) use to slowly fading channels only. With arrangements of pooling pilot powers, like the common channels in WCDMA, this difficulty can be overcome to some extent.

- Capacity-reaching distributions are vector modulations with Gaussian inputs, which provide no Tx diversity. This seems to indicate that diversity and rate increase are complementary concepts. However, as we shall see in Chapters 7 and 9, for matrix modulation with fixed discrete modulation constellations, *full Tx diversity and full rate increase may be realized simultaneously*. In addition, with maximum likelihood detection, or some simplified non-linear version of it, full Rx diversity may always be achieved. These claims, again, are contradicted by the results of [85], where a fundamental tradeoff between Tx diversity and rate increase is observed. This latter argument is based on constellation sizes that scale with SNR. The solution of this apparent paradox lies in the order of taking limits. If the constellation size is chosen to realize near capacity throughput at a given SNR, it may be constructed to have full Tx-diversity using matrix modulation and the prescriptions in [86] and Chapter 7, and full Rx-diversity with a sufficiently complex receiver. The diversity gains would, however, only yield significant performance improvement at a larger SNR, where to reach near capacity throughput, one should use a larger constellation, as prescribed in [85], etc. In a realistic cellular communication system, however, the set of possible modulation alphabets would be restricted, and the operation range would have regions where it is beneficial to design the constellations to provide Tx-diversity, and the receiver to provide Rx-diversity, irrespective of the number of parallel streams employed.

2.3.2 Channel State Information at Transmitter

The MIMO channel capacity with perfect channel state information at the transmitter and receiver is given in [16]. In this section, we shall not distinguish between various methods to acquire CSI at the Tx, and refer to any capacity expression with channel state information at Tx as "closed-loop capacity," with the understanding that the previous section discussed "open-loop capacity".

The idea is that by using Equation (2.22), the channel may be concretely diagonalized to the parallel channels, using transmit beam-forming matrix $\mathbf{W}$ and receive beam-forming matrix $\mathbf{V}^\dagger$. With this multibeam-forming, $\min(N_t, N_r)$ orthogonal and non-interfering parallel channels are constructed. The parallel channels have generally a different gain, i.e. the corresponding eigenvalues are different. Thus with side information at the transmitter, the transmitter may choose the covariance of the transmitted symbols to be

$$\mathbf{Q} = \tilde{\mathbf{W}}^\dagger \, \mathbf{P} \, \tilde{\mathbf{W}} \, , \tag{2.28}$$

where $\tilde{\mathbf{W}}$ is a generalized beam-forming matrix, which constructs altogether N_t orthogonal beams. With perfect CSI, the beam-forming matrix is taken to be the component $\mathbf{W}$ of the SVD (2.22). The power allocation matrix $\mathbf{P}$ is a diagonal matrix that may be chosen to exploit the differences of the eigenvalues of the eigenbeams.

With perfect generalized Tx and Rx beam-forming applying the beam-forming matrix $\mathbf{W}$, the signal model (2.19) becomes diagonal,

$$\tilde{\mathbf{y}} = \mathbf{y} \, \mathbf{V}^\dagger = \mathbf{x} \, \mathbf{A} \, \Sigma + \mathbf{n} \, . \tag{2.29}$$

Here $\mathbf{A}$ is the diagonal amplitude matrix, which satisfies $\mathbf{A}^2 = \mathbf{P}$. The transmitted signal is $\hat{\mathbf{x}} = \mathbf{x}\,\mathbf{A}\,\mathbf{W}$. Note that the relevant ingredients $\mathbf{V}$ and $\mathbf{W}$ of the singular value decomposition can be determined as the eigenvectors of the channel correlation matrices

$$\mathbf{R}_{\mathrm{Rx}} = \mathbf{H}^{\dagger}\mathbf{H} \quad \text{and} \quad \mathbf{R}_{\mathrm{Tx}} = \mathbf{H}\mathbf{H}^{\dagger}. \qquad (2.30)$$

With perfect knowledge of the eigenbeams, the capacity becomes

$$\mathcal{C} = \max_{\sum_i P_i = P} \sum_{i=1}^{\min(N_t, N_r)} \mathrm{E}\left\langle \log\left(1 + P_i/\sigma^2\,\lambda_i\right)\right\rangle_H, \qquad (2.31)$$

where the diagonal elements of the power allocation matrix are P_i.

With perfect CSI at the transmitter, optimal power allocation strategies may be found in closed form. Beams with stronger gain, i.e. a larger eigenvalue, should be allocated more power. In [16] it was shown that the capacity of a multi-antenna channel (in bits per dimension) is reached with a water-filling power-allocation policy. The power allocated to the i'th row of the $\mathbf{W}$ matrix in the SVD, corresponding to the eigenvalue λ_i, is

$$P_i = \sigma^2\left[\mu - \lambda_i^{-1}\right]_+. \qquad (2.32)$$

The variable μ is defined by the total power constraint: $\sum_i P_i \leq P$. The function $[\;]_+$ sets negative numbers to zero. The resulting capacity expression is

$$\mathcal{C} = \sum_{i=1}^{\min(N_t, N_r)} \mathrm{E}\left\langle [\log \mu\lambda_i]_+\right\rangle. \qquad (2.33)$$

The generalized beam-forming matrix $\mathbf{W}$ above assumes that the receiver experiences additive white Gaussian noise. The results are, however, extendible to cases with coloured noise and to particular receiver structures, as shown e.g. in [87, 88].

In equation (2.33), *instantaneous* CSI is assumed at Tx—for each transmission, the channel is known perfectly. In addition, averaging over channel realizations assumes that a coding scheme is applied which extends over multiple channel coherence times. These two assumptions are contradictory in most realistic wireless environments. More realistic capacity expressions may be considered by omitting the averaging over channel states, leading to non-ergodic outage capacity expressions [16], or by assuming imperfect CSI. An example of the latter is long-term beam-forming, considered in [89–93] and suggested for MIMO long-term beams in [94, 95]. The idea is to exploit the possible structure in the channel correlation matrix $\mathbf{H}\mathbf{H}^{\dagger}$. The correlation matrix is calculated by filtering over a number of instantaneous channel realizations. Power allocation between long-term beams, optimally water-filling [92] is applied to increase capacity.

Example: $N_r = 1$. To get an insight in the open- and closed-loop capacity expressions (2.24) and (2.33) it is illuminating to consider one receive antenna case. With $N_r = 1$, the one row of the $N_t \times N_t$ matrix $\mathbf{W}$ is the single optimal eigenbeam w, whereas the $N_t - 1$ remaining rows are beams that span the noise subspace. The open-loop capacity (2.24) is

$$\mathcal{C}^{\mathrm{OL}}_{N_r=1} = \mathrm{E}\left\langle \log\left(1 + \eta/N_t\,\lambda\right)\right\rangle_{\mathbf{h}}, \qquad (2.34)$$

where the single eigenvalue is $\lambda = |h|^2$. Note that vector modulation reaches open-loop capacity, as observed in [96].

With perfect CSI, and one Rx antenna, choosing the optimal power allocation is trivial. All power is allocated to the single eigenbeam. This intuitive result follows from (2.32). The closed-loop capacity (2.33) becomes

$$\mathcal{C}^{\mathrm{CL}}_{N_{\mathrm{r}}=1} = \mathrm{E} \left\langle \log \left(1 + \eta \, \lambda \right) \right\rangle_{\mathbf{h}} . \tag{2.35}$$

In both expressions (2.34) and (2.35), the eigenvalue λ is the same, and is given by (2.22). The difference is that in the open-loop case, transmit power is evenly divided among antennas with no phase shifts, i.e. evenly among the N_{t} vectors in $\mathbf{W}$, $(N_{\mathrm{t}} - 1)$ of which span the noise subspace. Conversely, in the closed-loop case, transmit power is still divided among antennas, but phase shifts and weightings have now been applied so that the total power is transmitted to the best eigenbeam and no power is wasted on the noise subspace. The difference is exactly the N_{t}-fold beam-forming (or SNR) gain discussed in the previous sections, i.e $\eta/N_{\mathrm{t}} \mapsto \eta$.

Example: $N_{\mathrm{r}} = N_{\mathrm{t}}$. For further insight, the square matrix channel $N_{\mathrm{r}} = N_{\mathrm{t}}$ may be considered. In this case, all rows in $\mathbf{W}$ represent true eigenbeams that have an associated eigenvalue. If in the closed-loop scenario, we construct concrete N_{t} eigenbeams, and transmit *with equal power* to these beams, the analysis above shows that *there is no capacity gain from using generalized beam-forming.* In the square matrix case, transmit beam-forming compared to vector modulation is just a change of basis at the transmitter. Only if power allocation between the parallel beams is applied may one gain in capacity over an open loop transmission. As can be observed from the results in [87], the capacity gain due to optimal power allocation is small compared to the capacity gain due to increased parallel streams.

From a communication point of view, this seeming uselessness of generalized beam-forming reveals a blind point of using capacity arguments based on continuous input. The complexity required to reach near-capacity throughput would be much diminished if beam-forming is used to decompose the matrix channel to independent parallel channels.

2.4   MIMO CAPACITY IN CORRELATED CHANNELS

The linear capacity growth phenomenon was observed above for i.i.d. channels. In realistic systems, there are always correlations between the channels. Capacity aspects of correlated MIMO channels have been discussed, e.g. in [72,77]. Here we shall assess MIMO capacities without CSI at the transmitter in wireless scenarios corresponding to the cases discussed in section 2.1.3. The downlink direction is assumed, in uplink, the transmitter and receiver parameters change place, and the capacities are the same.

In all cases, both BS and MS have a uniform linear array with half wavelength spacing. The MS is surrounded by scattering clutter and has an angular spread of $360°$. The correlation matrix is of the type (2.11), extended to 16 antennas with the model in [65]. The base station is equipped with up to 16 antennas, with the correlation matrix similarly extended. The considered scenarios are

- A macro cell, where the BS has angular spread $5°$. The correlation matrix is of the type (2.9).

- A micro cell, where the BS has angular spread $15°$. The correlation matrix is of the type (2.10).

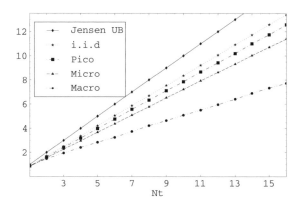

Fig. 2.9: $N_{\mathrm{r}} = N_{\mathrm{t}}$. Capacities (2.21) in bps/Hz of four wireless MIMO environments with different correlations, as well as the Jensen upper bound (2.27), for SNR=0 dB, $\mathrm{E}\left\langle |h_{kl}|^2 \right\rangle = 1$.

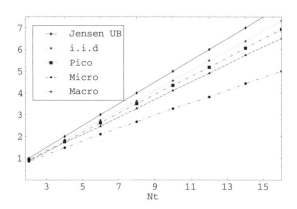

Fig. 2.10: $N_{\mathrm{r}} = N_{\mathrm{t}}/2$. Capacities (2.21) in bps/Hz of four wireless MIMO environments with different correlations, as well as the Jensen upper bound (2.27), for SNR=0 dB, $\mathrm{E}\left\langle |h_{kl}|^2 \right\rangle = 1$.

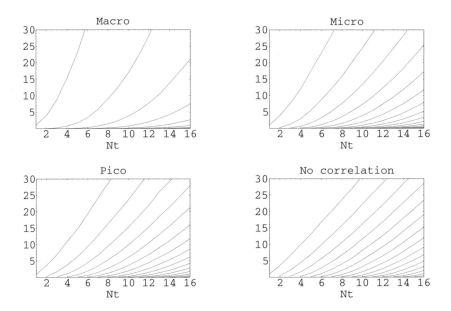

Fig. 2.11: $N_{\mathrm{r}} = N_{\mathrm{t}}$. Average values of some of the largest eigenvalues of $\mathbf{H}^{\dagger}\mathbf{H}$ in four cellular MIMO scenarios, $\mathrm{E}\left\langle |h_{kl}|^{2} \right\rangle = 1$.

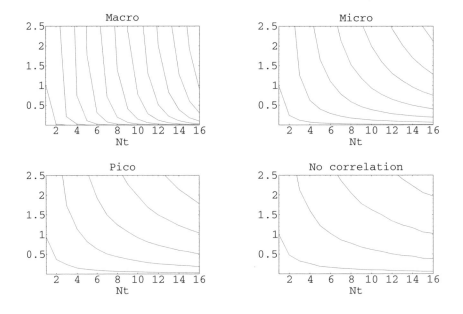

Fig. 2.12: $N_{\mathrm{r}} = N_{\mathrm{t}}$. Average values of some of the smallest eigenvalues of $\mathbf{H}^{\dagger}\mathbf{H}$ in four cellular MIMO scenarios, $\mathrm{E}\left\langle |h_{kl}|^{2} \right\rangle = 1$.

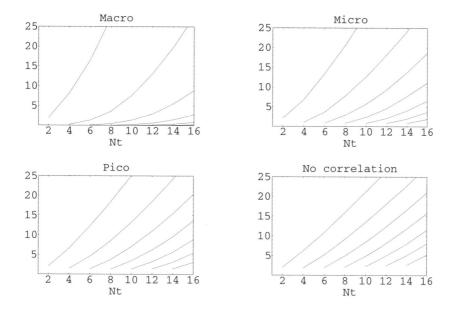

Fig. 2.13: $N_r = N_t/2$. Average values of some of the largest eigenvalues of $\mathbf{H}^{\dagger}\mathbf{H}$ in four cellular MIMO scenarios, $E\left\langle |h_{kl}|^2 \right\rangle = 1$.

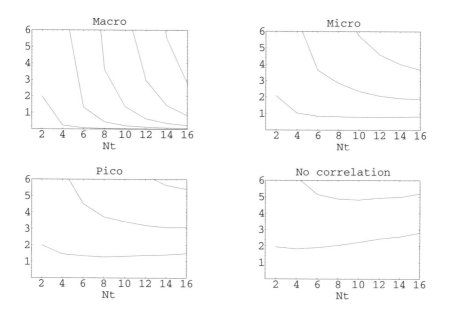

Fig. 2.14: $N_r = N_t/2$. Average values of some of the smallest eigenvalues of $\mathbf{H}^{\dagger}\mathbf{H}$ in four cellular MIMO scenarios, $E\left\langle |h_{kl}|^2 \right\rangle = 1$.

Table 2.4: Number of large and small eigenvalues in the example environments, for $N_t = N_r = 16$ and $N_t = 2N_r = 16$. Altogether 16 and 8 eigenvalues, respectively

| | $N_t = N_r = 16$ | | | $N_t = 2N_r = 16$ | | |
	Large	Medium	Small	Large	Medium	Small
Macro	4	1	11	3	0	5
Micro	7	4	5	5	2	1
Pico	9	4	3	6	2	0
No correlation	11	3	2	7	1	0

- A pico cell, where the BS has angular spread $360°$. The correlation matrix is of the type (2.11).

- An uncorrelated array, where all channels are i.i.d.

In all cases, the square matrix channel with $N_r = N_t$ will be considered, along with an asymmetric channel $N_r = N_t/2$ (or or equivalently $N_r = 2N_t$).

For the case of $N_r = N_t$, capacity results can be found in Figure 2.9. The decrease in capacity with increasing correlation is explained by Figure 2.11, where the average values of some of the largest eigenvalues are plotted. It can be seen that in a strongly correlated channel, there are few strong eigenmodes. With decreasing correlation, the largest eigenvalues become smaller, whereas the smaller eigenvalues become larger. The relevant difference between the less correlated cellular MIMO scenarios is visible in Figure 2.12, where it can be seen how the number of essentially vanishing eigenvalues increases with correlation. In the macro cell scenario, half of the eigenvalues are always almost zero, whereas in the uncorrelated case, on the average, only one eigenvalue is almost vanishing.

The same plots for $N_t = 2N_r$, can be found in Figures 2.10, 2.13 and 2.14. It is notable how the smallest eigenvalues do not vanish, as for the square matrix channel case. Also, for the stronger correlations, it is clearly visible that capacity is not proportional to $\min(N_t, N_r)$. Table 2.4 summarizes the results regarding large and small eigenvalues, with an arbitrary threshold that an eigenvalue is large if it is $> 5\, E\left\langle|h|^2\right\rangle$, whereas it is small if it is $< E\left\langle|h|^2\right\rangle$. Here $E\left\langle|h|^2\right\rangle$ is the expectation value of a individual channel, which is 1 in the plots. Note that with this normalization, the total expected channel power is $E\left\langle \mathrm{Tr}\, \mathbf{H}^\dagger \mathbf{H}\right\rangle = E\left\langle\sum_{i=1}^{N_r} \lambda_i\right\rangle = N_t N_r$, irrespective of the correlation.

The essential observations regarding correlated MIMO capacity are

- With strong correlations, there are only a few dominant beams. These have large beam-forming gain (large eigenvalues), as these few beams collect most of the channel power.

- With weak correlation, large beam-forming gains are changed to multiple parallel streams—the eigenvalue spectrum becomes more even. These parallel streams are amenable for rate increase, which is visible in an increased capacity.

- With strong correlations, most of the eigenvalues are practically zero.

- With no correlations and a $N_r = N_t$ channel, the smallest eigenvalue is practically zero.

Prank: The results of Table 2.4 indicate that often the rank of a MIMO channel for all practical purposes, i.e. the practical rank or Prank, is $< \min(N_t, N_r)$. This should be taken into account when choosing (open loop) signalling schemes for specific environments. Thus it is intuitively clear that e.g. in a macro cell of the type discussed above, with a ULA consisting of $N_t = 8$ antennas, one would not be advised to transmit more than two parallel streams. Eigenvalues that are of the order of the noise power should not be relied on when separating streams from each other. In a more exact analysis, it becomes clear that the ratios of the various singular values should be considered.

As a simple example, consider transmitting one stream of 16-QAM symbols, as opposed to two streams of QPSK symbols, over a 2×2 matrix channel, with left SVD matrix

$$\mathbf{W} = \left[\begin{array}{cc} w_{11} & w_{12} \\ w_{21} & w_{22} \end{array} \right]$$

In the latter case, the received signal after matched filtering with $\mathbf{V}$ to the receive eigenmodes will consist of

$$(x_1 w_{11} + x_2 w_{21})\, \sigma_1\ , \quad (x_1 w_{12} + x_2 w_{22})\, \sigma_2$$

received over the first and second eigenmode, respectively, plus noise. That is, two (generically different) sums of the two QPSK symbols, rotated and scaled by respective elements in $\mathbf{W}$, are received over the eigenmodes. In each sum there are 16 possible points, corresponding to the two QPSK symbols. If one eigenmode is much weaker than the other, detection is dominated by just one mode. In that case, it is preferable to transmit a symbol with 16 possibilities so that the minimum distance is not affected by the channel, e.g. a 16-QAM symbol.

In the 2×2 case, the appropriate measure is the relative condition number of the channel matrix (i.e. the ratio of the two singular values). In a channel with more antennas, this is extended to the concept of Prank, which can be defined as the number of singular values that are larger than a given fraction of the largest singular value, resulting in an integer which is $\leq$ rank $\mathbf{H}$. The choice of threshold depends on the actual signalling schemes considered. This allows a smooth interpolation between rate $R_s = 1$ in macro cells and rate $R_s = N_t$ in pico cells with lots of scattering. A simple example of using Prank is discussed in section 12.5.

The concept of Prank is intuitively clear in the closed-loop MIMO setting, where the power allocation (Equation 2.32) automatically discards channels with weak eigenvalues. Here, we want to argue that such thresholds should be applied in open-loop cases as well.

In [97], a selection between parallel stream and Tx-diversity use of a MIMO system was discussed, based on Demmel condition number, i.e. the square root of the ratio of the total channel power to the smallest eigenvalue. Prank differs from this in that the whole spectrum of schemes with symbol rate $1, \ldots, \min(N_r, N_t)$ may be considered, and that the relative condition number is used. The reasons for this become clear in Chapter 9, especially section 9.2. There we shall see that irrespective of the number of parallel streams, it is possible to design MIMO signalling schemes which enjoy full diversity, and have a matched filter bound characterized by the total channel power.

2.5   PERFORMANCE MEASURES FOR CLOSED-LOOP TRANSMIT DIVERSITY

When the transmit antennas are fully uncorrelated both open-loop and closed-loop provide diversity gain. Bit-error probability (BEP) $P(e)$ is a frequently used performance measure, which is conventionally calculated as

$$P(e) = \int_{\mathbb{R}} p(\xi)g(\xi)d\xi, \tag{2.36}$$

where $g(\cdot)$ is the error rate of the modulation in terms of the received SNR and $p(\cdot)$ is the probability density function of the SNR. However, in case of practical closed-loop transmit diversity schemes employing low bit-rate quantization of the feedback messages, the emerging integrals are difficult to solve in closed form. Therefore, we mainly illustrate the performance of closed-loop algorithms using the expected SNR gain γ as a performance measure. This quantity is also often referred to as antenna gain or aperture gain in beam-forming literature. The SNR gain illustrates the performance of the MISO system well and also provides means to analyse different closed-loop algorithms using explicit quantization schemes of the feedback signal. However, in MIMO antenna processing, examining the SNR gain is not enough, because the system can allocate multiple parallel channels to the user when the rank of the channel correlation matrix is larger than one, and thereby increase the data rate of the system.

Another frequently used performance measure is the pairwise codeword error probability. However, when signal design is independent of transmit weights, minimizing pairwise codeword error probability is the same as maximizing SNR. In this case, given two codewords $\mathbf{c}$ and $\mathbf{e}$, $P(\mathbf{cWH} \to \mathbf{eWH})$ is minimized when SNR gain γ is maximized, if only the beam-forming matrix $\mathbf{W}$ is optimized.

Consider a MISO system with N_t transmit antennas in the base station and a single receive antenna in the mobile station. In the case of scalar coding, the mobile station receives the signal from the dedicated channel in a multipath environment as

$$\mathbf{y} = x(\mathbf{wH}) + \mathbf{n}, \tag{2.37}$$

where x is the transmitted symbol, $\mathbf{n}$ is zero-mean complex Gaussian noise, $\mathbf{w} = [w_1, w_2, \cdots, w_{N_t}]$ consists of transmit weights signalled from the mobile station, and components of the channel matrix $\mathbf{H} = [\mathbf{h}_1, \mathbf{h}_2, \cdots, \mathbf{h}_{N_t}]^T$ are distributed according to some cumulative distribution function. The components of $\mathbf{y} = [y_1, \cdots, y_L]$ correspond to the different propagation paths of the channel. We assume that all channel impulse response vectors $\mathbf{h}_m$ are identically distributed and $E \langle |h_{m,l}|^2 \rangle = 2\sigma_l^2$, where $l \in \{1, 2, \cdots, L\}$ corresponds to the path number. Moreover, we normalize channels by assuming that $E \langle \|\mathbf{h}_m\|^2 \rangle = 1$ for all $m = 1, 2, \cdots, N_t$. Finally, when analysing SNR gains of different closed-loop methods in Chapter 11 we assume that impulse responses $\mathbf{h}_m$ are uncorrelated if not otherwise stated.

In a single path channel, $\mathbf{h}_m$ is a complex scalar rather than vector, and we denote h_m instead of $\mathbf{h}_m$ and $\mathbf{h} = [h_1, h_2, \cdots, h_{N_t}]^T$ instead of $\mathbf{H} = [\mathbf{h}_1, \mathbf{h}_2, \cdots, \mathbf{h}_{N_t}]^T$. The mobile station receives the signal from the dedicated channel as

$$y = x(\mathbf{w\,h}) + n, \tag{2.38}$$

where components of the channel vector $\mathbf{h}$ are independent identically distributed random variables.

The expected SNR gain is defined by

$$\gamma = \frac{\mathrm{E}\langle\xi\rangle}{\mathrm{E}\langle\xi_0\rangle}, \tag{2.39}$$

where $\mathrm{E}\langle\xi_0\rangle = \mathrm{E}\left\langle\frac{P}{\sigma^2}\frac{\|\mathbf{h}\|}{N_t}\right\rangle$ is the expected received SNR without channel state information in the transmitter and $\mathrm{E}\langle\xi\rangle = \mathrm{E}\left\langle\frac{P}{\sigma^2}|\mathbf{wh}|^2\right\rangle_\mathbf{h}$ is the expected received SNR with CSI. Without feedback information, $\gamma = 1$. Thus γ indicates the relative gain from the coherent combining of signals. A gain in received SNR affects system performance in a straightforward manner. Moreover, the link capacity increase of a feedback method can be estimated from the SNR gain. Using Jensen upperbounding similar to (2.27)

$$\mathcal{C} = \mathrm{E}\langle\log_2(1+\xi)\rangle \leq \log_2(1+\gamma\cdot\mathrm{E}\langle\xi_0\rangle), \tag{2.40}$$

where $\mathcal{C}$ is the capacity expressed as bits/s/Hz assuming an optimal rate adaptation to the channel fading with a constant transmit power [98]. Hence, the SNR gain γ provides an upper bound for the system capacity improvement in a fading channel. On the other hand, when $\mathbf{w} = \mathbf{h}$ and the transmitter knows the channel completely all the time, which corresponds to transmit beam-forming, capacity and γ are maximized simultaneously as discussed in Example 1 on page 37.

There are two important facts worth mentioning. First, in FDD WCDMA dedicated downlink channels fast power control can be used on top of the closed-loop transmit diversity modes. However, near the cell boundary the dynamics of the fast power control is limited because the intercell interference must be carefully controlled, so the capacity formula (2.40) is valid for our purposes. Furthermore, from the network point of view it is advantageous if better channel quality can be offered to mobiles near the cell boundary without increasing the base station transmission power. Second, if the expectation of ξ_0 is low in (2.40), then

$$\mathcal{C} \approx \log_2(e)\gamma\cdot\mathrm{E}\langle\xi_0\rangle. \tag{2.41}$$

From a FDD WCDMA point of view, this approximation is useful since the operating SNR is low, especially near the cell boundary.

2.6 SUMMARY

Both open-loop and closed-loop techniques improve link performance using different approaches. Hybrid open-loop and closed-loop techniques are also possible, and are further examined in Chapter 12. In the following we summarize some important differences in the approaches affecting the link performance.

- *Feedback imperfections.* The performance of closed-loop transmit diversity schemes deteriorates with high mobile velocities due to the feedback latency. Furthermore, errors in signalling the feedback command reduce the asymptotic diversity order to 1, as further discussed in section 11.7. However, in case of feedback errors, closed-loop methods are still able to provide SNR gain, and it depends on the operation point of the system whether the loss of diversity is considered important.

- *Channel estimation in WCDMA.* Practical low-complexity closed-loop techniques require a dedicated pilot channel for coherent detection, while open-loop techniques can utilize common pilot channels only.

- *Intra-cell interference.* When closed-loop transmit diversity is applied, different intra-cell users assume different channels, and therefore single-user interference cancellation techniques in multipath downlink channels are not as effective as in SISO systems. On the other hand, when open-loop techniques and vector coding are applied in the transmitter, the receiver should be able to equalize several channels simultaneously, which is difficult+ in MISO systems.

- *Correlated channel environment.* Both open-loop and closed-loop techniques lose diversity gain, but closed-loop schemes can still provide SNR gain while the SNR gain of open-loop schemes is equal to one.

In addition, a MIMO channel may be exploited by transmitting parallel streams. The most promising use of such channels is a combination of parallel streams and Tx-diversity/beam-forming/closed loop methods. The reasons for this are the following

- Even with a ULA with half-wavelength spacing in rural macro cells, typical for beam-forming and smart antenna applications, the MIMO channel is rich enough to support multiple parallel streams when there are more than about 6 Tx antenna elements, and at least 2 Rx elements.

- Even in a completely uncorrelated MIMO channel, the smallest eigenvalue of the channel matrix is almost zero.

- In a square matrix MIMO channel, generalized beam-forming without power allocation does not increase capacity; it only reduces complexity of the receiver.

Part II

Open-loop Methods

3

Open-loop Concepts: Background

In open-loop transmit diversity multiple transmit antennas are used to convey information to the receiver. In contrast to closed-loop systems, studied later, here the transmitted signal is independent of the prevailing channel state. However, significant performance gains can be obtained even in the absence of channel state information in the transmitter. When in single-antenna modulation and coding schemes the signal matrix is defined in time, in multi-antenna concepts the transmitted signal is defined in both space and time dimensions. Thus, multiple antenna elements allow the expansion of the dimension of the signal space, and well-designed space–time modulation schemes utilize these new dimensions in an efficient manner. In the recently developed space–time coding concepts, multiple transmit antenna are used to increase the diversity order of the received signal.

In this chapter we review a number of open-loop transmit diversity solutions, and discuss their properties, applicability and applications in current and future wireless systems. Among other things, we review the principles of the transmit diversity solutions defined currently for WCDMA and cdma2000 systems.

3.1 DELAY DIVERSITY

With delay diversity [99, 100] the base station transmits delayed copies of the same signal from different transmit antennas. Without channel state information, the transmit power is divided evenly between the antennas. The usage of delay diversity is transparent to the receiver, which simply experiences a longer delay spread. Therefore delay diversity can be deployed without modifications to existing standards. Transmission schemes of this kind providing diversity benefits in a manner that is transparent to the receiver are called *implicit transmit diversity schemes*.

In order to generate two uncorrelated signal branches with two transmit antennas in a flat fading environment, the delay should be at least of the order of the symbol length T_s. In general, to obtain full diversity benefit from the delay diversity when $N_t = 2$, the transmit delay between the two antennas should be larger than the delay spread of the channel. Delay diversity can also be interpreted as a simple space–time trellis code [19], which provides only diversity gain, no coding gain. Furthermore, delay diversity can be combined with phase-hopping and frequency offset schemes.

The resulting multipath channel requires an equalizer or RAKE at the receiving end. Due to delay diversity there are more channel parameters to estimate than in a single antenna transmission, and therefore the additional diversity may not always be able to compensate for the performance loss caused by the compromised channel estimation. In practice, the receiver can only handle a certain number of different channel taps, and if the delay spread is already large enough, the receiver is not able to exploit the additional diversity at all. In a flat fading environment the WCDMA downlink has no intra-cell interference, due to the orthogonal channelization codes. Delay diversity would decrease the orthogonality factor of the WCDMA downlink, and therefore 3GPP standardization efforts have concentrated on other open-loop transmit diversity schemes.

3.2 IMPLICIT DIVERSITY VIA PHASE MODULATION

The simplest transmitter solutions offering implicit transmit diversity are based on phase modulation. In phase modulation solutions, the signals transmitted from different antennas are subjected to intentional frequency offset, frequency sweep or to phase-hopping. This increases the time-selectivity of the channel by reducing channel coherence time. Time-selectivity can be exploited by a channel code. In contrast with space–time coding, phase modulation-based solutions do not improve performance at symbol level. However, performance improvement can be significant when open-loop systems are concatenated with an outer code. We summarize the main approaches below.

Phase Sweep: Phase Sweep Transmit Diversity (PSTD) [23] is an efficient and simple linear processing transmit diversity method. Consider a multi-antenna transmitter with N_t antenna elements and a single-path channel. At time t antennas $n = 1, .., N_t$ transmit signals $\{\exp(j\,\delta(n-1)t)x[t]\}_{n=1}^{N_t}$ and the receiver obtains

$$y[t] = \frac{1}{\sqrt{N_t}} \sum_{n=1}^{N_t} h_n[t] \exp(j\,\delta(n-1)t)x[t] + n[t], \tag{3.1}$$

where $\delta > 0$ determines the phase sweep step size and h_n is the complex channel amplitude between transmit antenna n and the receiver antenna. The receiver experiences a linear combination of the channels, which changes in time to:

$$h[t] = \frac{1}{\sqrt{N_t}} \sum_{n=1}^{N_t} h_n[t] \exp(j\,\delta(n-1)t). \tag{3.2}$$

Frequency Offset: With PSTD a discretized model was assumed, with a time-varying antenna-specific phase modulation at symbol level. Alternatively, a continuous (antenna specific) frequency offset f_0 can be applied at each transmit antenna. In this case, the complex baseband representation of the signal transmitted from antenna n is

$$x[t] \exp(j\,2\pi f_o t(\frac{n}{N_t} - 1)),$$

where f_o determines the frequency offset.

Transmit diversity solutions applying continuous phase offset can be analysed straight-forwardly in a Rayleigh fading channel. In particular, the effect of frequency offset can be seen by examining the channel autocorrelation function. As an example, in a Rayleigh fading channel the channel autocorrelation function $R(\tau) = E(h[t]h[t - \tau]^*)$ with different numbers of transmit antennas is given by [101]

$$R(\tau) = 2\sigma_s^2 J_0(2\pi f_D \tau) \text{ if } N_t = 1 \qquad (3.3)$$

$$R(\tau) = \frac{2}{3}\sigma_s^2 J_0(2\pi f_D \tau)\cos(\pi f_0 \tau) \text{ if } N_t = 2 \qquad (3.4)$$

$$R(\tau) = \frac{2}{3}\sigma_s^2 J_0(2\pi f_D \tau)(1 + 2\cos(\pi f_0 \tau)) \text{ if } N_t = 3 \qquad (3.5)$$

$$R(\tau) = \frac{1}{4}\sigma_s^2 J_0(2\pi f_D \tau)(2\cos(\pi f_0 \tau) + 2\cos(\pi f_0 \tau/3)) \text{ if } N_t = 4 \qquad (3.6)$$

where J_0 is the Bessel function of order zero, f_D is the maximum Doppler spread for each channel, and f_0 the controllable frequency offset or sweep.

It is instructive to consider some special cases. If $f_0 = 0$ Hz the autocorrelation function is identical regardless of the number of transmit antennas. Thus, there is no performance benefit in using more than one transmit antenna. However, when frequency offset deviates from zero, the channel autocorrelation function has a different functional form. In particular, Figure 3.1 depicts the autocorrelation functions when $f_0 = 35$ Hz, assuming Doppler spread $f_D = 10$ Hz and $f_D = 50$ Hz. With one transmit antenna the channel is uncorrelated when $T_s = 1.25$ s. However, with two transmit antennas the first zero crossing occurs when $f_0 T_s$ is below 0.5. Hence, time-selectivity or time-diversity, measured loosely as the number of independently fading symbols in a given time interval, is increased. Essentially, when frequency offset is applied, a time-varying channel arises, even in stationary environments where $f_D = 0$.

Phase Hopping: In phase hopping the phase difference for consecutive symbols is pseudo random. Alternatively, it can follow some predetermined phasor sequence $\{\exp(j\delta_t)\}_{n=1}^{N_t}$ that is known to both the transmitter and the receiver. As a rule of thumb, with two transmit antennas, δ_t should sample $(0, 2\pi]$ uniformly such that the distance between $\exp(j\delta_t)$ and $\exp(j\delta_{t+1})$ is maximized. This reduces the channel correlations between successive information symbols, and potentially allows a reduction in latency due to bit-interleaving. As an example, δ_t can repeat the sequence $(0, \pi, \pi/2, 3\pi/2, \pi/4, 5\pi/4, 3\pi/4, 7\pi/4)$, hopping within different states in an 8-PSK constellation. In CDMA systems, with antenna specific pilots the effective channel can be constructed as in Equation (3.7).

Channel Estimation Issues: Despite the benefits of phase modulation diversity, there are also some disadvantages. The induced time-varying channel can exacerbate the channel estimation problem at the receiver. However, the acquired diversity gains often outweigh the loss induced by increased channel estimation errors, as analysed in [101]. In phase sweep transmit diversity, the receiver estimates a single rapidly time-varying impulse response or a channel coefficient. In one implementation, each of the N_t antennas can have an associated training sequence or a pilot channel. If these pilot signals are transmitted without the phase

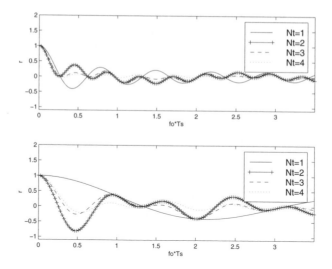

Fig. 3.1: Normalized channel autocorrelation function $R(\tau)$ for $f_D = 50$ Hz, $f_o = 35$ Hz (top) and $f_D = 10$ Hz, $f_o = 35$ Hz (bottom) with $N_t = 1, .., 4$ transmit antennas.

sweep, the receiver can obtain an estimate for $h_n[t], n = 1, ..., N_t$, and construct

$$\hat{h}[t] = \frac{1}{\sqrt{N_t}} \sum_{n=1}^{N_t} \hat{h}_n[t] \exp(\mathrm{j}\,\delta(n-1)t). \tag{3.7}$$

provided that $\exp(\mathrm{j}\,\delta(n-1)t)$ is known *a priori* in the receiver. However, the latter option is not transparent to the receiver and it needs to be standardized.

Comparison and Applicability: Phase sweep, with sufficiently small step size, is in practice transparent to the receiver. It is applicable e.g. in a CDMA system without explicit standardization, assuming that the step size is small enough to enable efficient channel estimation. Frequency offset is another transparent solution for CDMA systems. Frequency offset transmit diversity can be used as long as the offset remains within the implementation margins allowed in WCDMA or cdma2000 specification.

In TDMA systems, such as GSM, a random phase can be selected for different GSM bursts. In GSM the impulse response for each burst is estimated independently and therefore phase modulation has no effect on channel estimation performance.

In contrast to delay diversity, phase modulation concepts retain downlink orthogonality and impulse response length. Hence, phase modulation is simpler from the receiver viewpoint, and appropriate for both CDMA and TDMA systems. Clearly, the hopping sequences and sweeps should be tailored to the given service to attain maximal gains, as will be explained later. A unified framework for the linear filtering approach, proposed in [102], subsumes as a special case most of the aforementioned techniques, and many of those proposed by Wittneben [103]. Analysis and simulations in [104] provide additional insight into relative performance between TDTD, CDTD (described below), and phase sweep transmit diversity schemes.

3.3 CODE AND TIME DIVISION TRANSMIT DIVERSITY

In the most rudimentary transmit diversity method, each symbol is repeated N_t times and the repeated symbols are transmitted simultaneously via N_t antennas using quasi-orthogonal signalling channels. Quasi-orthogonal or orthogonal signalling channels can be constructed via different means. In Code Division Transmit Diversity (CDTD) orthogonal signalling channels are constructed using orthogonal channelization codes (e.g. Hadamard codes) [39]. In Frequency Division Transmit Diversity (FDTD) they are constructed using different transmit frequencies. Analogously, in Time Division Transmit Diversity (TDTD), different time intervals [23, 40] are used. If the same symbols are repeated in each signalling channel the aforementioned techniques reduce bandwidth efficiency by a factor of N_t.

Repetition coded transmit diversity via CDTD or TDTD is not attractive for 3G systems due to inherent bandwidth inefficiency. However, bandwidth efficiency in CDTD and TDTD can be restored by sending different symbols via different antennas. Here, the transmitted frame (sequence of symbols) is divided into N_t parts, and each signal part is transmitted from a different antenna using e.g. code division or time division multiplexing [103, 105, 106]. For example, in TDTD the received baseband signal after matched filtering is

$$y[t] = h_{t \ (\mathrm{mod} \ N_t)+1}[t]x[t] + n[t], \tag{3.8}$$

where $x[t]$ is the symbol transmitted at time t and $n[t]$ denotes channel noise. In a corresponding CDTD case, also known as Orthogonal Transmit Diversity (OTD) in the cdma2000 system [39], different (orthogonal) Walsh codes are used to spread the signal in each transmit antenna. CDTD and TDTD convert spatial diversity into time diversity. Performance is improved only if channel coding is able to exploit time diversity. Basically, this requires that the encoding rate is sufficiently low, inversely proportional to the number of transmit antennas [107]. The performance of TDTD and CDTD can be further improved by applying well-defined symbol constellations, as will seen in the following section.

3.4 DIVERSITY TRANSFORM *~ preceding*

Fading resistant modulation methods using diversity transforms, or symbol precoders that output rotated constellations [108–112] are effective in increasing system performance without compromising decoding delay or bandwidth efficiency. Fading resistant multidimensional constellations can be constructed via linear transforms at the transmitter, without channel side information. The transform matrix often needs to be manually designed for the given channel, or for a given input constellation. In fading resistant modulation the coordinates of a rotated multidimensional signal are distributed over multiple signal dimensions, defined e.g. in space, time or frequency. Ideally, the rotation is such that the original information stream can be retrieved unless all signal coordinates fade simultaneously. Orthogonal or unitary precoders maintain the Euclidean distances between the signal states at precoder input and output. Some related techniques, such as Trellis coded modulation [113], also increase the robustness to fading, but they tend to impose a performance loss in a Gaussian channel.

If a D-dimensional symbol constellation before the transform is $\mathbb{A}$, the set

$$\{\mathbb{A}_{tr} | \mathbf{U}\mathbf{x}, \mathbf{x} \in \mathbb{A}\}$$

defines the states of the transformed signal constellation. The transformation matrix $\mathbf{U}$, also called the precoding matrix, is typically a unitary (rotation) matrix. The target is to pin down $\mathbf{U}$ so that the rotated constellation $\mathbb{A}_{tr}$ is more immune to fading. The design criteria that apply here include (i) rank criteria, discussed in the following section, (ii) cutoff rate [109, 114, 115], and (iii) product distance.

Product distance is commonly used when developing full diversity modulation alphabets [111]. The minimum product distance of constellation $\mathbb{A}$ is given by

$$d_{min} = \min_{\boldsymbol{\Delta}=\mathbf{x}^{(c)}-\mathbf{x}^{(e)},\ \mathbf{x}^{(c)}\neq\mathbf{x}^{(e)}\in\mathbb{A}} \prod_{i=1}^{D} |\Delta_i| \tag{3.9}$$

Here $\mathbf{x}^{(c)}$ is the transmitted symbol vector, and $\mathbf{x}^{(e)}$ is an erroneously detected symbol vector. The symbol difference vector is $\boldsymbol{\Delta}$. The constellation rotation $\mathbf{U}$ can be optimized by maximizing the minimum product distance:

$$\mathbf{U} = \arg\max_{\mathbf{U}} \min_{\boldsymbol{\Delta}=\mathbf{x}^{(c)}-\mathbf{x}^{(e)},\ \mathbf{x}^{(c)}\neq\mathbf{x}^{(e)}\in\mathbb{A}} \prod_{i=1}^{D} |(\mathbf{U}\boldsymbol{\Delta})_i| \tag{3.10}$$

where $\mathbf{U}$ is a unitary (rotation) matrix. In addition, it is desirable to minimize the number of constellation vectors that meet the minimum product distance.

It is apparent that in a conventional D-dimensional QPSK constellation the minimum product distance is zero, since all coordinate constellations are identical. The purpose of constellation rotation is to provide modulation diversity, or signal space diversity [111], such that the coordinate constellations are different.

Example for $N_t = 2$: In order to appreciate the benefits of symbol precoding for spatial diversity consider the following illustrative example in the spirit of [111, 115]. The presentation below differs from ones in the literature in that the method is described explicitly in terms of the received signal model and the ensuing correlations.

We assume that 2 QPSK symbols are transmitted orthogonally over a fading channel corrupted by Gaussian noise, during two symbol epochs. The rotation matrix is of dimension 2×2. The QPSK vector with two complex dimensions is precoded, and the coordinates of the rotated symbol vector at the output of the precoder are transmitted via two orthogonal signalling channels. Orthogonal signalling is implemented e.g. with CDTD, such that the first complex coordinate is spread with code $\mathbf{s}_1$ transmitted from antenna 1, whereas the second coordinate is spread with code $\mathbf{s}_2$ and transmitted from antenna 2. Hence, we use a code matrix $\mathbf{S} = [\mathbf{s}_1 \ \mathbf{s}_2]$.

In a flat fading channel the most transparent received signal model is given by

$$\mathbf{y} = \mathbf{S}\mathbf{H}\mathbf{U}\mathbf{x} + \mathbf{n} ,$$

where the diagonal elements of $\mathbf{H} = \mathrm{diag}(h_1, h_2)$, designate the complex channel amplitudes between transmit antennas 1 and 2 and the receive antenna, $\mathbf{x}$ is a 2-dimensional QPSK symbol vector, and $\mathbf{n}$ is Gaussian noise. For comparison with other linear space–time modulation schemes, the signal model may be written according to Equation (1.1) as

$$\mathbf{y} = \mathbf{X}\mathbf{h} + \mathbf{n} , \tag{3.11}$$

where $\mathbf{h} = [h_1 \ h_2]^T$ is the channel vector. Denoting the rotated symbols by $\tilde{\mathbf{x}} = \mathbf{U}\mathbf{x}$, the matrix modulation is

$$\mathbf{X} = \mathbf{S} \begin{bmatrix} \tilde{x}_1 & 0 \\ 0 & \tilde{x}_2 \end{bmatrix}. \tag{3.12}$$

The symbol precoding matrix $\mathbf{U}$ assumes the following generic form

$$\mathbf{U}(\mu, \nu) = \begin{bmatrix} \mu & \nu \\ -\nu^* & \mu^* \end{bmatrix}, \tag{3.13}$$

with power constraint $|\mu|^2 + |\nu|^2 = 1$. It is a unitary 2×2 matrix with unit determinant. The received signal after matched filtering is

$$\mathbf{z} = \mathbf{U}^\dagger \mathbf{H}^\dagger \mathbf{S}^\dagger \mathbf{S} \mathbf{H} \mathbf{U} \mathbf{x} + \mathbf{n}. \tag{3.14}$$

Clearly, with arbitrary orthonormal signalling channels, $\mathbf{S}^\dagger \mathbf{S} = \mathbf{I}_2$. The effect of the transform is manifested in the correlation matrix

$$\mathbf{U}^\dagger \mathbf{H}^\dagger \mathbf{H} \mathbf{U} = \begin{bmatrix} a_1 & b \\ b^* & a_2 \end{bmatrix} \tag{3.15}$$

where

$$a_1 = |h_1|^2|\mu|^2 + |h_2|^2|\nu|^2 \tag{3.16}$$

$$a_2 = |h_2|^2|\mu|^2 + |h_1|^2|\nu|^2 \tag{3.17}$$

$$b = (|h_2|^2 - |h_1|^2)\mu^*\nu. \tag{3.18}$$

Note that the code is orthogonal if $\nu = 0$ or $\mu = 0$, or if $|h_1| = |h_2|$. If $|\mu| = |\nu|$ the diagonal elements of the code correlation matrix are all identical, and the energy of both symbols is distributed evenly over the two antennas. Intuitively it is clear that this provides the best diversity. This feature will be discussed at length in Chapter 4 as the concept of Maximal Symbolwise Diversity (MSD).

With Δ_k the symbol errors in the two original QPSK symbols, the product distance is

$$d(\mu, \nu, \mathbf{\Delta}) = \left| \nu\mu^* \Delta_2^2 - \mu\nu^* \Delta_1^2 + \left(|\mu|^2 - |\nu|^2 \right) \Delta_1 \Delta_2 \right|. \tag{3.19}$$

With the MSD choice $|\mu| = |\nu|$, there is one essential parameter left in choosing $\mathbf{U}$, namely the phase difference of μ and ν. Denoting this by ϕ the product distance becomes

$$d(\phi, \mathbf{\Delta}) = \tfrac{1}{2} \left| e^{j\phi} \Delta_2^2 - e^{-j\phi} \Delta_1^2 \right| = \tfrac{1}{2} \left| \Delta_1^2 - \left(e^{j\phi} \Delta_2 \right)^2 \right|. \tag{3.20}$$

Thus the phase difference should be chosen so that the distance between squared differences in a QPSK constellation and a rotated QPSK constellation is maximized with this angle. For QPSK symbols, the possible non-trivial symbol differences are $\bar{A} = \sqrt{2}\{\pm 1, \pm j, \pm 1 \pm j\}$. The minimum product distance as a function of ϕ is plotted in Figure 3.2. The problem of choosing the optimal ϕ will be encountered in exactly the same form in a different setting in section 7.3.1. More details, including solutions for closed form optima can be found there. Any ϕ in the interval $(0.15\pi, 0.35\pi)$ provides ample diversity protection. The rotation $\phi = \pi/4$, although not the ultimate optimum, is easy to implement, because the resulting constellations is similar to 8-PSK modulation alphabet.

Figures 3.3 and 3.4 show all possible signal states and related signal labels for the first and second elements of $\tilde{x}$, when $\mu = 1/\sqrt{2}$ and $\nu = \exp(j\pi/4)/\sqrt{2}$. Note that to maintain orthogonal signalling, two symbol epochs are needed to transmit 4 bits. Therefore, the spectral efficiency remains at 2 bps/Hz, which is equivalent to the original QPSK constellation.

It is apparent from Figures 3.3 and 3.4 that a properly selected rotation extends each symbol over multiple transmission intervals or more generally over multiple dimensions defined in time, space, code or frequency domain. As a result, each bit can be recovered unless channel states in both dimensions vanish simultaneously due to fading. On the other

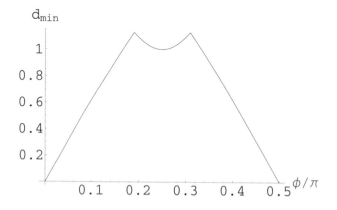

Fig. 3.2: Product distance for diversity transform (Equation (3.13)) as a function of $\phi = \arg \nu/\mu$, for $|\mu| = |\nu|$.

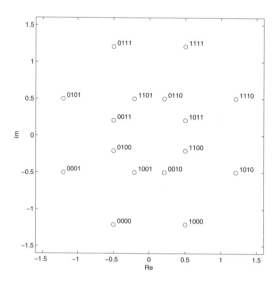

Fig. 3.3: Symbol alphabet and symbol labels for the first dimension of a two-dimensional rotated constellation.

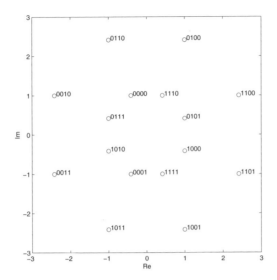

Fig. 3.4: Symbol alphabet and symbol labels for the second dimension of a two-dimensional rotated constellation.

hand, if both symbols are sampled from QPSK constellation, both channel realizations need to be of sufficient quality to recover all four bits. Here, symbol precoding improves the diversity order at symbol level, and performance gain is not dependent on some well-designed outer code. Hence diversity transforms fit in well within the modular system design principle advocated in the introduction.

In the previous example the code matrix was two-dimensional. In the presence of more than two transmit antennas, full modulation diversity cannot be achieved, unless the dimension of the precoding matrix is also increased. The optimal precoding matrix is difficult to find when the matrix size is increased. Often the matrix is found by a numerical search over some parameterized family of rotation matrices [109, 111].

The diversity transform example above provided us with first insights to many concepts discussed more thoroughly in coming chapters. Thus we noted that some concepts relating to explicit channel realizations are easier to understand using a correlation matrix description, whereas others, like explicit rotations, are better discussed with the channels integrated out, using e.g. the product distance. In Chapter 4, extensive use will be made of diversity transforms to improve high SNR performance of linear space–time modulation schemes.

3.5 SPACE–TIME CODING

Following a well-known classification in coding theory, space–time coding approaches have been divided into space–time trellis codes [19] and space–time block codes [10, 18].

3.5.1 Space–Time Trellis Codes

The concept of space–time trellis coding was introduced by Tarokh, Seshadri and Calderbank in their seminal work [19]. space–time trellis coding extends the Ungerböck paradigm to the spatial domain. Instead of set-partitioning large modulation constellations, STTCs set-partition various spatial transmission constellations, i.e. constrain the vector modulation transmitted at a given time using a trellis structure.

As an example, consider the space–time trellis in Figure 3.5. This trellis operates on a vector modulation consisting of two QPSK symbols. At each symbol period, two bits are transmitted. These two bits select the state transition. The QPSK symbol transmitted from antenna 2 is chosen according to the previous state, and the QPSK symbol transmitted from antenna 1 is chosen according to the new state. The selection constrains the *a priori* 16 alternatives of a vector modulation to four alternative combinations during each symbol period.

State 1: 00 01 02 03

State 2: 10 11 12 13

State 3: 20 21 22 23

State 4: 30 31 32 33

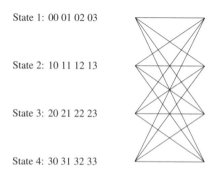

Fig. 3.5: A fully connected four-state QPSK STTC considered in [19].

Space–time codes differ from the diversity schemes discussed above, in that in addition to diversity, STTCs provide coding gain by increasing the minimum distance between code sequences. Thus they aim at joint design of coding and modulation in the MIMO/MISO transmission chain depicted in Figure 1.1.

Unlike rate 1 (2 bps/Hz) orthogonal schemes, rate 1 space–time trellis codes are able to reach full diversity when the number of transmit antennas is larger than two, but their decoding is more difficult than that of orthogonal schemes, requiring a trellis search. The trellis complexity of a full diversity code is at least M^{N_t-1}, where M and N_t refer to the cardinality of the modulation alphabet, and the number of transmit antennas, respectively. Criteria for designing full diversity space–time codes employing PSK modulations have been investigated in [116].

Space–Time Trellis Codes with Space–Time Block Code Alphabet: space–time trellis codes seek to minimize pairwise error probabilities between coded symbol sequences at high SNR. In a very interesting recent development, concatenating space–time trellis and block codes is considered [117–119]. This is done in the true Ungerböck spirit. The modulation alphabets consist of sets of equivalent orthogonal space–time block codes instead of vector modulation with QAM or PSK symbols as in conventional STTC. More exactly, a trellis code selects among the set of orthogonal block codes, and uncoded bits are transmitted with the selected space–time block code. These schemes have been called "coset" or "super-orthogonal" space–time codes.

Concatenated Schemes: Concatenating STTC with an outer code is discussed in [19]. Trellis decoding causes error bursts, and it was suggested that concatenating with a Reed–Solomon outer code is suitable for correcting the burst errors. This approach is referred to as smart–greedy codes. For using STTC as a space–time module in e.g. WCDMA, concatenating with an outer convolutional or turbo code should be discussed. The error bursts deteriorate the performance of an outer convolutional code [120]. WCDMA provides several different services and quality classes so that flexible generation of different data rates is important. Therefore, introducing space–time codes to WCDMA is a challenging task requiring a redesign of error-correcting codes.

Another strategy to increase the minimum distance of the code symbols is to introduce multidimensional modulation symbols [121], and let the trellis operate on these. Additional dimensions may be provided by, for example, sending a code symbol at multiple time instants, decreasing the rate of the space–time code.

Space–Time and the Turbo Principle: The first suggestions for space–time turbo coding [122, 123] multiplexed directly the output of turbo decoders to a vector modulation. In an alternative development [124], space–time block codes were concatenated with an outer turbo trellis-coded modulation or a binary turbo code. More involved space–time turbo codes have been discussed in [125–127]. In these works, the recursivity of the constituent STTCs were stressed. In addition, the importance of puncturing the constituent STTCs, and applying bit-interleaving, were stressed in [126]. Recently, in [128] space–time trellis codes and low-density parity-check codes were concatenated, claiming that this facilitates using different data rates.

3.5.2 Space–Time Code Design Criteria

In [19, 129, 130], a number of design criteria for optimizing space–time code performance in slowly fading channels were developed. These criteria apply irrespective of the underlying space–time coding philosophy, be it based on trellises, block matrices or some non-linear scheme.

A collection of bits transmitted in a space–time code matrix constitutes a codeword, characterized by the transmission matrix $\mathbf{X}$. The set of all codewords is $\{\mathbf{X}\}$. If there are Q symbols, and each symbol is in a Mary modulation (each symbol encodes $\log_2 M$ bits), the maximum cardinality of the set of codewords is $|\{\mathbf{X}\}| = M^Q$. The total error probability P_c for a transmission of a given space–time codeword $\mathbf{X}^{(c)}$ is upper-bounded by the union bound, i.e. the sum over all codewords $\mathbf{X}^{(e)} \in \{\mathbf{X}\}$ of the pairwise error probabilities $P_{c \mapsto e}$:

$$P_c \leq P_{\mathrm{UB}c} = \sum_{\mathbf{X}^{(e)} \in \{\mathbf{X}\}} P_{c \mapsto e} \, . \tag{3.21}$$

The average error rate assuming equiprobable codewords is

$$P_{\mathrm{ave}} = 1/M^Q \sum_{X^{(c)} \in \{\mathbf{X}\}} P_c \, . \tag{3.22}$$

A crucial assumption enabling space–time performance analysis is that the union bound is tight, so that properties of the average error rate can be deduced from the pairwise error probabilities $P_{c \mapsto e}$. This restricts the validity of all space–time code performance criteria to a loosely defined region of "high" signal-to-noise ratio (SNR).

In quasi-static i.i.d. Rayleigh fading, minimizing the pairwise error probability of deciding in favour of an erroneous space–time transmission matrix $\mathbf{X}^{(e)}$ when transmitting $\mathbf{X}^{(c)}$ leads to the well-known rank [129] and determinant [19] criteria. In addition, in [130] the trace criterion was suggested, which yields heuristic insight into the structure of non-orthogonal space–time block codes. For these criteria it is important that the coherence time of the channel is longer than the space–time code block. They are expressed in terms of properties of the codeword difference matrix

$$\mathbf{D}^{(ce)} = \mathbf{X}^{(c)} - \mathbf{X}^{(e)} , \tag{3.23}$$

or its Hermitian square, the distance matrix. The design criteria are based on a Chernoff bound of the pairwise error probabilities:

$$P_{c \mapsto e} \leq \det \left(\mathbf{I}_{N_t} + \rho \mathbf{D}^{(ce)\,\dagger} \mathbf{D}^{(ce)} \right)^{-N_r} . \tag{3.24}$$

This bound should be minimized for optimal performance. At high SNR ρ, this leads to the following criteria:

- *The rank criterion* [129]: The transmit diversity of a multiple transmitter scheme is

$$\min_{e \neq c} \text{Rank}[\mathbf{D}^{(ce)}] \leq \min[T, N_t] . \tag{3.25}$$

  To achieve maximal diversity, $\mathbf{D}^{(ce)}$ should have full rank for all non-vanishing code word pairs.

- *The determinant criterion (MAX-MIN-DET)* [19]: To optimize performance in a (Rayleigh) fading environment, the code should be designed to maximize

$$\min_{e \neq c} \det' \left[\mathbf{D}^{(ce)\,\dagger} \mathbf{D}^{(ce)} \right] . \tag{3.26}$$

  The prime in the determinant indicates that zero eigenvalues should be left out from the product of eigenvalues when computing the determinant. The determinant criterion is a generalization of the product distance in Equation (3.9) to the space–time domain. Indeed, when diversity transforms are applied in the spatial direction, as in section 3.4, the product distance (3.19) is nothing but the square root of the MAX-MIN-DET of the corresponding space–time transmission matrix (3.12).

Apart from these, a trace (Euclidean distance) criterion may be derived from the determinant criterion [130, 131]. Following [130], $\text{Tr}\left[\mathbf{D}^{(ce)\,\dagger}\mathbf{D}^{(ce)}\right]$ plays the role of Euclidean distance between codeword pairs, and it should be maximized. Moreover, to optimize performance in Rayleigh fading, $\mathbf{X}$ should be designed so that the eigenvalues of $\mathbf{D}^{(ce)\,\dagger}\mathbf{D}^{(ce)}$ are as close as possible to each other and to $\text{Tr}\left[\mathbf{D}^{(ce)\,\dagger}\mathbf{D}^{(ce)}\right]/N$. Also the row-wise sum of the absolute values of the off-diagonal elements should be minimized. This criterion is a consequence of the determinant criterion and Hadamard's determinant inequality (A.14).

For linear schemes, the transmission matrix is a function of symbols, and the codeword difference matrix simplifies to

$$\mathbf{D}^{(ce)} = \mathbf{X}\left(\mathbf{x}^{(c)}\right) - \mathbf{X}\left(\mathbf{x}^{(e)}\right) = \mathbf{X}\left(\mathbf{x}^{(c)} - \mathbf{x}^{(e)}\right). \tag{3.27}$$

This leads to a number of design criteria for optimizing linear space–time codes.

In [132], it was argued that the total self-interference power (non-orthogonality) should be minimized, and that each modulation matrix should be unitary (maximal symbolwise diversity). In [133], maximizing mutual information was considered. This is a very natural requirement, given the bit-interleaved coded space–time modulation paradigm, i.e. when relying on an effective outer code to exploit the information provided by the space–time modulation. Further, in [134], the tight frame criterion was proposed. In [51, 135], the additional criteria of Frobenius orthogonality (traceless self-interference) and symbol homogeneity were suggested. These criteria can be heuristically understood from the trace criterion [51]. Below, these criteria will be derived, and their interrelations discussed. Some criteria can be seen as steps to maximizing the mutual information, while some are directly performance optimization criteria.

3.5.3 Space–Time Trellis Versus Space–Time Block Coding

A characteristic difference between space–time trellis codes and space–time block codes is that typical space–time block codes are linear codes over the field of complex numbers. The complex numbers in question are symbols in some two-dimensional modulation alphabet with M constellation points. The only non-linearity is the possible non-linearity of the mapping of $\log_2 M$ bits to the symbols, if higher-order modulations ($M > 4$) are used. As functions of the encoded bits, space–time trellis codes may be linear codes over some finite field, but they are not linear over complex numbers. A more thorough discussion of space–time block codes, and more general linear space–time modulation schemes, is postponed to the following sections. Here, we shall concentrate on the fundamental differences between space–time trellis and block coding principles.

Operation Point: From a performance point of view, a significant difference between space–time block and trellis codes is in their performance optimization and typical operation point. For interference limited systems, like WCDMA and possible future wideband systems, lowering the operation point is of primary interest. Thus for a typical WCDMA link, the bit energy per noise + interference power is below 5 dB. space–time trellis codes are designed with a high SNR performance in mind. In this book we shall see how space–time block codes and more generic matrix modulation schemes may be designed for optimum performance at low and medium SNR. In this regime, full diversity, although improving performance, is only of secondary interest.

Two Tx Antenna Schemes: The bulk of space–time coding work has concentrated on the case with $N_t = 2$ transmit antennas and $N_r = 1$ receive antenna. Optimal space–time trellis codes have been discussed in e.g. [136]. The optimal space–time block code was presented by Alamouti [10]. One example of the Alamouti code encodes the $Q = 2$ complex symbols x_1, x_2 to be transmitted during $T = 2$ symbol periods, in the form

$$\mathbf{X}_{\text{Ala}}(x_1, x_2) = \begin{bmatrix} x_1 & x_2 \\ -x_2^* & x_1^* \end{bmatrix}. \tag{3.28}$$

For transmission according to signal model (1.1), each row is transmitted during the same symbol period from a different antenna. For reliable detection of (3.28), the channel should be constant at least during the $T = 2$ symbol periods that (3.28) is transmitted over. This code is the open-loop transmit diversity mode of WCDMA, and it is known under the acronym STTD (Space–Time Transmit Diversity).[1]

Concatenated vs. Joint, Complexity vs. Performance: It has been shown that the Alamouti code received with a single receive antenna achieves capacity without channel state information in the transmitter [37]. Furthermore, linear detection implies compatibility with concatenated outer codes, leading to the conclusion that there is not much room for joint coding and modulation design of the STTC type for $N_t = 2, N_r = 1$.

For two transmit antennas, a complexity vs. performance tradeoff comparison between space–time trellis codes and space–time block codes concatenated with binary codes has been performed in [137]. It was found that if (3.28) is concatenated with trellis-coded modulation, the performance is systematically better than for a space–time trellis code with the same number of states. This observation has been further confirmed in [138] which simulates (3.28) with several channel codes, binary and non-binary, and concludes that the combination of the Alamouti code and forward error correcting code outperforms other combinations. Whether this property extends to multiple Tx antennas employing possibly non-orthogonal space–time modulation remains to be seen.

ISI Channel: To retain orthogonality, space–time block codes require a channel with negligible inter-symbol interference, although this can be alleviated by applying the space–time codes to blocks instead of symbols [47] on the condition that the channel can be considered a block fading one. On the other hand, space–time trellis codes can combine channel equalization and decoding, which makes them suitable to channels subject to inter-symbol interference.

Since there is no general agreement on 4G air interface, space–time trellis codes should not be overlooked as a study item for future cellular communication systems. However, in case of OFDM the simulations in [138] show that the performance of Alamouti code with forward error correcting coding is superior to the smart–greedy codes of [19].

3.6   SPACE–TIME BLOCK CODES

For space–time block codes, the essential criteria are the provided transmit (Tx) diversity, the (symbol) rate R_s of the code, and the delay. The degree of Tx diversity is characterized by the number of independently detectable channels that each symbol is transmitted over. For full diversity it equals the number of transmit antennas. If multiple receive (Rx) antennas are deployed, the total diversity degree is the product of the Tx and Rx diversity degrees. The (symbol) rate of the code is the number of symbols transmitted by per symbol period. The delay is the length of the space–time code block T. Typically the goal for designing space–time block codes is to maximize rate and minimize delay, keeping full diversity.

---

[1]To be exact, a transposed version (1.3) of (3.28) is accepted by the WCDMA standard. A transposition does not change the properties of the scheme.

3.6.1 Two Tx Antennas: STTD

For $N_r = 1$ receive antenna, the optimal space–time block code for two transmit antennas is STTD (3.28), with symbol rate $R_s = Q/T = 1$. Optimality is seen in many different ways. In flat fading, the received signal is

$$\mathbf{y} = \tfrac{1}{\sqrt{2}} \mathbf{Xh} + \text{noise} = \tfrac{1}{\sqrt{2}} \begin{bmatrix} x_1 h_1 + x_2 h_2 \\ x_1^* h_2 - x_2^* h_1 \end{bmatrix} + \text{noise} . \tag{3.29}$$

Here, the $\mathbf{X}$ is normalized by $1/\sqrt{2}$ so that the transmit power is $\operatorname{Tr} \mathbf{X^\dagger X} = |x_1|^2 + |x_2|^2$, i.e. the same power is used for STTD transmission as would be used if the symbols were independently transmitted from one antenna. Conjugating the received signal during the second symbol period, the received signal may be written in terms of an equivalent signal model

$$\begin{bmatrix} y_1 \\ y_2^* \end{bmatrix} = \mathcal{H} \begin{bmatrix} x_1 \\ x_2 \end{bmatrix} + \text{noise} , \tag{3.30}$$

where the "equivalent channel matrix" is

$$\mathcal{H} = \tfrac{1}{\sqrt{2}} \begin{bmatrix} h_1 & h_2 \\ h_2^* & -h_1^* \end{bmatrix} . \tag{3.31}$$

Now space–time matched filtering of (3.28) proceeds simply by applying the Hermitian conjugate of the equivalent channel matrix on the received signal (3.30):

$$\mathcal{H}^\dagger \begin{bmatrix} y_1 \\ y_2^* \end{bmatrix} == \tfrac{1}{2} (|h_1|^2 + |h_2|^2) \begin{bmatrix} x_1 \\ x_2 \end{bmatrix} + \text{noise} . \tag{3.32}$$

The matched filter gave a result where both symbols have been transmitted at half-power over both channels, and have been maximum ratio combined (MRC) at the receiver. This is a consequence of the fact that the equivalent channel matrix is proportional to a unitary matrix. Compared to MRC at the receiver, with two receive antennas, there is a 3 dB loss due to splitting the transmit power in two between the antennas.

With one receive antenna STTD is an optimal linear open-loop scheme, assuming that the receiver has sufficient channel state information. It provides full diversity, with linear matched filter detection. Also, it reaches channel capacity [37, 133].

3.6.2 More Than Two Tx Antennas

It is likely that the number of antennas at a base station will be extended to $N_t > 2$. Typically orthogonally polarized antennas provide independently fading channels. Thus two separate antennas, each with two polarizations, already counts as $N_t = 4$ transmit antennas. This is conceivable in future wireless systems and, as discussed in Chapter 1, is currently a work item in WCDMA standardization. Further in the future, larger numbers of Tx antennas are possible as well.

STTD is the optimal space–time block code for $N_t = 2$. Constructing optimal schemes for $N_t > 2$, however, is a non-trivial task. In [18], space–time block coding was extended to more than two transmit antennas. The underlying characteristic of (3.28), explaining its excellent performance, was found to be the unitarity of the code matrix. For STTD, this is simply expressed as

$$\mathbf{X^\dagger X} = \tfrac{1}{2} \left(|x_1|^2 + |x_2|^2 \right) \mathbf{I}_2 , \tag{3.33}$$

where $\mathbf{I}_2$ is the two-dimensional identity matrix. Orthogonality/unitarity leads to the theory of orthogonal designs, for real or complex modulation symbols. These have a simple linear detection scheme with optimal MRC performance, generalizing (3.32).

Based on the principle of orthogonality/unitarity, the problem of designing rate 1, full diversity space–time block codes was solved in [18].

Real Modulation: Rate $R_s = 1$ real codes were found for any number of Tx antennas. For up to 8 antennas, these were constructed from orthogonal designs in 2, 4 and 8 dimensions. An orthogonal design is an orthogonal matrix with entries $\pm c_j$, where c_j are real symbols. For $R_s = 1$, the number of symbols equals the matrix dimension. For example, the four-antenna real block code constructed in [18] is based on the orthogonal design

$$\mathbf{X}(c_1, c_2, c_3, c_4) = \begin{bmatrix} c_1 & c_2 & c_3 & c_4 \\ -c_2 & c_1 & -c_4 & c_3 \\ -c_3 & c_4 & c_1 & -c_2 \\ -c_4 & -c_3 & c_2 & c_1 \end{bmatrix} \tag{3.34}$$

Here c_j are real valued modulation symbols, as opposed to complex valued symbols x_j.

If the number of Tx antennas N_t is not equal to the orthogonal design dimensions 2, 4 or 8, delay optimal codes for $N_t < 8$ can be constructed by deleting a column from a higher-dimensional orthogonal design, i.e. by switching off antennas. Thus, for three transmit antennas, the optimal code can be written by deleting a column from the 4×4 matrix above, and optimal 5, 6, or 7-antenna schemes can be constructed from an 8×8 orthogonal design. For more than 8 Tx antennas, generalized orthogonal designs were constructed in [18] that have $R_s = 1$ for any number of Tx antennas. For these, the delay grows exponentially with N_t:

$$T_{\min} = 16^{\lfloor (N_t - 1)/8 \rfloor} 2^{\lceil \log_2(1 + (N_t - 1) \bmod 8) \rceil} . \tag{3.35}$$

Here $\lfloor r \rfloor$ indicates the greatest integer $\leq r$, and $\lceil r \rceil$ the smallest integer $\geq r$. For 9 antennas, the delay is 16, for 10 antennas 32, and for 11 to 16 antennas 64.

Complex Modulation: Allowing complex signal constellations severely restricts the number of Tx antennas for which $R_s = 1$ codes exist. Indeed, $R_s = 1$ complex modulation space–time block codes exist only for two transmit antennas [18]. This code is the STTD code of (3.28).

The series of $R_s = 1$ real modulation schemes was extended in [18] to a series of $R_s = 1/2$ complex modulation schemes. In these schemes, first a $R_s = 1$ orthogonal design is transmitted with complex symbols instead of real, followed by transmitting the same orthogonal design with the symbols complex-conjugated. This halves the rate and doubles the delay. For example, the rate 1/2 code for four transmit antennas, corresponding to the rate 1 real code for 4 antennas (3.34), is

$$\mathbf{X}(x_1, x_2, x_3, x_4) = \begin{bmatrix} x_1 & x_2 & x_3 & x_4 \\ -x_2 & x_1 & -x_4 & x_3 \\ -x_3 & x_4 & x_1 & -x_2 \\ -x_4 & -x_3 & x_2 & x_1 \\ x_1^* & x_2^* & x_3^* & x_4^* \\ -x_2^* & x_1^* & -x_4^* & x_3^* \\ -x_3^* & x_4^* & x_1^* & -x_2^* \\ -x_4^* & -x_3^* & x_2^* & x_1^* \end{bmatrix} . \tag{3.36}$$

Similarly, one would get a $R_s = 1/2$ complex modulation scheme for $N_t = 8$ Tx antennas with delay $T = 16$ from the corresponding 8×8 orthogonal design. Such $R_s = 1/2$ schemes can be constructed for any number of Tx antennas, with an exponentially growing delay which is twice (3.35).

Apart from the rate-halving schemes discussed above, a $R_s = 3/4$ square matrix scheme for four transmit antennas was presented in [18]

$$
\mathbf{X} =
\begin{bmatrix}
x_1 & x_2 & \frac{1}{\sqrt{2}} x_3 & \frac{1}{\sqrt{2}} x_3 \\
-x_2^* & x_1^* & \frac{1}{\sqrt{2}} x_3 & -\frac{1}{\sqrt{2}} x_3 \\
\frac{1}{\sqrt{2}} x_3^* & \frac{1}{\sqrt{2}} x_3^* & -\operatorname{Re}[x_1] + j \operatorname{Im}[x_2] & j \operatorname{Im}[x_1] - \operatorname{Re}[x_2] \\
\frac{1}{\sqrt{2}} x_3^* & -\frac{1}{\sqrt{2}} x_3^* & j \operatorname{Im}[x_1] + \operatorname{Re}[x_2] & -\operatorname{Re}[x_1] - j \operatorname{Im}[x_2]
\end{bmatrix}
\tag{3.37}
$$

A three-antenna scheme may again be constructed from (3.37) by deleting one column. In this scheme, a disadvantage of square matrix space–time block codes with rate less than one becomes visible. Even for equal power constellations (M-PSK), they are power-unbalanced in that the power transmitted from any given antenna fluctuates in time. For non-equal power constellations (M-QAM etc.), this property of the schemes makes power fluctuations worse. These (amplified) power fluctuations may be problematic in view of power-amplifier design, as the region of linear amplification has correspondingly to be extended.

The theory of square matrix space–time block codes was completed in [139–141]. These codes have minimum delay, but restricted rate. In [141], the principles of linearity as functions over complex numbers, and unitarity (3.33) were investigated, in the setting of square matrices with complex entries. It was found that the maximum achievable rates fall off inversely with the number of Tx antennas:

$$
R_{s\ \max} = \frac{Q_{\max}}{T} = \frac{\lceil \log_2 N_t \rceil + 1}{2 \lceil \log_2 N_t \rceil} .
\tag{3.38}
$$

This is in contrast to rate-halving codes, for which R_s is fixed at 1/2, and the minimal delay increases exponentially.

Minimal delays/maximal rates for rate-halving and square matrix orthogonal designs for some N_t can be found in Table 3.1. For 3–8 Tx antennas, removing antennas from a square matrix design is preferable. The existence of generalized orthogonal designs (rectangular designs that are not restrictions of a square design) with rate $> 1/2$ for generic N_t is still an open problem.

The simplest forms of square matrix space–time block codes are complex orthogonal designs, i.e. square matrices with elements either $\pm$ a symbol or its complex conjugate, or 0. STTD (3.28) is an orthogonal design. For $N_t = 4$ Tx antennas, an example of an orthogonal design corresponding to (3.37) is [139, 140]

$$
\mathbf{X}_{3/4}(x_1, x_2, x_3) =
\begin{bmatrix}
x_1 & x_2 & x_3 & 0 \\
-x_2^* & x_1^* & 0 & -x_3 \\
-x_3^* & 0 & x_1^* & x_2 \\
0 & x_3^* & -x_2^* & x_1
\end{bmatrix}
\tag{3.39}
$$

It should be noted that the design above has two versions of the Alamouti code on the block diagonal. Compared to (3.37), this is a simpler form. This scheme will be used as a starting point when constructing layered non-orthogonal schemes in the coming chapters.

Table 3.1: Rates and delays for full-diversity linear space–time block codes with complex symbols.

Scheme	N_t	T	Q	$R_s = Q/T$
Rate-halving:	2	4	2	1/2
	3-4	8	4	1/2
	5-8	16	8	1/2
	9	32	16	1/2
	10	64	32	1/2
	11-16	128	64	1/2
Square matrix:	1	1	1	1
	2	2	2	1
	3 to 4	4	3	3/4
	5 to 8	8	4	1/2
	9 to 16	16	5	5/16
	$2^{K-2}+1$ to 2^{K-1}	2^{K-1}	K	$K/2^{K-1}$

STBCs and Link Capacity: In [37, 133] it was shown that complex orthogonal space–time block codes are capacity suboptimal if more than one Rx antenna is used. Also, for one Rx antenna, they are capacity suboptimal if they have rate < 1. The only capacity achieving orthogonal design is STTD, when received with one antenna. With increasing N_r and/or N_t, orthogonal space–time block codes become increasingly capacity suboptimal. In the light of the result of [96] that unconstrained vector modulation reaches capacity for $N_r = 1$, but other diversity schemes (delay diversity, orthogonal transmit diversity) are capacity-suboptimal, it is remarkable that (3.28) reaches capacity even with one Rx antenna.

3.6.3 Space–Time Block Coding Terminology

The characteristics of a space–time block code are compared to the single antenna transmission of the same symbol stream. The ratio of the transmission rate of the block code to this "uncoded" scheme, $R_s = Q/T$, is called the (symbol) rate of the block code. It should be noted that this rate is a relative modulation rate; as observed in [18], for an orthogonal scheme, a rate less than one does not imply any coding, nor does it imply any increase in Euclidean distance. It is just a measure of the (in)efficiency of the use of the antenna resource. In the next chapters we shall see that for non-orthogonal linear schemes too, there is no coding gain. Thus the use of the word "code" is misleading when discussing space–time block coding.

When it comes to linear schemes with a rate above the maximum allowed by orthogonality, the number of various space–time block coding concepts is daunting, and numerous appellations have been used for schemes with different properties. These include "non-orthogonal" and "layered" space-time block codes" [132], "quasi-orthogonal space-time block codes" [142], "linear dispersion codes" [133], "multi-stratum space-time codes" [143], "Khatri–Rao space-time codes" [53], "diagonal algebraic" [52] and "threaded algebraic space-time codes" [144]. In this book, we adopt the term "matrix modulation" for all matrix valued space–time and MIMO transmission methods where there is not an underlying trellis structure. The term "space–time block code" is reserved for orthogonal schemes, in line with the original introduction of the term in [18]. The reason for the term "matrix modulation" is immediately visible in the form of the STTD modulation (3.28), and the generalizations above. Various kinds of space–time block codes are typically space–time modulation schemes that are linear (in terms of complex symbols) over the field of complex numbers, and will accordingly be called "linear matrix modulation". With the same terminology, examples of non-linear matrix modulation would be the unitary space–time modulation schemes discussed in [22], and the space–time block codes based on matrix symbol sets discussed in [145]. Similarly, the space–time modulation used in space–time trellis codes, and in basic MIMO systems of the BLAST type [20, 21], shall be called a "vector modulation". In these, $T = 1$. A vector modulation for $N_t = 2$ Tx antennas would look like

$$\mathbf{x} = \begin{bmatrix} x_1 & x_2 \end{bmatrix},$$ (3.40)

The signal model (1.1), omitting beam-forming, gives

$$\mathbf{y} = \mathbf{xH} + \mathbf{n},$$ (3.41)

for (3.40), and thus a vector modulation is a row vector of symbols, not a column vector.

A generic matrix modulation is transmitted over T symbol periods from N_t antennas. For reliable detection, the channel coherence time should typically be $> T$. The underlying space–time coding philosophy is that of bit-interleaved coded space–time modulation; the bits coming out from a binary encoder are interleaved and mapped to the space–time modulation matrix $\mathbf{X}$. This is the way that STTD is embedded into the modular design of WCDMA Rel'99. With suitable design of interleaver periodicity, a multilevel coding concept may be used as well, where orthogonal parts of the matrix modulation are encoded at different levels.

3.7 NON-LINEAR MATRIX MODULATION

Non-linear schemes may be split into three categories.

1. The first non-linear space–time modulation scheme discussed were the unitary matrix modulations of [22]. These are based on non-linear mappings from the symbol set to space–time transmission matrices. The idea is to choose a set of constellation points that lie on the hyper-surface of a unitary group manifold. In the first works [22, 146], the motivation was that of differential space–time modulation. The main argument for differential signalling is that for increasing numbers of transmit antennas, the ratio of pilot power needed for reliable channel estimation grows to the degree that channel estimation overheads becomes excessive [84]. With PSK modulation, orthogonal designs are unitary linear space–time modulations, and can be used for differential signalling, as discussed in [147].

2. Another set of non-linear space–time block codes is based on sets of linearly independent modulation matrices [145]. In these sets, one has M matrices which can in principle be used to transmit $\log_2 M$ bits. As opposed to linear schemes, these modulation matrices do not have a simple interpretation in terms of bits.

3. Some schemes may be understood as binary codes or coded modulation schemes concatenated with a linear matrix or vector modulation. Thus e.g. the non-linear

space–time block codes discussed in [148] can be understood as a concatenation of a linear, orthogonal space–time block code with CRC (cyclic redundancy check).

Non-linear schemes are probably optimal with many transmit antennas (differential encoding) and high SNR. These will, however, not be discussed in this book due to the following reasons.

- Mapping bits into non-linear schemes is hard. In non-linear schemes, constellation points are mapped on to a non-linear manifold. These points do not form regular lattices where bits are easy to map. This leads to suboptimal Euclidean distance properties for the bits. As we shall see, Euclidean distance dominates low SNR performance.

- Detection of non-linear schemes is hard, as there is no inherent linear structure which would simplify detection. An exception to this are schemes based on the Cayley transform [149], for which linearized maximum likelihood detectors of the sphere decoder type may be used.

- The problem of channel estimation may be solvable without relying on differential encoding. In a cellular multiuser system, it is likely that more antennas are deployed at the base station than in mobile units. Thus in the uplink direction, channel estimation overhead may thus not be big enough to justify the use of non-coherent schemes. In the downlink direction, at least in CDMA-based multiuser systems, the pilot powers of all users may be pooled to yield improved channel estimates, as in WCDMA CPICH common pilot channels. Also, according to the results in [150], for four Tx antennas, the performance of a non-coherent scheme is better than that of a coherent scheme only at very high SNR. At lower SNR, it is preferable to dedicate power to pilot signalling. Combining CPICH ideas with the results of [150] leads to the conclusion that for power-efficient signalling with realistic numbers of Tx antennas, non-linear modulation needs not be considered, at least in CDMA.

3.8   SUMMARY

Transmit diversity concepts come in many different flavours. Implicit transmit diversity solutions are attractive due to their simplicity. They can often be used to provide diversity for systems that were initially designed for a SISO channel. Symbol rotations can be used in addition to capture diversity from a time-varying or a multi-antenna wireless channel. The rotated symbols can be transmitted using TDTD or CDTD/OTD, for example. However, current wireless standards do not support rotated constellations. Thus, in order to benefit from symbol rotations (via diversity transforms), the system specification would need to be modified.

If we go about modifying system specification, or defining a new system, we cannot avoid stepping into the territory of space–time codes. Proper space–time code design requires that space–time coding criteria are taken into account when designing matrix-valued modulation schemes. The criteria were summarized here, and several codes designed using the criteria were given. However, the inherent rate limitations in orthogonal space–time designs are seen to be restrictive for future high rate systems. This, however, is only part of the story. We will see in the following chapters that a conscious concatenation and design of the different diversity concepts described in this chapter pave the way for extremely efficient multi-antenna modulation techniques that offer ample diversity, rate and performance.

4

Matrix Modulation: Low SNR Aspects

In this chapter, linear matrix modulators will be discussed, concentrating on issues related to performance at low and medium SNR. As we shall see, at low SNR, performance optimality is related to achieving link capacity. That is, the mutual information of the modulator should be maximized. Of two modulators with the same rate, the one providing more information performs better at low SNR.

Section 3.6 discussed how requiring simultaneous linearity and orthogonality leads to a decrease in rate. This leads to inefficiency in the use of multiple antennas, a dilution of maximal symbol rates and waste of channel capacity. Simultaneously, the inefficient use of antennas gives rise to fluctuating transmit powers, i.e. the power-unbalance problem.

Increase the Symbol Rate: If one wishes to increase the symbol rate of a space–time block code above the maximal allowed by orthogonality, there are three options.

- The diversity gain may be sacrificed. This was considered in [49], where diversity 2, rate 1 orthogonal space–time block codes were proposed for three and four antennas, constructed by concatenating STTD (3.28) with two-fold orthogonal transmit diversity. With concatenated channel coding, the remaining spatial diversity can partly be transformed to temporal diversity.

- Linearity may be sacrificed, keeping diversity gain and unitarity. This was considered in [22, 146] in the context of unitary space–time modulators.

- Orthogonality may be sacrificed, keeping the linearity and (the possibility for) full diversity gain.

In this chapter, the third method will be investigated. Methods will be developed to design high-rate non-orthogonal matrix modulators with the non-orthogonality under control. In the open literature, such schemes have been discussed in [116, 132–134, 142, 151], under a variety of names. These include non-orthogonal space–time block codes [132], quasi-orthogonal STBCs [142], linear dispersion codes [133]. In [116, 132, 142, 151], methods to increase the symbol rate of a space–time block code above the upper limit allowed by orthogonality were discussed. In [133], a more general MIMO setting was considered.

The intuition underlying the approach of this chapter comes from multiuser detection. Non-orthogonality gives rise to self-interference, which characterizes layers of a matrix modulator. Symbol streams that do not interfere with each other are considered to be in the same layer. Such "quasi-orthogonal" layers are thus analogous to users in a multiuser setting, just in the same way as symbol streams in vector modulation are treated as layers in BLAST-receiving algorithms [20, 21]. As for vector modulators, various interference cancellation (IC) methods can be used for detecting layered matrix modulators. This will be discussed in more detail in Chapter 6.

The methods discussed in this chapter apply equally well in MIMO and MISO settings. For concreteness, the properties of matrix modulation discussed in this chapter will be demonstrated in terms of a set of examples for $N_t = 2$ and $N_t = 4$ Tx antennas.

The approach of this chapter is geared towards communication and information theory. The reader interested in concrete open-loop designs is referred to Chapters 8 and 9.

4.1   LINEAR MATRIX MODULATION

A matrix modulator employing N_t Tx antennas and T symbol periods is described by a $T \times N_t$ modulation matrix $\mathbf{X}$. For reliable detection, it is essential that the channel coherence time is greater than T, so that the channel is essentially constant during the transmission of the matrix. Reliable equalization of multipath components is also crucial. Matrix modulation is especially suitable for channels with negligible inter-symbol interference (ISI). Examples of this are WCDMA with large spreading factors, and OFDM. In ISI channels, extended block schemes of the type discussed in [152] and [153] may be generalized to an arbitrary modulation matrix to make it amenable for equalization. For simplicity, we shall consider single-path channels only. Also, in this chapter, no beam-forming is assumed. Thus the signal model is (1.1) with $\mathbf{W} = \mathbf{I}_{N_t}$. With N_r receive antennas, it reads

$$\underset{T \times N_r}{\mathbf{Y}} = \underset{T \times N_t}{\mathbf{X}} \quad \underset{N_t \times N_r}{\mathbf{H}} + \underset{T \times N_r}{\text{noise}} . \tag{4.1}$$

Here $\mathbf{Y}$ is the $T \times N_r$ matrix of received signals, and $\mathbf{H}$ is the $N_t \times N_r$ channel matrix. In concrete calculations, the elements of $\mathbf{H}$ are assumed i.i.d., with zero mean complex Gaussian (Rayleigh) distribution. The additive noise is Gaussian. Also, perfect channel state information (CSI) is assumed at the receiver (Rx), and no CSI, except information of the fading statistics, is assumed at the transmitter (Tx).

It should be noted that in space–time and MIMO signalling, the spatial dimension, divided into N_t parts, is non-orthogonal, whereas the (properly equalized) temporal dimension, divided into T parts, is essentially orthogonal. Instead of using time as the orthogonal complementary dimension, any other substantially orthogonal dimension (frequency, spreading code, subcarrier) can be used, and correspondingly one might discuss space–frequency code, space–code codes etc. Also, instead of spatially dislocated antennas as the spatial transmission resource, different polarizations may be used.

4.1.1 Basis Matrices

The modulation matrix $\mathbf{X}$ is a linear function of the Q (complex modulation) symbols $x_k, k = 1 \ldots Q$ to be transmitted. Thus the symbol rate, the number of complex symbols transmitted per channel use, is $R_\mathrm{s} = Q/T$. As a consequence of linearity, a linear modulation matrix may be expanded as

$$\mathbf{X} = \sum_{k=1}^{Q} \mathbf{X}_k(x_k, x_k^*) \,, \tag{4.2}$$

where each $\mathbf{X}_k$ is a $T \times N_\mathrm{t}$ matrix with complex entries, which are linear functions of the symbol x_k and its complex conjugate.

Often it is preferable to deal separately with symbols and complex conjugates [141], or equivalently, with the real and imaginary parts of the symbols [133]. For this, the $\mathbf{X}_k$ may be expanded in terms of $2Q$ constant matrix coefficients, $\left\{\mathbf{B}^{(k\pm)}\right\}_{k=1}^{Q}$ or $\left\{\mathbf{B}^{(k)}\right\}_{k=1}^{2Q}$, as

$$\mathbf{X}_k(x_k, x_k^*) = x_k\,\mathbf{B}^{(k-)} + x_k^*\,\mathbf{B}^{(k+)} = c_{2k-1}\,\mathbf{B}^{(2k-1)} + c_{2k}\,\mathbf{B}^{(2k)} \,. \tag{4.3}$$

Here c_{2k-1}, c_{2k} are the real and imaginary part of the symbol x_k, so that

$$\mathbf{X} = \sum_{k=1}^{2Q} c_k\,\mathbf{B}^{(k)} \,. \tag{4.4}$$

Some modulators may be expressed without complex conjugations. In such cases, $\mathbf{B}^{(k+)} = 0$, and the most transparent description is

$$\mathbf{X} = \sum_{k=1}^{Q} x_k\,\mathbf{B}^{(k-)} \,. \tag{4.5}$$

The real and imaginary parts of the complex symbols will often be considered as independent real symbols. For QPSK and M-QAM modulations, they indeed are such.

The set of matrices $\mathbf{B}^{(k)}$ will be called a *basis* of the matrix modulation, and each individual $\mathbf{B}^{(k)}$, encoding an independent real symbol, will be called a *basis matrix*. The reason for this is that often the basis matrices constitute a part of a linearly independent basis for generic matrices in the matrix modulation dimensions. For an example, see section B.2 in Appendix B. The basis matrices may be considered as space–time generalizations of I and Q- branches of 2-D modulation alphabets.

An alternative way to describe linear matrix modulation was used in most of [18] and in [116]. There, a matrix modulation was described through a set of $N_\mathrm{t} \times Q$ generator matrices $\{G_t\}_{t=1}^{T}$. The codeword transmitted during a given symbol period t is constructed from the vector $\mathbf{x}$ of symbols to be transmitted as $G_t\mathbf{x}$. Here Equation (4.4) is used as it yields a more transparent description of the diversity and self-interference properties of a matrix modulator.

4.1.2 Diversity and Self-interference

As discussed in section 3.5.2, performance of a space–time code is determined in terms of unitary invariants of the codeword difference matrix. These are the rank, the trace and the determinant of the distance matrix (the squared codeword difference matrix). More exactly, in section 7.1.2 we shall see that the maximum likelihood metric, as well as the bit error probability, can be written in terms of unitary invariants.

In this chapter we shall see that for linear modulators, the unitary invariants of the modulation matrix $\mathbf{X}$ itself play a crucial role for the mutual information, and indeed for performance.

A most fundamental characteristic of linear matrix modulation is that properties of the codeword difference matrices and properties of the modulation matrix itself are directly related. This is a consequence of linearity. The codeword difference matrix, crucial for space–time code performance analysis, can be expressed in terms of the modulation matrix as

$$\mathbf{D}^{(ce)} = \mathbf{X}\left(\mathbf{x}^{(c)}\right) - \mathbf{X}\left(\mathbf{x}^{(e)}\right) = \mathbf{X}\left(\mathbf{\Delta}\right) \tag{4.6}$$

where $\Delta_k = x_k^{(c)} - x_k^{(e)}$ are the symbol differences between the transmitted codeword $\mathbf{x}^{(c)}$ and the possibly erroneous detected codeword $\mathbf{x}^{(e)}$. The distance matrix (the Hermitian square of the codeword difference matrix) is nothing but the Hermitian square of the modulation matrix, with symbols replaced by symbol differences.

Generic properties of unitary invariants are discussed in Appendix A. The essential feature is that all unitarily invariant properties may be described in terms of traces of powers of $\mathbf{X}^\dagger\mathbf{X}$. These "trace invariants" are of the form

$$t_n\left(\mathbf{X}^\dagger\mathbf{X}\right) = \mathrm{Tr}\left[\left(\mathbf{X}^\dagger\mathbf{X}\right)^n\right] . \tag{4.7}$$

In Appendix A it is shown that for a $N \times N$ matrix, the determinant can be expressed in terms of the N lowest trace invariants. From the cyclicity of trace it follows that $t_n\left(\mathbf{X}^\dagger\mathbf{X}\right) = t_n\left(\mathbf{X}\,\mathbf{X}^\dagger\right)$, i.e. $\mathbf{X}^\dagger\mathbf{X}$ and $\mathbf{X}\,\mathbf{X}^\dagger$ have the same unitary invariants. The former has dimensions $N_t \times N_t$, the latter $T \times T$. Accordingly, $\mathbf{X}^\dagger\mathbf{X}$ gives explicit information about the spatial properties of the code, and implicit information about the temporal properties. If $T < N_t$, $\mathbf{X}^\dagger\mathbf{X}$ carries spurious information. For $\mathbf{X}\,\mathbf{X}^\dagger$ it is vice versa. Due to cyclicity of trace, the unitary invariants are the same regardless of which Hermitian square is used, and the least spurious may be used. In this chapter, the case $T \geq N_t$ will mostly be considered, and thus $\mathbf{X}^\dagger\mathbf{X}$ will be discussed in most explicit calculations. The (spatial) diversity and self-interference matrices are defined as

$$\mathbf{X}^\dagger\mathbf{X} = \mathcal{D} + \mathcal{S} , \tag{4.8}$$

$$\mathcal{D} = \tfrac{1}{2}\sum_{k=1}^{2Q} c_k^2\, \mathcal{S}^{(kk)} \tag{4.9}$$

$$\mathcal{S} = \sum_{i=1}^{2Q}\sum_{k=i+1}^{2Q} c_i\, c_k\, \mathcal{S}^{(i,k)} \tag{4.10}$$

where interference of two basis matrices is given by

$$\mathcal{S}^{(i,k)} = \mathbf{B}^{(i)\,\dagger}\,\mathbf{B}^{(k)} + \mathbf{B}^{(k)\,\dagger}\,\mathbf{B}^{(i)} . \tag{4.11}$$

As these matrices are essential in investigating the Radon–Hurwitz structure of orthogonal designs, we shall call them Radon–Hurwitz matrices. Two symbols with vanishing Radon–Hurwitz matrix are said to be Radon–Hurwitz orthogonal (RHO). If $\mathcal{S} \neq \prime$, the matrix modulator is non-orthogonal.

Diversity Matrix: The diversity matrix $\mathcal{D}$ is the part of $\mathbf{X}^\dagger \mathbf{X}$ which depends on real symbol powers c_k^2. It describes the distribution of transmit power over the multiple transmit antennas, and thus measures how well the code exploits the inherent transmit diversity of the system.

The concept of *symbolwise diversity degree* of a matrix modulation is defined by the diversity matrices of individual symbols as

$$D_{\text{symbol}} = \min_k \text{ rank } \mathcal{S}^{(kk)} . \tag{4.12}$$

This is the diversity degree captured by a matched filter bound, where the interference from other symbols is supposed to be completely removed. If each $\mathcal{S}^{(kk)}$ has rank N_t, the transmission of each symbol is distributed over all antennas and the modulator is said to have full symbolwise diversity.

Self-interference Matrix: With self-interference, we mean the ISI induced by the space–time transmission, as opposed to the ISI induced by the multipath structure of the channel. The self-interference (non-orthogonality) induced by the matrix modulation is described by the matrix $\mathcal{S}$. It measures to what extent the encoded symbols interfere with each other, and affects the capacity and performance of the code.

The temporal diversity and self-interference are defined as

$$\mathbf{X}\,\mathbf{X}^\dagger = \tfrac{1}{2} \sum_{k=1}^{2Q} \sum_{i=1}^{2Q} c_k c_i \, \mathcal{S}_T^{(ki)} = \mathcal{D}_T + \mathcal{S}_T , \tag{4.13}$$

where the temporal Radon–Hurwitz matrices are

$$\mathcal{S}_T^{(i,k)} = \mathbf{B}^{(i)} \, \mathbf{B}^{(k)\,\dagger} + \mathbf{B}^{(k)} \, \mathbf{B}^{(i)\,\dagger} . \tag{4.14}$$

To get an intuitive understanding of the relation of the diversity and self-interference matrices to the diversity properties of the code, some space–time modulators presented in the literature are reconsidered in the perspective of (4.2–4.13).

- *Vector modulation:* Capacity achieving MISO [96] and MIMO [16,20,21] modulators were constructed by transmitting independent signals from each antenna during each symbol interval. In [96] it was shown that this is also capacity-optimal in a MISO channel. $\mathbf{X}$ is a row-vector encoding $Q = N_t$ symbols;[1]

$$\mathbf{X} = \begin{bmatrix} x_1 & x_2 & \cdots & x_{N_t} \end{bmatrix} . \tag{4.15}$$

The basis vectors $\mathbf{B}^{(k)}$ are simply row-vectors with one non-zero element, either 1 or j. This leads to

$$\mathcal{D} = \text{Diagonal matrix } \begin{bmatrix} |x_1|^2, |x_2|^2, \ldots |x_{N_t}|^2 \end{bmatrix} \tag{4.16}$$

$$\mathcal{S} = \begin{bmatrix} 0 & x_1^* x_2 & \cdots & x_1^* x_{N_t} \\ x_2^* x_1 & 0 & \cdots & x_2^* x_{N_t} \\ \vdots & \vdots & \ddots & \vdots \\ x_{N_t}^* x_1 & x_{N_t}^* x_2 & \cdots & 0 \end{bmatrix} . \tag{4.17}$$

---

[1]Note that the signal model (4.1) makes $\mathbf{X}$ a row vector, in contrast to the usage in the majority of the literature.

Thus on the level of space–time modulation, there is no transmit diversity; each symbol is transmitted from just one antenna. Also, spatially all symbols interfere with each other. When analysing performance properties, the spurious information in $\mathbf{X}^\dagger\mathbf{X}$ may be omitted by investigating the temporal diversity and self-interference (4.13), which are 1×1 matrices:

$$\mathcal{D}_T = \sum_{k=1}^{Q} |x_k|^2 \qquad (4.18)$$

$$\mathcal{S}_T = 0 .$$

These reveal that vector modulation provides full temporal diversity. The modulator covers only one symbol period, so the possible degree of temporal diversity is indeed one. Similarly, as $T = 1$, there is no room for temporal self-interference.

- *Unitary (complex orthogonal) designs:* The basic design rule [18] is that the diversity matrix is proportional to identity, and that there is *no self-interference* (the code is RHO),

$$\mathcal{D} = \sum_{k=1}^{Q} |x_k|^2 \, \mathbf{I}_{N_t} , \qquad \mathcal{S} = 0 . \qquad (4.19)$$

The former condition requires each individual symbol to be encoded by a unitary matrix[2] which guarantees optimal diversity gain. Optimality is intuitively clear, and will be discussed in section 5.1.1. The set of basis matrices forming $N_t = 2$ STTD, the "STTD basis", is a very special set of matrices. Thus we shall reserve a special notation for them, they will be denoted by

$$\mathbf{T}^{(1)} = \mathbf{I}_2; \ \ \mathbf{T}^{(2)} = \begin{bmatrix} j & 0 \\ 0 & -j \end{bmatrix}; \ \ \mathbf{T}^{(3)} = \begin{bmatrix} 0 & 1 \\ -1 & 0 \end{bmatrix}; \ \ \mathbf{T}^{(4)} = \begin{bmatrix} 0 & j \\ j & 0 \end{bmatrix} \quad (4.20)$$

The diagonal matrices $\mathbf{T}^{(1)}$ and $\mathbf{T}^{(2)}$ encode the real and imaginary parts of x_1, respectively, whereas the anti-diagonal matrices $\mathbf{T}^{(3)}$ and $\mathbf{T}^{(4)}$ encode x_2.

The two examples discussed above may be summarized as follows. Vector modulation is rate- and capacity-optimal, but diversity and performance sub-optimal. Orthogonal designs are diversity-optimal, but rate and capacity sub-optimal. The obvious way to improve these schemes is to construct something in between. This is the realm of generic (linear) matrix modulation.

4.2 EXAMPLES

In this chapter, a number of examples with symbol rate $R_s = 1$ for $N_t = 4$ antennas and $T = 4$ symbol periods will be used to demonstrate the principles discussed. The examples include one orthogonal modulator with sub-optimal diversity, and some non-orthogonal modulators

---

[2]Here we use the term "unitary" loosely, to indicate a matrix for which the inverse is *proportional to* its Hermitian conjugate.

with full symbolwise diversity. Most of these examples are not relevant for the future development of WCDMA, or for possible 4G systems. They are only considered as simple examples to hone the design tools on. The ultimate use of the methods developed in this chapter is the design of high-rate information and performance-optimal matrix modulators for MIMO channels discussed in Chapter 9. Robust and practical developments with rate $R_s = 1$ are discussed in Chapter 8.

Most examples considered are *quasi-orthogonal*; they have a structure consisting of multiple RH-orthogonal layers. The layering structure will be denoted by a sum. If the scheme has four complex symbols, of which two pairs are mutually RHO, it is said to be 2+2 layered.

Orthogonal Schemes: As an example of a linear orthogonal modulator with sub-optimal diversity, consider the STTD-OTD scheme of [49] concatenating STTD with orthogonal transmit diversity. In the simplest form this means transmitting two copies of (3.28) on the block-diagonal:

$$\mathbf{X}_{\text{STTD}-\text{OTD}} = \begin{bmatrix} \mathbf{X}_A(x_1, x_2) & \mathbf{0}_2 \\ \mathbf{0}_2 & \mathbf{X}_B(x_3, x_4) \end{bmatrix}. \tag{4.21}$$

Here we have introduced a notation often used in this book, where $\mathbf{X}_A, \mathbf{X}_B, \mathbf{X}_C$ etc. denote STTD blocks with independent symbols. $\mathbf{0}_2$ denotes a 2×2 matrix of zeros. A power-balanced version reads

$$\mathbf{X}_{\text{STTD}-\text{OTD}} = \begin{bmatrix} \mathbf{X}_A & \mathbf{X}_B \\ \mathbf{X}_A & -\mathbf{X}_B \end{bmatrix}. \tag{4.22}$$

Note that power is normalized only when numerical comparisons are performed between schemes. For (4.21) and (4.22), the diversity and self-interference matrices are

$$\mathcal{D} = \begin{bmatrix} \left(|x_1|^2 + |x_2|^2\right)\mathbf{I}_2 & \mathbf{0}_2 \\ \mathbf{0}_2 & \left(|x_3|^2 + |x_4|^2\right)\mathbf{I}_2 \end{bmatrix} \tag{4.23}$$

$$\mathcal{S} = 0 \tag{4.24}$$

The two-dimensional identity matrix is $\mathbf{I}_2$. Each symbol enjoys double transmit diversity, and the modulator is RH-orthogonal with no self-interference.

Non-orthogonal Schemes: Non-orthogonal matrix modulation has been considered in [26, 116, 132, 133, 135, 142, 151, 154, 155]. They have the characteristic property that the symbol rate is higher than the one allowed by orthogonality (see e.g. Table 3.1), and $\mathcal{S} \neq 0$. Except for the first example provided in [116], they typically have the property that $\mathcal{D}$, expressed either on the symbol-level [132, 135, 142, 151, 155] or the bit-level [133] is proportional to the identity matrix.

A Full Rank Scheme: An example of a full rank matrix modulator is the 4×4 scheme generalizing the 3×3 non-orthogonal $R_s = 1$ space–time block code of [116], of the form

$$\mathbf{X}_{\text{PSK}-\text{rank}} = \begin{bmatrix} x_1 & x_2 & x_3 & x_4 \\ x_2 & x_3 & x_4 & x_1 + x_2 \\ x_3 & x_4 & x_1 + x_2 & x_2 + x_3 \\ x_4 & x_1 + x_2 & x_2 + x_3 & x_3 + x_4 \end{bmatrix}. \tag{4.25}$$

This matrix fulfils the PSK-rank criterion of [116], i.e. it fulfils the rank criterion for any PSK-modulation. It may be described in terms of (4.5) with the four basis matrices

$$
\mathbf{B}^{(1-)} = \begin{bmatrix} 1 & 0 & 0 & 0 \\ 0 & 0 & 0 & 1 \\ 0 & 0 & 1 & 0 \\ 0 & 1 & 0 & 0 \end{bmatrix} ; \quad
\mathbf{B}^{(2-)} = \begin{bmatrix} 0 & 1 & 0 & 0 \\ 1 & 0 & 0 & 1 \\ 0 & 0 & 1 & 1 \\ 0 & 1 & 1 & 0 \end{bmatrix} ;
$$

$$
\mathbf{B}^{(3-)} = \begin{bmatrix} 0 & 0 & 1 & 0 \\ 0 & 1 & 0 & 0 \\ 1 & 0 & 0 & 1 \\ 0 & 0 & 1 & 1 \end{bmatrix} ; \quad
\mathbf{B}^{(4-)} = \begin{bmatrix} 0 & 0 & 0 & 1 \\ 0 & 0 & 1 & 0 \\ 0 & 1 & 0 & 0 \\ 1 & 0 & 0 & 1 \end{bmatrix} . \tag{4.26}
$$

The diversity matrix reads

$$
\mathcal{D} = \sum_{k=1}^{4} |x_k|^2 \begin{bmatrix} 1 & 0 & 0 & 1 \\ 0 & 1 & 0 & 0 \\ 0 & 0 & 1 & 0 \\ 1 & 0 & 0 & 2 \end{bmatrix}
$$
$$
+ \begin{bmatrix} 0 & 0 & 0 & -|x_1|^2 \\ 0 & |x_2|^2 & |x_2|^2 & 0 \\ 0 & |x_2|^2 & |x_2|^2 + |x_3|^2 & |x_2|^2 + |x_3|^2 \\ -|x_1|^2 & 0 & |x_2|^2 + |x_3|^2 & -|x_1|^2 \end{bmatrix} \tag{4.27}
$$

The self-interference matrix is non-transparent. For further use, however, we shall note the diagonal part of the self-interference, which is

$$
2\,\mathrm{Re} \begin{bmatrix} 0 & x_1 x_2^* & x_1 x_2^* + x_2 x_3^* & x_1 x_2^* + x_2 x_3^* + x_3 x_4^* \end{bmatrix} . \tag{4.28}
$$

Only x_1 is encoded with a unitary basis matrix; the remaining symbols are encoded non-unitarily. Furthermore, all symbols interfere with each other. The scheme is thus 1+1+1+1 layered.

A Cyclic Scheme: Another 1+1+1+1 layered scheme is the cyclically shifted matrix modulation of the form

$$
\mathbf{X}_{\mathrm{cyclic}} = \begin{bmatrix} x_1 & x_2 & x_3 & x_4 \\ x_4 & x_1 & x_2 & x_3 \\ x_3 & x_4 & x_1 & x_2 \\ x_2 & x_3 & x_4 & x_1 \end{bmatrix} . \tag{4.29}
$$

Each symbol is encoded by a cyclically shifted identity matrix, which is a unitary matrix. Accordingly, the diversity matrix is

$$
\mathcal{D} = \sum_{k=1}^{4} |x_k|^2 \, \mathbf{I}_4 . \tag{4.30}
$$

The self-interference consists of cyclic permutations of the vector

$$
\begin{bmatrix} 0 & \sum_{k=1}^{4} x_k x_{k-1}^* & 2\,\mathrm{Re}\,[x_1 x_3^* + x_2 x_4^*] & \sum_{k=1}^{4} x_k x_{k+1}^* \end{bmatrix} , \tag{4.31}
$$

so that all diagonal elements are 0. In the sums, the indexes wrap around: $x_5 \equiv x_1$ etc.

A 3+1 Layered Quasi-orthogonal Scheme: This was constructed in [156] by encoding an extra symbol on the anti-diagonal of the rate 3/4 orthogonal design (3.39):

$$
\mathbf{X}_{3+1} = \begin{bmatrix} x_1 & x_2 & x_3 & x_4 \\ -x_2^* & x_1^* & x_4 & -x_3 \\ -x_3^* & -x_4 & x_1^* & x_2 \\ -x_4 & x_3^* & -x_2^* & x_1 \end{bmatrix}.
\tag{4.32}
$$

The basis matrices are

$$
\mathbf{B}^{(1)} = \begin{bmatrix} 1 & 0 & 0 & 0 \\ 0 & 1 & 0 & 0 \\ 0 & 0 & 1 & 0 \\ 0 & 0 & 0 & 1 \end{bmatrix} \quad \mathbf{B}^{(2)} = \begin{bmatrix} j & 0 & 0 & 0 \\ 0 & -j & 0 & 0 \\ 0 & 0 & -j & 0 \\ 0 & 0 & 0 & j \end{bmatrix}
$$

$$
\mathbf{B}^{(3)} = \begin{bmatrix} 0 & 1 & 0 & 0 \\ -1 & 0 & 0 & 0 \\ 0 & 0 & 0 & 1 \\ 0 & 0 & -1 & 0 \end{bmatrix} \quad \mathbf{B}^{(4)} = \begin{bmatrix} 0 & j & 0 & 0 \\ j & 0 & 0 & 0 \\ 0 & 0 & 0 & j \\ 0 & 0 & j & 0 \end{bmatrix}
\tag{4.33}
$$

$$
\mathbf{B}^{(5)} = \begin{bmatrix} 0 & 0 & 1 & 0 \\ 0 & 0 & 0 & -1 \\ -1 & 0 & 0 & 0 \\ 0 & 1 & 0 & 0 \end{bmatrix} \quad \mathbf{B}^{(6)} = \begin{bmatrix} 0 & 0 & j & 0 \\ 0 & 0 & 0 & -j \\ j & 0 & 0 & 0 \\ 0 & -j & 0 & 0 \end{bmatrix}
$$

$$
\mathbf{B}^{(7)} = \begin{bmatrix} 0 & 0 & 0 & 1 \\ 0 & 0 & 1 & 0 \\ 0 & -1 & 0 & 0 \\ -1 & 0 & 0 & 0 \end{bmatrix} \quad \mathbf{B}^{(8)} = j\,\mathbf{B}^{(7)}.
$$

These basis matrices may be expressed in a more condensed format as *tensor (Kronecker) products* of elements in the STTD basis (4.20). The tensor product of a 2×2 matrix $A = \begin{bmatrix} a_{11} & a_{12} \\ a_{21} & a_{22} \end{bmatrix}$ with a $m \times m$ matrix B is a $2m \times 2m$ matrix $A \otimes B$ which reads in block form $A \otimes B = \begin{bmatrix} a_{11}B & a_{12}B \\ a_{21}B & a_{22}B \end{bmatrix}$. With this notation, the basis above may be written as

$$
\begin{aligned}
\mathbf{B}^{(1)} &= \mathbf{I}_4 & \mathbf{B}^{(2)} &= -j\,\mathbf{T}^{(2)} \otimes \mathbf{T}^{(2)} \\
\mathbf{B}^{(3)} &= \mathbf{I}_2 \otimes \mathbf{T}^{(3)} & \mathbf{B}^{(4)} &= \mathbf{I}_2 \otimes \mathbf{T}^{(4)} \\
\mathbf{B}^{(5)} &= -j\,\mathbf{T}^{(3)} \otimes \mathbf{T}^{(2)} & \mathbf{B}^{(6)} &= -j\,\mathbf{T}^{(4)} \otimes \mathbf{T}^{(2)} \\
\mathbf{B}^{(7)} &= -j\,\mathbf{T}^{(3)} \otimes \mathbf{T}^{(4)} & \mathbf{B}^{(8)} &= \mathbf{T}^{(3)} \otimes \mathbf{T}^{(4)}
\end{aligned}
$$

Each symbol is encoded with a unitary matrix, so the diversity matrix is (4.30). The self-interference is

$$
\mathcal{S} = 2\,\mathrm{Im} \begin{bmatrix} 0 & x_3 x_4 & -x_2 x_4 & -x_1 x_4^* \\ -x_3 x_4 & 0 & x_1 x_4 & -x_2 x_4^* \\ x_2 x_4 & -x_1 x_4 & 0 & -x_3 x_4^* \\ x_1 x_4^* & x_2 x_4^* & x_3 x_4^* & 0 \end{bmatrix}
\tag{4.34}
$$

which immediately shows that all self-interference is between x_4 and the other symbols.

A 2+2 Layered Quasi-orthogonal Scheme: 2+2 layered rate $R_s = 1$ matrix modulators for $N_t = 4$ Tx antennas have been discussed in [132, 142, 151]. Here, the simplest form, the

"ABBA" of [132] is considered:

$$
\mathbf{X}_{\mathrm{ABBA}}(x_1, x_2, x_3, x_4) = \begin{bmatrix} x_1 & x_2 & x_3 & x_4 \\ -x_2^* & x_1^* & -x_4^* & x_3^* \\ x_3 & x_4 & x_1 & x_2 \\ -x_4^* & x_3^* & -x_2^* & x_1^* \end{bmatrix} \equiv \begin{bmatrix} \mathbf{X}_A & \mathbf{X}_B \\ \mathbf{X}_B & \mathbf{X}_A \end{bmatrix}. \tag{4.35}
$$

The basis matrices are easy to write in terms of the STTD basis (4.20). The last equality in (4.35) indicates that simple tensor products may be used. Indeed,

$$
\begin{array}{llll}
\mathbf{B}^{(1)} = & \mathbf{I}_4 & \mathbf{B}^{(2)} = & \mathbf{I}_2 \otimes \mathbf{T}^{(2)} \\
\mathbf{B}^{(3)} = & \mathbf{I}_2 \otimes \mathbf{T}^{(3)} & \mathbf{B}^{(4)} = & \mathbf{I}_2 \otimes \mathbf{T}^{(4)} \\
\mathbf{B}^{(5)} = -\mathrm{j}\ \mathbf{T}^{(4)} \otimes \mathbf{I}_2 & \mathbf{B}^{(6)} = -\mathrm{j}\ \mathbf{T}^{(4)} \otimes \mathbf{T}^{(2)} \\
\mathbf{B}^{(7)} = -\mathrm{j}\ \mathbf{T}^{(4)} \otimes \mathbf{T}^{(3)} & \mathbf{B}^{(8)} = -\mathrm{j}\ \mathbf{T}^{(4)} \otimes \mathbf{T}^{(4)}
\end{array} \tag{4.36}
$$

The block structure makes analysing ABBA especially simple. The diversity matrix is again (4.30), and the self-interference is

$$
\mathcal{S} = 2\,\mathrm{Re}\,[x_1 x_3^* + x_2 x_4^*] \begin{bmatrix} 0 & 0 & 1 & 0 \\ 0 & 0 & 0 & 1 \\ 1 & 0 & 0 & 0 \\ 0 & 1 & 0 & 0 \end{bmatrix} = -\mathrm{j}\ \mathbf{T}^{(4)} \otimes \left(\mathbf{X}_A^\dagger \mathbf{X}_B + \mathbf{X}_B^\dagger \mathbf{X}_A \right). \tag{4.37}
$$

Only symbols x_1, x_3 and x_2, x_4 interfere.

4.3   HEURISTIC DESIGN RULES AT LOW SNR

At high SNR, the performance of a matrix modulator is dictated by the rank and the determinant of the codeword difference matrix, just as for any space–time code. At low and medium SNR, reaching full diversity protection for all possible codeword pairs is not of primary interest. It is more important that the most frequently occurring codeword errors have full diversity protection. Also error events that have a small number of bit errors should have higher diversity protection than error events with multiple bit errors. This intuition follows from the linearity of the scheme, combined with the interpretation of the trace of the distance matrix as an Euclidean distance, in the spirit of the trace criterion discussed in section 3.5.2. The importance of the trace may be seen if the spirit of the union bound in section 3.5.2 is violated, and the Chernoff bound (3.24) is considered at low SNR. In the Chernoff bound, the determinant $\det\left(\mathbf{I} + \eta \mathbf{X}^\dagger \mathbf{X}(\boldsymbol{\Delta})\right)$ appears. Here, the distance matrix $\mathbf{D}^{(\mathrm{ce})\dagger}\mathbf{D}^{(\mathrm{ce})}$ has been replaced by $\mathbf{X}^\dagger \mathbf{X}(\boldsymbol{\Delta})$ due to the linearity of $\mathbf{X}$. The error vector is the vector of differences of transmitted symbols and detected symbols: $\boldsymbol{\Delta} = \mathbf{x}^{(c)} - \mathbf{x}^{(e)}$. Now we use the following matrix identity:

$$
\det \mathbf{M} = \exp \mathrm{Tr} \log \mathbf{M}, \tag{4.38}
$$

as well as the series expansion

$$
\log(1 + x) = x - \tfrac{1}{2} x^2 + \tfrac{1}{3} x^3 - \dots \tag{4.39}
$$

Using these we have

$$
\det \left(\mathbf{I} + \eta\, \mathbf{X}^\dagger \mathbf{X}\right) = \exp \left\{ \eta\ \mathrm{Tr}\, \mathbf{X}^\dagger \mathbf{X} - \tfrac{1}{2}\eta^2\, \mathrm{Tr} \left(\mathbf{X}^\dagger \mathbf{X}\right)^2 + \dots \right\}
$$

$$= 1 + \eta \, \text{Tr} \, \mathbf{X}^{\dagger}\mathbf{X} - \tfrac{1}{2} \eta^2 \left(\text{Tr} \, (\mathbf{X}^{\dagger}\mathbf{X})^2 - (\text{Tr} \, \mathbf{X}^{\dagger}\mathbf{X})^2 \right) + \; .. \, (4.40)$$

which immediately indicates that for optimizing performance at low SNR, the Euclidean distance

$$\text{Tr} \, \mathbf{X}^{\dagger}\mathbf{X}(\boldsymbol{\Delta}) = \text{Tr} \, \mathcal{D}(\boldsymbol{\Delta}) + \text{Tr} \, \mathcal{S}(\boldsymbol{\Delta}) \tag{4.41}$$

should be maximized. In addition, the second trace invariant appears, in combination with $\left(\text{Tr} \, \mathbf{X}^{\dagger}\mathbf{X} \right)^2$, which is of the form of the modified trace invariants discussed in Appendix A. A number of heuristic rules may be deduced from (4.40).

4.3.1 Frobenius Orthogonality

Due to linearity, the distance matrix $\mathbf{X}^{\dagger}\mathbf{X}(\boldsymbol{\Delta})$ is a quadratic function of the symbol differences $\boldsymbol{\Delta}$. This is true for the self-interference part $\mathcal{S}(\boldsymbol{\Delta})$ of the distance matrix as well, and for the part of the Euclidean distance arising from non-orthogonality, namely $\text{Tr} \, \mathcal{S}$. From the definition (4.10) of $\mathcal{S}$ it follows that its diagonal entries are real functions of pairs of symbol differences, i.e. functions of $\text{Re} \, [\Delta_k \Delta_l]$ and/or $\text{Im} \, [\Delta_k \Delta_l]$ for $k \neq l$.

In a generic complex modulation alphabet, a number of Δ_k:s may be related by rotations in the complex plane. For example, a $90°$ rotation maps QPSK modulation points to each other, and accordingly also possible QPSK errors. If $\text{Tr} \, \mathcal{S} \neq 0$, the value of the trace, and thus the Euclidean distance, changes by rotating Δ_k, even if the other Δ_l are averaged over. This means that the Euclidean distance between a symbol and its nearest neighbours differs from the distance between an equivalent rotated symbol and its nearest neighbours. Thus the modulation points are not homogeneously situated in modulation matrix space, and the matrix modulation cannot be optimal.

The requirement of homogeneity comes from the convexity of error functions. The total power used for transmitting a modulation matrix is $\text{Tr} \, \mathbf{X}^{\dagger}\mathbf{X} = \text{Tr} \, \mathcal{D} + \text{Tr} \, \mathcal{S}$. If $\text{Tr} \, \mathcal{S} \neq 0$, the total power used for transmitting a set of symbols, and the same set of symbols with one symbol rotated to another constellation point, would differ. From the convexity of error functions (e.g. (3.24)) it follows that the union bound of error is minimized if the transmit power does not depend on the actual bits to be transmitted, to the extent to which that it is possible.

This indicates that the self-interference should be traceless [51, 135],

$$\text{Tr} \, \mathcal{S} = 0 \, , \tag{4.42}$$

which further indicates that

$$\text{Tr} \, \mathcal{S}^{(kl)} = \text{Tr} \, \left(\mathbf{B}^{(k)\dagger}\mathbf{B}^{(l)} + \mathbf{B}^{(l)\dagger}\mathbf{B}^{(k)} \right) = 0 \, , \quad k \neq l \tag{4.43}$$

i.e. the basis matrices $\mathbf{B}^{(k)}$ and $\mathbf{B}^{(l)}$ used for different (real) symbols should be orthogonal with respect to the Frobenius norm. The squared Frobenius norm of the difference of two basis matrices is

$$\begin{aligned} \|\mathbf{B}^{(k)} - \mathbf{B}^{(l)}\|_F^2 &= \text{Tr} \left[\left(\mathbf{B}^{(k)} - \mathbf{B}^{(l)} \right)^{\dagger} \left(\mathbf{B}^{(k)} - \mathbf{B}^{(l)} \right) \right] \\ &= \|\mathbf{B}^{(k)}\|_F^2 + \|\mathbf{B}^{(l)}\|_F^2 - \text{Tr} \, \mathcal{S}^{(kl)} \, . \end{aligned} \tag{4.44}$$

The last term defines the Frobenius inner product of $\mathbf{B}^{(k)}$ and $\mathbf{B}^{(l)}$, and Frobenius orthogonality means that it vanishes.

When considering Frobenius orthogonality, a hidden assumption regarding the concept of "symbol" and "symbol rate" in this book should be discussed. Often a set of constellation points, say a 4-PAM symbol, can be interpreted as a linear combination of symbols with fewer constellation points, say two BPSK symbols (with different power). The concept of "symbol" is defined modulo such trivial reinterpretations. A real or complex symbol is a real or complex degree of freedom, irrespective of whether bits are encoded linearly (as when interpreting a 4-PAM symbol as a sum of two QPSK symbols with different power) or non-linearly (as in a Gray-encoded 4-PAM symbol). This means that the basis matrices used to transmit independent real symbols should be linearly independent when considered as elements of a vector space of matrices with real coefficients. Frobenius orthogonality fortifies such linear independence to strict orthogonality. A trivial example of Frobenius orthogonality is that it is better to arrange two BPSK symbols into a QPSK symbol than into a linearly dependent 4-PAM symbol, if signal space dimensionality allows.

Of the examples in section 4.2, the "PSK-rank" scheme (4.25) has $\mathrm{Tr}\mathcal{S} \neq 0$, which is immediately seen from (4.28). All other schemes have $\mathrm{Tr}\mathcal{S} = 0$, which is a trivial statement, as the transmission matrices (4.21), (4.29), (4.32) and (4.35) do not transmit two symbols in overlapping matrix elements. For high-rate schemes, discussed in Chapter 9, this becomes a non-trivial statement.

4.3.2   Minimal Self-interference

The preceding criterion dealt with maximizing the first-order term in (4.40), the Euclidean distance (4.41). To minimize the Chernoff bound at low SNR, the second-order term $\mathrm{Tr}\,(\mathbf{X}^\dagger\mathbf{X})^2 - (\mathrm{Tr}\,\mathbf{X}^\dagger\mathbf{X})^2$ should also be minimized. This term carries parts that are related to the diversity matrix, the trace of which was maximized above when optimizing the Euclidean distance. According to the principle of Frobenius orthogonality, $\mathcal{S}$ should not contribute to the Euclidean distance. Its contribution to the second-order term may thus be independently minimized. This is given by

$$\mathcal{N} = \tfrac{1}{4} \sum_{k,l\,:\,k\neq l} \mathrm{Tr}\left(\mathcal{S}^{(kl)}\right)^2 , \qquad (4.45)$$

which is the total self-interference power. The self-explanatory and intuitively clear principle of minimal self-interference [132] requires that $\mathcal{N}$ is minimized.

When comparing different schemes, total self-interference should be normalized. The total basis matrix power may be normalized to

$$\sum_{k=1}^{2Q} \mathrm{Tr}\,\mathcal{S}^{(kk)} = 4Q , \qquad (4.46)$$

implying that the average of $\mathrm{Tr}\,\mathbf{B}^{(k)\,\dagger}\mathbf{B}^{(k)}$ over real symbols in one. The normalized self-interference per transmitted real symbol is then

$$\widetilde{\mathcal{N}} = \frac{1}{2Q}\,\frac{\mathcal{N}}{\left(\sum_k \mathrm{Tr}\,\mathcal{S}^{(kk)}/4Q\right)^2} = \frac{1}{2Q}\,\mathcal{N} . \qquad (4.47)$$

Note that the measure (4.47) differs slightly from the measure of non-orthogonality introduced in [132]. For the examples, the self-interferences $\widetilde{\mathcal{N}}$ may be found in Table 4.1 on page 95.

4.3.3 Symbol Homogeneity

With a given non-vanishing self-interference, the division of self-interference between different symbols becomes an issue. Another principle that follows directly from requiring homogeneity in constellation space is symbol homogeneity (SH) [51, 135]. This means that all symbols should experience the same average interference from the other symbols. The ISI experienced by symbol c_k is

$$\mathcal{N}_k = \sum_{l\,:\,l\neq k} \mathrm{Tr}\left(\mathcal{S}^{(kl)\ 2}\right) . \tag{4.48}$$

Using a homogeneity argument as above, based on convexity of error functions, it follows that for a well designed matrix modulator,

$$\mathcal{N}_k = \mathcal{N}/2Q , \ \forall\, k . \tag{4.49}$$

Of the examples, "3+1" (4.32) and "PSK-rank" are not SH. For the former this is a direct consequence of the layering structure; all interference is between the three orthogonally encoded symbols of (3.39) and the fourth symbol x_4. For the latter this follows from the fact that different symbols are transmitted with different power.

4.3.4 Maximal Symbolwise Diversity

For a linear scheme, the diversity matrix is diagonal, or diagonally dominated, with a matrix Euclidean distance which is proportional to the sum of the individual symbol Euclidean distances. This is clearly visible in the examples discussed in the previous section. Thus the larger the sum $E_\Delta = \sum_{k=1}^{Q} |\Delta_k|^2$ of symbol Euclidean distances corresponding to an error event Δ is, the better the protection against error. From (4.40), it is heuristically clear that the smaller E_Δ is, the more important the higher-order expansion terms are. Thus a medium SNR design rule is to use the rank and determinant criteria for error events with small sum Euclidean distance. The first step towards this is to require Maximal Symbolwise Diversity (MSD) [132, 133, 135], each individual symbol should be encoded by a unitary matrix:

$$\mathbf{B}^{(k)\dagger}\mathbf{B}^{(k)} = \mathbf{I}_{N_t} \ \forall\, k . \tag{4.50}$$

This guarantees full diversity protection for one-symbol error events, which have the smallest Euclidean distances. Intuitively it is clear that if the matrix modulation does not provide full diversity protection against one-symbol error events, it cannot provide full diversity protection against all error events. This is the reason why MSD was called "full raw diversity" in [132]. Exploiting this raw diversity to construct true diversity, essential for high SNR performance, will be discussed in Chapter 7.

Of the examples, STTD-OTD (4.21) and "PSK-rank" (4.25) do not fulfil MSD, whereas the other schemes do. For the MSD modulators, the diversity matrices are of the form (4.30).

4.3.5 Maximizing Mutual Information

This self-explanatory design principle was suggested in [133]. In the following sections we shall see that minimal self-interference, Frobenius orthogonality and symbol homogeneity are all prerequisites for maximizing the mutual information.

4.4   MATCHED FILTERING AND MAXIMUM LIKELIHOOD METRIC

In order to be able to consider the information provided by a matrix modulation, the equivalent channel matrix should be discussed, along with the equivalent signal model. These concepts bear a near relationship to matched filtering and to the maximum likelihood detection metric.

4.4.1   Equivalent Channel Matrix

The received signals in (4.1) may be reconsidered as a $(TN_r) \times 1$ vector $\mathbf{y}$; the vectors corresponding to a given receive antenna are concatenated. Due to linearity of the modulation matrix $\mathbf{X}$ in the real symbols c_k, this vector may be expressed in the form

$$\mathbf{y} = \widetilde{\mathcal{H}}\,\mathbf{c} + \mathbf{n}\,. \tag{4.51}$$

The equivalent channel matrix $\widetilde{\mathcal{H}}$ is a $TN_r \times 2Q$ matrix which depends on the structure of the code and the channel. Additive noise is denoted by $\mathbf{n}$. Note that both $\widetilde{\mathcal{H}}$ and $\mathbf{y}$ have complex entries. The equivalent channel may be expressed as a column vector of N_r different $T \times 2Q$ equivalent channel matrices $\widetilde{\mathcal{H}}^{(n)}$ for a given receive antenna n,

$$\widetilde{\mathcal{H}} = \begin{bmatrix} \widetilde{\mathcal{H}}^{(1)} \\ \widetilde{\mathcal{H}}^{(2)} \\ \vdots \\ \widetilde{\mathcal{H}}^{(N_r)} \end{bmatrix}. \tag{4.52}$$

For many schemes, splitting the equivalent signal model (4.51) into parts related to the I and Q branches (real symbols) of the individual symbols is not necessary. For these, the equivalent channel matrix is a $TN_r \times Q$ matrix, and the equivalent signal model reads

$$\mathbf{y} = \mathcal{H}\,\mathbf{x} + \mathbf{n}\,. \tag{4.53}$$

Examples of such are all schemes with a modulation matrix $\mathbf{X}$ including no complex conjugations of symbols, e.g. "PSK-rank" (4.25) and "cyclic" (4.29). If during one symbol period, only symbols or complex conjugate symbols are transmitted, the signal model may also be written in the form (4.53). This includes STTD (3.28) and ABBA (4.35). Schemes that cannot be written in the form (4.53) include complex orthogonal designs for $N_t > 2$, e.g. (3.39) and their generalizations, e.g. the 3+1 layered scheme (4.32). Here, the more general real signal model (4.51) will be discussed. It is straightforward to restrict the results to the complex signal model (4.53).

A third signal model is possible for schemes allowing (4.53). The complex signal model may be split into real and imaginary parts as

$$\bar{\mathbf{y}} = \begin{bmatrix} \mathrm{Re}\ \mathbf{y} \\ \mathrm{Im}\ \mathbf{y} \end{bmatrix} = \begin{bmatrix} \mathrm{Re}\ \mathcal{H} & -\,\mathrm{Im}\ \mathcal{H} \\ \mathrm{Im}\ \mathcal{H} & \mathrm{Re}\ \mathcal{H} \end{bmatrix} \begin{bmatrix} \mathrm{Re}\ \mathbf{x} \\ \mathrm{Im}\ \mathbf{x} \end{bmatrix} + \begin{bmatrix} \mathrm{Re}\ \mathbf{n} \\ \mathrm{Im}\ \mathbf{n} \end{bmatrix} \equiv \bar{\mathcal{H}}\bar{\mathbf{x}} + \bar{\mathbf{n}}\,. \tag{4.54}$$

Here $\bar{\mathcal{H}}$ is a $2TN_r \times 2Q$ matrix, operating on real symbols, with a *real vector* as output. The bars adorning objects indicate that we are considering signal model with real input and output. The notation $\bar{\mathbf{x}}$ stresses the fact that the $2Q$ real symbols are real and imaginary parts of the Q complex symbols, not an arbitrary collection of $2Q$ real symbols. This signal model will be used when analysing detection in Chapter 6.

With perfect or partial channel-state information at the receiver, matched filtering may be performed by acting on the received signal with the estimate of the Hermitian conjugate (complex conjugate transpose) of $\widetilde{\mathcal{H}}$. As the transmitted symbols are decomposed into real symbols, one should consider real-valued matched filtering,

$$\bar{\mathbf{z}} = \text{Re}\left[\widetilde{\mathcal{H}}^\dagger \mathbf{y}\right] = \tfrac{1}{2}\left[\widetilde{\mathcal{H}}^\dagger \ \widetilde{\mathcal{H}}^T\right] \left[\begin{array}{c} \mathbf{y} \\ \mathbf{y}^* \end{array}\right] \equiv \bar{\mathcal{R}}\,\mathbf{c} + \text{noise}\,, \tag{4.55}$$

where the $2Q \times 2Q$ matrix

$$\bar{\mathcal{R}} = \text{Re}\left[\widetilde{\mathcal{H}}^\dagger \widetilde{\mathcal{H}}\right] = \sum_{n=1}^{N_r} \text{Re}\left[\widetilde{\mathcal{H}}^{(n)\,\dagger}\widetilde{\mathcal{H}}^{(n)}\right] \equiv \sum_{n=1}^{T} \bar{\mathcal{R}}^{(n)} \tag{4.56}$$

is the (real valued) matched filter correlation matrix. The last equality expresses the correlation matrix as a sum of correlation matrices $\bar{\mathcal{R}}^{(n)}$ for one receive antenna.

Different detectors are different ways to invert $\bar{\mathcal{R}}$ in (4.55). It is intuitively clear that the ratio of the power of the off-diagonal elements to the diagonal element of the k:th row of $\bar{\mathcal{R}}$ describes the power of the interference caused by other symbols on c_k. From a multiuser detection point of view, it is natural to demand the expected values of these ratios to be as small as possible. This is the essence of the minimal self-interference criterion.

For a linear space–time modulation of the form (4.4), the matrix elements of $\widetilde{\mathcal{H}}^{(n)}$ may be expressed in terms of matrix elements of the basis matrices $\mathbf{B}^{(k)}$ and the channel matrix $\mathbf{H}$ as

$$\widetilde{\mathcal{H}}^{(n)}_{tk} = \sum_m b^{(k)}_{tm} h_{mn} = \left(\mathbf{B}^{(k)}\,\mathbf{H}\right)_{tn}. \tag{4.57}$$

The matrix elements of the correlation matrix are given in terms of traces, and may be expressed in terms of the RH matrices (4.11) as

$$\bar{\mathcal{R}}_{jk} = \tfrac{1}{2}\,\text{Tr}\left[\mathcal{S}^{(j,k)}\,\mathbf{H}\,\mathbf{H}^\dagger\right] \tag{4.58}$$

Written in terms of objects in signal model (4.1), the matched filter outputs read

$$\bar{z}_k = \text{Re}\,\text{Tr}\left(\mathbf{H}^\dagger\,\mathbf{B}^{(k)\dagger}\,\mathbf{Y}\right). \tag{4.59}$$

Mathematically, these may be be understood as coordinates of projections of the received signals to the basis matrix direction of the k'th encoded symbol. The real projection operators corresponding to $\bar{z}_k$ are constructed from the k'th column of the equivalent channel matrix. They are $2TN_r \times 2TN_r$ matrices with elements

$$\mathcal{P}^{(k)}_{tt'} = \frac{1}{2\bar{\mathcal{R}}_{kk}}\left[\begin{array}{c} \widetilde{\mathcal{H}}_{tk} \\ \widetilde{\mathcal{H}}^*_{tk} \end{array}\right]\left[\ \left(\widetilde{\mathcal{H}}^\dagger\right)_{kt'}\ \ \left(\widetilde{\mathcal{H}}^T\right)_{kt'}\ \right], \tag{4.60}$$

acting on vector $[\mathbf{y}^T\ \mathbf{y}^\dagger]^T$. It follows from the definition of $\bar{\mathcal{R}}$ in (4.56) that these are indeed projection operators, $\mathcal{P}^{(k)\,2} = \mathcal{P}^{(k)}$.

Complex valued matched filter outputs, corresponding to the complex modulation symbols x_k in (4.4), are

$$z_k = \bar{z}_{2k-1} + \text{j}\,\bar{z}_{2k}. \tag{4.61}$$

These can be written in terms of the basis matrices for the complex symbols and their conjugate, $\mathbf{B}^{(k\pm)}$, resulting in expressions of the kind discussed in [141].

For schemes allowing complex valued matched filtering based on signal model (4.53), the complex valued matched filter outputs (4.61) may be constructed directly from (4.53). For these models, the correlation matrix $\mathcal{R}$ is a $Q \times Q$ matrix of the form

$$\mathcal{R} = \mathcal{H}^\dagger \mathcal{H} \ . \tag{4.62}$$

Now the real symbol correlation matrix corresponding to $\bar{\mathcal{H}}$ in (4.54) can be written in terms of the complex symbol correlations as

$$\bar{\mathcal{R}} = \bar{\mathcal{H}}^\dagger \bar{\mathcal{H}} = \begin{bmatrix} \operatorname{Re} \mathcal{R} & -\operatorname{Im} \mathcal{R} \\ \operatorname{Im} \mathcal{R} & \operatorname{Re} \mathcal{R} \end{bmatrix}, \tag{4.63}$$

and it coincides with the real symbol correlation matrix obtained from $\widetilde{\mathcal{H}}$ for schemes allowing complex valued matched filtering.

Examples of equivalent channel matrices and correlation matrices may be found in section 6.7.

4.4.2   Maximum Likelihood Detection Metric

From the signal models (4.1) and (4.51), a maximum likelihood (ML) detection metric of a linear matrix modulation can be constructed. Written in terms of the matched filter outputs $\bar{z}_k$ of (4.55), the detection metric is

$$\begin{aligned} \Omega &= \operatorname{Tr} (\mathbf{Y} - \hat{\mathbf{X}}\hat{\mathcal{H}})^\dagger (\mathbf{Y} - \hat{\mathbf{X}}\hat{\mathcal{H}}) = (\mathbf{y} - \hat{\tilde{\mathcal{H}}}\,\hat{\mathbf{c}})^\dagger (\mathbf{y} - \hat{\tilde{\mathcal{H}}}\,\hat{\mathbf{c}}) \\ &\equiv \Omega_0 + \sum_k \Omega_k + \Omega_{\mathrm{no}} \end{aligned} \tag{4.64}$$

$$\Omega_0 = \operatorname{Tr}\left(\mathbf{Y}^\dagger \mathbf{Y}\right) - \sum_k \bar{z}_k^2 / \bar{\mathcal{R}}_{kk}$$

$$\Omega_k = \left(\bar{z}_k - \bar{\mathcal{R}}_{kk}\,\hat{c}_k\right)^2 / \bar{\mathcal{R}}_{kk} \tag{4.65}$$

$$\Omega_{\mathrm{no}} = \operatorname{Tr}\left(\mathcal{S}\,\hat{\mathbf{H}}\,\hat{\mathbf{H}}^\dagger\right) \tag{4.66}$$

Here the ˆs indicate estimates, and will be suppressed below. The different channels that the symbol c_k was transmitted over are diversity-combined in $\bar{\mathcal{R}}_{kk}$, the diagonal element of the correlation matrix (4.58), which is a function of the part of the diversity matrix (4.9) pertaining to c_k.

The possible non-orthogonality of the modulator shows up in Ω_{no}, which depends on the self-interference $\mathcal{S}$ of the code. Ω_0 is an irrelevant normalization factor. The most important part of the metric is the symbolwise part $\sum_k \Omega_k$, expressing the connection of the matched filter projections $\bar{z}_k$ and the symbol estimates $\hat{c}_k$.

If the diversity protection is the same for the real and imaginary parts of the original complex symbols, the metric may be written in terms of complex symbols and the complex projections of (4.61). This would restore the symbol-wise detection metric discussed in [141] in the absence of self-interference.

4.4.3   Design Criteria and ML Metric

Here, the heuristic design criteria of section 4.3 are reassessed on an intuitive level in terms of the metric (4.64) and the matched filter projections. MSD can be understood in terms of the linear part (4.65) of the metric. This part expresses the protection of symbols against noise, omitting self-interference. The detection power of symbol s_k in (4.65) is given by the corresponding diagonal element $\bar{\mathcal{R}}_{kk}$ of the correlation matrix. The values of $\bar{\mathcal{R}}_{kk}$ for different k are coupled by the overall power constraint. Due to the convexity of the Q function, it is beneficial that each symbol is detected with the same power; the $\bar{\mathcal{R}}_{kk}$ should be as equal as possible. MSD (4.50) ensures that $\bar{\mathcal{R}}_{kk}$ are equal for all symbols, so that the detection of each symbol in (4.65) is supported by a diversity combination of all channels used for transmission.

Symbol homogeneity (4.49) continues the same intuition to the non-orthogonal term (4.66) in the metric, and ensures that the interference experienced by different symbols is the same on average.

To understand Frobenius orthogonality(4.43), notice that the received signal R lives in $2TN_r$ real dimensions, and can be seen as a result of $2TN_r$ independent measurements of a real quantity. As long as $K \leq T\min(N_r, N_t)$, (4.43) indicates that the projection operators $\mathcal{P}^{(k)}$ of (4.60) project out linearly independent subspaces of the received signal space for generic channel realizations. That is, with $K \leq T\min(N_r, N_t)$, one independent real measurement may be performed for each real symbol, which dominantly affects this symbol, and affects the other symbols only through the non-orthogonal part (4.66) of the metric. The result of this measurement is used to detect the symbol according to (4.64). For non-generic channel realizations, the channel coefficients conspire so that some matched filter projections are linearly dependent; the information-carrying dimensionality of the received signal matrix is lowered. *A priori*, such channel realizations form a zero-measure set and need not be considered. With added noise, they have a finite probability.

For an orthogonal design, self-interference vanishes. Thus the interference part of the metric (4.66) vanishes, and the linear part (4.65) provides a Euclidean metric in the the space of projections of the received signals to the basis matrix directions. In other words, all channel realizations are generic.

As an example, consider STTD (3.28), received with one antenna. A Euclidean metric in the received signal space is given by the vector norm in two complex dimensions, $\mathbf{Y}^\dagger\mathbf{Y}$. The fact that (3.28) can be linearly detected shows up in the fact that the complex matched filter projections (4.61) form another basis in the received signal space, which is orthogonal with respect to an Euclidean metric:

$$\mathbf{Y}^\dagger\mathbf{Y} = \sum_{k=1}^{2} |z_k|^2 / \mathbf{H}^\dagger\mathbf{H} . \tag{4.67}$$

In other words, (4.65) gives a Euclidean metric in the space of matched filter outputs, which is directly inherited from the metric in received signal space.

With non-vanishing $\mathcal{S}$, (4.66) pinches the manifold of projected received signals so that some points far from each other in terms of the unpinched Euclidean metric become neighbours. These points correspond to non-generic channel realizations. A crude measure of the probability of a channel realization to be in the vicinity of a pinching is provided by the measure of self-interference (4.47). Optimizing performance using the rank, determinant or other more refined criteria arising from (3.24) takes on the task of investigating concrete realizations of this pinching, and minimizing the contribution of the worst configurations to the selected error rate.

4.5 MUTUAL INFORMATION

In [133] it was suggested that the mutual information of a linear matrix modulation (4.4) should be maximized. This is a very intuitive design metric, which is especially pertinent for concatenated schemes, where the task to turn information into performance is left to the concatenated channel code. In this section we shall take a close look at the mutual information. We shall see that minimizing the self-interference and maximizing the mutual information are equivalent.

4.5.1 Information and Interference

Generalizing Equation (2.21), the mutual information provided by a linear space–time modulation $\mathbf{X}$ extending over T channel uses can be written in terms of the equivalent channel model as [133]

$$\mathcal{I} = \frac{1}{T} \, \mathrm{E}\Big\langle \log\det \Big(\mathbf{I}_{TN_{\mathrm{r}}} + \frac{1}{\sigma^2} \, \widetilde{\mathcal{H}} \mathbf{Q} \widetilde{\mathcal{H}}^\dagger \Big) \Big\rangle_{\mathbf{H}} . \tag{4.68}$$

Here $\mathbf{I}_{TN_{\mathrm{r}}}$ is the $TN_{\mathrm{r}} \times TN_{\mathrm{r}}$ identity matrix and $\mathrm{E}\langle\rangle_{\mathbf{H}}$ is the expectation value over channel realizations, with appropriate pdf for the channel states. The real symbols are assumed continuous Gaussian with covariance $\mathbf{Q} = \mathrm{E}\left\langle c\, c^{\mathrm{T}} \right\rangle$. The power constraint is $\mathrm{Tr}\,\mathbf{Q} = PT$. Note that here, information is measured in nats. With the covariance proportional to the identity, the mutual information becomes

$$\mathbf{Q} = \frac{TP}{2Q} \, \mathbf{I}_{2Q} , \tag{4.69}$$

the mutual information becomes

$$\mathcal{I} = \frac{1}{T} \, \mathrm{E}\Big\langle \log\det \Big(\mathbf{I}_{2Q} + \frac{\eta T}{2Q} \, \widetilde{\mathcal{H}}^\dagger \, \widetilde{\mathcal{H}} \Big) \Big\rangle_{\mathbf{H}} . \tag{4.70}$$

Here the SNR is defined as $\eta = P/\sigma^2$. Note that cyclicity of det was used to write the argument in real symbol space. The expression becomes more transparent if the argument is written in terms of the equivalent channel correlation matrix (4.56) as

$$\mathcal{I} = \frac{1}{2T} \, \mathrm{E}\Big\langle \log\det \Big(\mathbf{I}_{2Q} + \frac{\eta T}{Q} \, \mathcal{R} \Big) \Big\rangle_{\mathbf{H}} . \tag{4.71}$$

The definition of SNR and total transmission power P yield a normalization condition on the basis matrices. The average transmission power during the T channel uses of $\mathbf{X}$ is

$$P = \frac{1}{T} \, \mathrm{E}\left\langle \mathrm{Tr}\,\mathbf{X}^\dagger \mathbf{X} \right\rangle_c \tag{4.72}$$

$$= \frac{1}{2T} \sum_{k=1}^{2Q} \mathrm{E}\left\langle c_k^2 \right\rangle \mathrm{Tr}\, \mathcal{S}^{(kk)} = \frac{P}{4Q} \sum_{k=1}^{2Q} \mathrm{Tr}\, \mathcal{S}^{(kk)} .$$

The last equality follows from covariance (4.69), which indicates that $\mathrm{E}\left\langle c_k^2 \right\rangle = \frac{PT}{2Q}$ for all k. Consistency requires that the basis matrices are normalized as

$$\sum \mathrm{Tr}\, \mathcal{S}^{(kk)} = 4Q , \tag{4.73}$$

which, furthermore, is consistent with (4.46).

From (4.58) it is evident that RH-orthogonal modulators have diagonal $\bar{\mathcal{R}}$. The main idea of this section is to argue that well-designed non-orthogonal modulators, which provide as much information as possible, have dominantly diagonal $\bar{\mathcal{R}}$. For this, recall Hadamard's determinant inequality in (A.14). The upper bound of the determinant is reached if all off-diagonal elements vanish. Considering the mutual information in terms of the correlation matrix (4.71), one once more arrives at the intuitive understanding that self-interference, i.e. the off-diagonal elements in $\bar{\mathcal{R}}$, should be minimized. From (4.58) it is clear that the Frobenius norms of the off-diagonal RH matrices $\mathcal{S}^{(i,k)}$, $i < k$ should be minimized.

4.5.2 Expanding Information

As mutual information is generically a non-transparent object, either computer searches [133] or bounding tools [134] have to be used, when it is maximized. A different approach, exploiting series expansions for discrete input information, was proposed in [157, 158].

The capacity $\mathcal{C}$, as well as the mutual information $\mathcal{I}$ provided by a modulator, can be expanded in the SNR η,

$$\mathcal{C} = \sum_{n=1}^{\infty} \eta^n \mathcal{C}_n , \qquad \mathcal{I} = \sum_{n=1}^{\infty} \eta^n \mathcal{I}_n . \tag{4.74}$$

A modulator that reaches capacity has $\mathcal{I}_n = \mathcal{C}_n$ for all n. In [159] it was shown that these two first expansion coefficients of $\mathcal{I}$ determine the slope of the spectral efficiency curve (as a function of E_b/N_0) at minimum E_b/N_0.

The expansion coefficients for the capacity in i.i.d. Rayleigh fading were derived in [158]. SNR normalization is fixed by taking the matrix channel normalized as

$$\mathrm{E} \left\langle h^*_{m_1 n_1} h_{m_2 n_2} \right\rangle = \delta_{m_1 m_2} \, \delta_{n_1 n_2} , \tag{4.75}$$

Here δ_{mn} is Kronecker's δ. The two first coefficients of the capacity per symbol period are

$$\mathcal{C}_1 = N_{\mathrm{r}} \tag{4.76}$$

$$\mathcal{C}_2 = -\frac{N_{\mathrm{r}}(N_{\mathrm{r}} + N_{\mathrm{t}})}{2 N_{\mathrm{t}}} . \tag{4.77}$$

The linear growth of MIMO capacity in i.i.d. channels [16, 20] is visible already in these first coefficients; for $N_{\mathrm{r}} = N_{\mathrm{t}} = N$, both are linear in N.

Expectation Values and Wick's Theorem: To get insight into the calculations below, it is worthwhile to rederive the capacity coefficients (4.76) and (4.77). First one may use the variant $\log \det = \mathrm{Tr} \log$ of (4.38) in the capacity expression (2.21) to get

$$\mathcal{C} = \mathrm{E} \left\langle \mathrm{Tr} \log \left(\mathbf{I}_{N_r} + \eta/N_{\mathrm{t}} \, \mathbf{H}^\dagger \mathbf{H} \right) \right\rangle .$$

Using the expansion (4.39) for the matrix logarithm, one gets the two first terms

$$\mathcal{C} = \frac{\eta}{N_{\mathrm{t}}} \, \mathrm{E} \left\langle \mathrm{Tr} \, \mathbf{H}^\dagger \mathbf{H} \right\rangle - \frac{\eta^2}{2 N_{\mathrm{t}}^2} \, \mathrm{E} \left\langle \mathrm{Tr} \, \mathbf{H}^\dagger \mathbf{H} \, \mathbf{H}^\dagger \mathbf{H} \right\rangle + \cdots \tag{4.78}$$

Thus the calculation reduces to a calculation of expectation values of products of channel matrix elements. It is an elementary consequence of Wick's theorem (see e.g. [160]) that for complex Gaussian distributed channels, such calculations reduce to combinatorics. To be more exact, consider a set of N complex random variables h_n with distribution function $1/\pi^N \exp\langle-\sum_n h_n^* h_n\rangle$, which gives $\mathrm{E}\langle h_n^* h_m\rangle = \delta_{mn}$. From Wicks theorem it follows that

$$\mathrm{E}\left\langle \prod_{l=1}^{L} h_l^* \prod_{m=1}^{M} h_m \right\rangle = \delta_{LM} \sum_{P\in S_M} \prod_{m=1}^{M} \mathrm{E}\langle h_m^* h_{P(m)}\rangle$$

$$= \delta_{LM} \sum_{P\in S_M} \prod_{m=1}^{M} \delta_{m\,P(m)}, \qquad (4.79)$$

where the sum is over all permutations of M objects. That is, the expectation value of M holomorphic and M anti-holomorphic variables reduces to the product of M expectation values of a pair of a holomorphic and anti-holomorphic variable, summed over all such pairings. A pairing of this type is called a (Wick) *contraction*. This method extends directly to matrix expectation values of the kind in (4.78). The indexing of the random variables just becomes more involved.

The first-order term in (4.78) is straight forward to calculate, and evaluates to

$$\mathcal{C}_1 = \frac{1}{N_t}\,\mathrm{E}\left\langle \mathrm{Tr}\,\mathbf{H}^\dagger\mathbf{H}\right\rangle = \frac{1}{N_t}\sum_{n=1}^{N_r}\sum_{m=1}^{N_t}\mathrm{E}\langle h_{mn}^* h_{mn}\rangle = \frac{1}{N_t}\sum_{n=1}^{N_r}\sum_{m=1}^{N_t} 1 = N_r\,.$$

The second-order term is a sum of two contractions,

$$\mathcal{C}_2 = \frac{-1}{2N_t^2}\,\mathrm{E}\left\langle \mathrm{Tr}\,\mathbf{H}^\dagger\mathbf{H}\,\mathbf{H}^\dagger\mathbf{H}\right\rangle$$

$$= \frac{-1}{2N_t^2}\sum_{n_1,n_2=1}^{N_r}\sum_{m_1,m_2=1}^{N_t}\mathrm{E}\left\langle h_{m_1 n_1}^* h_{m_1 n_2} h_{m_2 n_2}^* h_{m_2 n_1}\right\rangle$$

$$= \frac{-1}{2N_t^2}\sum_{n_1,n_2=1}^{N_r}\sum_{m_1,m_2=1}^{N_t}\Big(\mathrm{E}\left\langle h_{m_1 n_1}^* h_{m_1 n_2}\right\rangle\mathrm{E}\left\langle h_{m_2 n_2}^* h_{m_2 n_1}\right\rangle$$

$$+\,\mathrm{E}\left\langle h_{m_1 n_1}^* h_{m_2 n_1}\right\rangle\mathrm{E}\left\langle h_{m_2 n_2}^* h_{m_1 n_2}\right\rangle\Big)$$

$$= \frac{-1}{2N_t^2}\sum_{n_1,n_2=1}^{N_r}\sum_{m_1,m_2=1}^{N_t}(\delta_{n_2 n_1}+\delta_{m_1 m_2}) = \frac{-1}{2N_t^2}\left(N_r N_t^2 + N_t N_r^2\right)\,.$$

Thus the results (4.76) and (4.77) are reproduced.

Convergence: With the Taylor series rigorously defined, the question of convergence becomes an issue. Actually, in fading channels, the capacity and information expansions are asymptotic, and their convergence radius is exactly 0. However, for physically meaningful quantities (such as capacity and information), asymptotic series often carry all information about the expanded functions. The capacity and mutual information expansions for fading channels are likely to be Borel resummable. After calculating explicit expansion coefficients, the series may be Borel resummed, leading to generically valid results. From this it follows that capacity and information may be compared on an order-to-order bases, despite the non-convergence of the series. In practice this means that even though the expansion coefficients cannot be used to approximate values of mutual information (except at very low SNR), they provide a full set of derivatives of the mutual information at the origin. Mutual information functions are monotonously increasing, with monotonously decreasing positive odd order derivative functions and monotonously increasing negative even order derivative functions. For such functions, the values of the derivatives at the origin determine much of the behavior of the function.

Second-order Mutual Information Coefficients: In [157, 158], the mutual information expansion was derived for discrete input, i.e. for quantized modulation alphabets. The argument for this is that realistic modulation alphabets (M-PSK, M-QAM) indeed are discrete, and rarely have a Gaussian distribution. In fact, if mutual information is to be used as a design criterion for matrix modulators, discrete input mutual information is the correct measure to use. The bit error rate of a discrete transmission scheme can be both upper and lower bounded by functions depending on the discrete input mutual information. The lower bound is Fano's inequality (see [161]), and the less known upper bound is proved in [162].

The expansion of mutual information, however, is much simpler to derive for continuous Gaussian input, simply by expanding (4.71) in η. The two first expansion coefficients for discrete and continuous mutual information coincide if the constellations are symmetric with respect to the origin. This is true for all realistic constellations, so the more transparent continuous information expansion based on (4.71) may be used. As for the capacity, (4.38) and (4.39) are used to get

$$\mathcal{I} = \frac{\eta}{2Q} \, \mathrm{E}\Big\langle \mathrm{Tr}_Q \bar{\mathcal{R}} \Big\rangle_{\mathbf{H}} - \frac{\eta^2 T}{4Q^2} \, \mathrm{E}\Big\langle \mathrm{Tr}_Q \bar{\mathcal{R}}^2 \Big\rangle_{\mathbf{H}} + \dots \qquad (4.80)$$

Here the notation Tr_Q stresses that the trace is over symbol indexes. In i.i.d. Rayleigh fading, the channels may be integrated out explicitly. Using the form (4.58) of correlation matrix elements, and the normalization (4.73), the linear term in η reads

$$\mathcal{I}_1 = \frac{1}{4Q} \sum_k \mathrm{E}\Big\langle \mathrm{Tr}\left[\mathcal{S}^{(kk)} \, \mathbf{H} \, \mathbf{H}^\dagger \right] \Big\rangle_{\mathbf{H}} = \frac{N_r}{4Q} \sum_k \mathrm{Tr}\, \mathcal{S}^{(kk)} = N_r \, . \qquad (4.81)$$

That is, the linear term is always proportional to the power constraint P. Using all of the available power, first-order capacity (4.76) is reached. Integrating over the channel proceeds much as above; using (4.75) leads directly to the result. In (4.81), $\mathrm{Tr}[\ ]$ without subscript is trace over transmit antenna space.

The second-order coefficient of discrete input information was constructed in [86, 157], and it is always negative. For i.i.d. Rayleigh fading, it is straightforward to integrate the second-order contribution of the continuous input expression (4.80) using the contraction rule (4.79). The result coincides with the discrete input one, and is

$$\mathcal{I}_2 = -\frac{T N_r}{16Q^2} \sum_{k,l} \left(\mathrm{Tr}\left(\mathcal{S}^{(kl)} \right)^2 + N_r \left(\mathrm{Tr} \mathcal{S}^{(kl)} \right)^2 \right) \qquad (4.82)$$

From the results of [158] it follows that typical equidistant modulation alphabets such as QPSK, M-QAM etc, cannot reach capacity in orders higher than $n = 2$. Thus, for such modulations, it becomes essential to maximize $\mathcal{I}_2$.

Heuristic Rules from Mutual Information: The similarities of the low SNR mutual information expansion (4.80) and the low SNR expansion of the Chernoff bound in section

4.3 are striking. Thus it is not surprising that the same criteria may be deduced from the second-order information as were deduced from expanding the Chernoff bound.

- *Minimal Self-interference.* In (4.82), the first term is proportional to the total self-interference, which should thus be minimized.

- *Equal Average Power for Symbols.* The second term may be split into two parts. The first, $\sum_k \left(\text{Tr} \mathcal{S}^{(kk)} \right)^2$, depends on the symbol transmit powers $\text{Tr} \, \mathcal{S}^{(kk)}$ and indicates that their squared sum should be minimized. The sum of the symbol powers is constrained by the the the power constraint and the first-order information. It is easy to show that to minimize the sum of the square of the quantities $\text{Tr} \, \mathcal{S}^{(kk)}$ with the constraint (4.81), all $\text{Tr} \, \mathcal{S}^{(kk)}$ should be equal.

- *Frobenius Orthogonality.* The remaining part of the second term depends on the Frobenius inner products $\text{Tr} \, \mathcal{S}^{(kl)}$ for $k \neq l$, which are not constrained. These should thus be minimized, and preferably vanish, implying Frobenius orthogonality. These two minimizations cannot always be performed independently.

The heuristic rule of symbol homogeneity is harder to deduce from (4.82), and is left for future work.

Examples: To see how essential features of the mutual information can be seen already from the second-order coefficient, consider the example of STTD (Alamouti code) and vector modulation for $N_t = 2$. First we note that the second-order capacity for $N_t = 2$ is $\mathcal{C}_2 = -N_r(2 + N_r)/4$.

As observed in (4.19), STTD has no self-interference; $\mathcal{S}^{(kl)} = 0$, and the $\mathcal{S}^{(kk)}$ are proportional to identity. To comply with the normalization (4.73), the STTD basis matrices (4.20) have to be divided by $\sqrt{2}$. This gives $\text{Tr} \, \mathcal{S}^{(kk)} = 2$ and $\text{Tr} \left(\mathcal{S}^{(kk)} \right)^2 = 2$. With $Q = T = 2$, Equation (4.82) gives $\mathcal{I}_2 = -N_r \left(1 + 2N_r \right)/4$. The information loss is

$$\mathcal{C}_2 - \mathcal{I}_2^{\text{STTD}} = N_r \left(N_r - 1 \right)/4 \,. \tag{4.83}$$

Thus for $N_r = 1$ STTD reaches capacity, whereas for $N_r > 1$ it does not. This most essential feature of STTD mutual information [37,133], discussed in section 3.6, is thus visible already in the second-order coefficient.

This should be contrasted to the vector modulation (3.40) for $N_t = 2$, with $T = 1$ and $Q = 2$. The basis vectors are

$$\mathbf{B}^{(1)} = [\, 1 \quad 0 \,]; \quad \mathbf{B}^{(2)} = [\, j \quad 0 \,]; \quad \mathbf{B}^{(3)} = [\, 0 \quad 1 \,]; \quad \mathbf{B}^{(4)} = [\, 0 \quad j \,] \,. \tag{4.84}$$

The diversity and self-interference matrices are of the form (4.16) and (4.17), respectively. More exactly, the squares $\mathcal{S}^{(kk)} = 2\mathbf{B}^{(k)\dagger} \mathbf{B}^k)$ of the basis matrices are

$$\mathcal{S}^{(11)} = \begin{bmatrix} 2 & 0 \\ 0 & 0 \end{bmatrix}; \; \mathcal{S}^{(22)} = \begin{bmatrix} 2 & 0 \\ 0 & 0 \end{bmatrix}; \; \mathcal{S}^{(33)} = \begin{bmatrix} 0 & 0 \\ 0 & 2 \end{bmatrix}; \; \mathcal{S}^{(44)} = \begin{bmatrix} 0 & 0 \\ 0 & 2 \end{bmatrix} \tag{4.85}$$

This gives $\text{Tr} \, \mathcal{S}^{(kk)} = 2$ and $\text{Tr} \left(\mathcal{S}^{(kk)} \right)^2 = 4$. The self-interference matrices involving $\mathbf{B}^{(1)}$ are

$$\mathcal{S}^{(12)} = 0; \quad \mathcal{S}^{(13)} = \begin{bmatrix} 0 & 1 \\ 1 & 0 \end{bmatrix}; \quad \mathcal{S}^{(14)} = \begin{bmatrix} 0 & -j \\ j & 0 \end{bmatrix} \,, \tag{4.86}$$

and for $\mathbf{B}^{(2)}, \mathbf{B}^{(3)}$ and $\mathbf{B}^{(4)}$ the same pattern repeats. From this it follows that $\operatorname{Tr} \mathcal{S}^{(kl)} = 0$ for $k \neq l$ (Frobenius orthogonality), leading to $\sum_{k,l} \left(\operatorname{Tr} \mathcal{S}^{(kl)} \right)^2 = 2Q \cdot 4 = 16$. Furthermore, $\operatorname{Tr} \left(\mathcal{S}^{(kl)} \right)^2 = 2$ for 8 of the 12 self-interference terms, and vanishes for the remaining four. Thus $\sum_{k,l} \left(\operatorname{Tr} \mathcal{S}^{(kl)} \right)^2 = 2Q \cdot 4 + 2 \cdot 2Q \cdot 2 = 32$. Equation (4.82) gives $\mathcal{I}_2 = -N_r(32 + 16N_r)/64$ which coincides with the second-order capacity for any N_r.

Lagrangian Extremization: The second-order mutual information is a complicated function of the $2QTN_t$ complex degrees of freedom in the set of basis matrices $\mathbf{B}^{(k)}$, constrained by the power constraint (4.72). Nevertheless it is possible to find an equation for the extrema of the second-order information with respect to the full set of independent variables. Depending on the eigenvalues of the Hessian, such extrema are either local minima, local maxima, or saddle points of the the second-order information as a function of the matrix modulator.

The power constraint (4.72) turns the linear manifold of the degrees of freedom in the $2Q$ basis matrices to a non-linear manifold. Differentiating with respect to the independent degrees of freedom on this non-linear manifold is much simplified by using the method of Lagrange multipliers. Before deriving extremality equations for the mutual information, this principle is revisited in the light of a simple example.

Lagrangian extremization works on constraint (hyper) surfaces. We have a function of many variables, with a parameter space constrained by a number of constraint equations. As an example, find the extrema of the function $f(x,y) = xy$ on the constraint surface $r^2 = x^2 + y^2$. The task is thus to find the minima and maxima of the product of coordinates on a one-dimensional circle. Lagrangian extremization proceeds by adding the constraint to the function, multiplied by a Lagrange multiplier λ: $f_L(x, y, \lambda) = xy + \lambda \left(x^2 + y^2 - r^2\right)$. This function is called a Lagrangian. Differentiating with respect to λ gives the constraint surface, and the joint extrema of f_L with respect to x, y, λ give the extrema of f on the constraint surface. At the extrema w.r.t. x, y we have

$$y + 2x\lambda = 0 , \quad x + 2y\lambda = 0$$

Note that there are two extremality equations for the two variables x, y. These equations have two solutions, $x = \pm y$ with $\lambda = \mp \frac{1}{2}$. The Hessian w.r.t. x, y is

$$\frac{\mathrm{d}^2 f_L}{(\mathrm{d}(x,y))^2} = \left[\begin{array}{cc} 2\lambda & 1 \\ 1 & 2\lambda \end{array} \right],$$

and the eigenvalues of the Hessian are $2\lambda \pm 1$. For the solution $x = y$, $\lambda = -1/2$, the eigenvalues are $[0, -1]$, whereas for $x = -y$, $\lambda = 1/2$, the eigenvalues are $[1, 0]$. Thus the first extremum is a maximum, whereas the second extremum is a minimum. On the constraint surface the solution $x = y$ corresponds to the global maxima $x = y = \pm r/\sqrt{2}$ and the solution $x = -y$ to the global minima $x = -y = \pm r/\sqrt{2}$.

In this simple case, extremization is easier to perform on the non-linear constraint surface itself. Changing variables to $x = r \sin \phi$ and $y = r \cos \phi$, the constraint equation is trivially fulfilled. Extremization of f means finding the extrema of $\sin 2\phi$. The maxima are at $\phi = \pi/4 + n\pi$, and the minima at $\phi = -\pi/4 + n\pi$.

Matrix Differentiation: Due to the matrix structure of the objects entering the second-order information (4.82), the extremality equations may be expressed in matrix form. Differentiating a trace with respect to matrix elements one gets

$$
\frac{\mathrm{d}}{\mathrm{d}\,a_{nm}} \operatorname{Tr} \mathbf{A}\,\mathbf{B} = \frac{\mathrm{d}}{\mathrm{d}\,a_{nm}} \sum_{kl} a_{kl}\,b_{lk} = \sum_{kl} \delta_{nk}\,\delta_{ml}\,b_{lk} = b_{mn}\;, \tag{4.87}
$$

This can be written in terms of the matrices as

$$
\frac{\mathrm{d}}{\mathrm{d}\,\mathbf{A}^{\mathrm{T}}} \operatorname{Tr}(\mathbf{A}\,\mathbf{B}) = \mathbf{B}\;, \tag{4.88}
$$

which gives a simple rule for matrix differentiation.

Closed Form Extremality Equation: To find extrema of the second-order information with respect to the basis matrices $\mathbf{B}^{(k)}$, the power constraint (4.72) is put into effect with a Lagrange multiplier λ, and the resulting Lagrangian is differentiated with respect to $\mathbf{B}^{(l)\mathrm{T}}$. The result is [86]

$$
\sum_{k=1}^{2Q} \left(N_{\mathrm{r}}\mathbf{B}^{(k)}\operatorname{Tr}\left(\mathcal{S}^{(kl)}\right) + \mathbf{B}^{(l)}\mathbf{B}^{(k)\dagger}\mathbf{B}^{(k)} + \mathbf{B}^{(k)}\mathbf{B}^{(l)\dagger}\mathbf{B}^{(k)}\right) = \lambda\mathbf{B}^{(l)} \tag{4.89}
$$

This is a collection of $2QTN_{\mathrm{t}}$ equations, which is exactly the number of equations that an extremum of a function of $2QTN_{\mathrm{t}}$ variables should fulfil. The dependence of the variables due to the power constraint gives rise to the λ term. At an extremum of the mutual information, each coefficient matrix $\mathbf{B}^{(l)}$ fulfils this equation for some real λ (the same for all l).

With a diversity optimal MSD basis, each coefficient matrix is unitary, so the second term in the left-hand side of (4.89) is trivially proportional to $\mathbf{B}_l$. Also, for Frobenius orthogonal (but not necessarily orthonormal) bases, we have

$$
\operatorname{Tr}\left(\mathcal{S}^{(kl)}\right) = 2p_k\delta_{kl}\;. \tag{4.90}
$$

The power of the basis matrix $\mathbf{B}_k$ is p_k. Typically all coefficient matrices have the same power, but in some cases optimal schemes do not have this property. With unitary coefficient matrices and (4.90), equation (4.89) reduces to

$$
\sum_{k=1}^{2Q} \mathbf{B}^{(k)}\,\mathbf{B}^{(l)\dagger}\,\mathbf{B}^{(k)} = (\tilde{\lambda} - 2N_{\mathrm{r}}\,p_l)\,\mathbf{B}^{(l)}\;,\quad l = 1,\ldots,2Q \tag{4.91}
$$

4.6 EXPANSION AROUND DIAGONAL DOMINANCE

In sections 4.3 and 4.5, low SNR properties of pairwise error probabilities and mutual information were considered, respectively, leading to a number of intuitive design criteria for linear matrix modulation. The validity of these criteria may be extended to all SNRs. The underlying principle is the diagonal dominance of positive semidefinite Hermitian matrices. To see the effects of this, a $N \times N$ matrix $\mathbf{M}$ is divided into two parts; a diagonal matrix depending on the trace, and a traceless part $\widetilde{\mathbf{M}}$:

$$
\mathbf{M} = \frac{1}{N}\operatorname{Tr}\mathbf{M}\;\mathbf{I}_N + \widetilde{\mathbf{M}}\;. \tag{4.92}
$$

Here, clearly

$$\operatorname{Tr} \widetilde{\mathbf{M}} = 0 . \tag{4.93}$$

Now in an expression of the type (3.24) or (4.71), a positive semidefinite Hermitian matrix (the distance matrix, the matched filter correlation matrix) appears within a determinant, combined with SNR and the identity matrix: $\det(\mathbf{I} + \eta\mathbf{M})$. Using (4.92), this may be written in the form

$$
\begin{aligned}
\det\left[\mathbf{I} + \eta\mathbf{M}\right] &= \det\left[\left(1 + \frac{\eta}{N}\operatorname{Tr}\mathbf{M}\right)\mathbf{I} + \eta\widetilde{\mathbf{M}}\right] \\
&= \left(1 + \frac{\eta}{N}\operatorname{Tr}\mathbf{M}\right)^N \det\left[\mathbf{I} + \frac{\eta}{1 + \frac{\eta}{N}\operatorname{Tr}\mathbf{M}}\widetilde{\mathbf{M}}\right] .
\end{aligned}
\tag{4.94}
$$

Thus the determinant may be expanded in small η, as was done in sections 4.3 and 4.5, or in large $\operatorname{Tr}\mathbf{M}$, or in both, and all these expansions are closely related to each other, as indicated by (4.94).

Due to the finite dimensionality of $\mathbf{M}$, the expansion in $\operatorname{Tr}\mathbf{M}$ is finite, and always converges. For positive-semidefinite Hermitian matrices, the expansion is dominated by a few first terms. Dominance by trace alone may be deduced from the well-known Hadamard inequality. The second-order term is related to the Schur inequality. These properties are discussed in Appendix A.

This means that for N_t Tx antennas, the expansion of (3.24) in the trace of the distance matrix $\operatorname{Tr}\mathbf{X}^\dagger\mathbf{X}(\Delta)$ leads to N_t design criteria, each of which relates to one unitary (trace) invariant of the distance matrix. The trace criterion in section 3.5.2 is the first of these criteria, and the other may be called higher trace criteria. The determinant (and rank) criterion is a combination of all N_t higher trace criteria.

The heuristic low SNR criteria of sections 4.3 and 4.5 may be derived from the mutual information and the pairwise error probability by considering the second-order trace invariants and the Schur inequality, and are valid at any SNR.

4.6.1 Diagonal Dominance in Information Measures

By simulations it can be seen that schemes with different second-order mutual information typically have different (continuous input) mutual information at all SNR. Generic statements can be derived by using the diagonal dominance of the correlation matrix $\mathcal{R}$ in the mutual information expression, based on the discussion in Appendix A. The following criteria are prerequisites for maximizing the mutual information at any SNR

- The tight frame criterion [134], which requires that the expectation value of the diversity matrix (4.9) is proportional to the identity:

$$\mathrm{E}\langle\mathcal{D}\rangle = \sum_k \mathcal{S}^{(kk)} \sim \mathbf{I}_{N_t} . \tag{4.95}$$

  This may be considered as an (over)completeness relation in Tx antenna space. That is, the modulator uses all transmit antennas equally much.

- Unitary equivalence of $\mathcal{S}^{(kk)}$. This means that all squared basis matrices $\mathcal{S}^{(kk)}$ should be possible to map on to each other with unitary transformations. For each $k \neq l$, a unitary matrix $\mathbf{U}^{(kl)}$ should exist that satisfies

$$\mathcal{S}^{(kk)} = \mathbf{U}^{(kl)\dagger} \mathcal{S}^{(ll)} \mathbf{U}^{(kl)} . \tag{4.96}$$

In particular, this means that each symbol is transmitted with the same power. This is an enhanced version of requiring the same average power for each symbol.

- Minimal self-interference, discussed above.

- Frobenius orthogonality, discussed above.

- Symbol homogeneity, discussed above.

Non-trivial examples of the two first criteria are provided by a vector modulation, and "PSK-rank" (4.25).

Consider e.g. the $N_t = 2$ vector modulation (3.40), with the $S^{(kk)}$ matrices (4.85). Their sum is

$$E\langle \mathcal{D} \rangle = \sum_k S^{(kk)} = 4\, \mathbf{I}_2 \, , \tag{4.97}$$

so the tight frame condition is fulfilled. Also, two $S^{(kk)}$s are either equal, or they may be mapped to each other with the unitary matrix $\mathbf{U}^{(13)} = \begin{bmatrix} 0 & 1 \\ 1 & 0 \end{bmatrix}$. Thus the $S^{(kk)}$s are unitarily equivalent.

For "PSK-rank" (4.25)

$$E\langle \mathcal{D} \rangle = \sum_k S^{(kk)} = \frac{1}{4} \begin{bmatrix} 4 & 0 & 0 & 3 \\ 0 & 5 & 1 & 0 \\ 0 & 1 & 6 & 2 \\ 3 & 0 & 2 & 7 \end{bmatrix} . \tag{4.98}$$

The frame is not tight. Nor are the $S^{(kk)}$ unitarily equivalent, which follows from the observation that different symbols are transmitted with different power.

All the other examples fulfil trivially the tight frame and unitary equivalence condition, as these are consequences of maximal symbolwise diversity.

4.6.2 Diagonal Dominance in Performance Measures

From the diagonal dominance of the distance matrix $\mathbf{X}^\dagger \mathbf{X}(\mathbf{\Delta})$ it follows that maximal symbolwise diversity is a prerequisite for optimal performance at any SNR, as long as the union bound is tight. Also, the information maximization criteria above may be derived from optimizing performance. These information maximization criteria are further restricted when optimizing performance. A prime example of this is MSD itself, which is a refinement of the tight frame and unitary equivalence of $S^{(kk)}$ conditions arising from information maximization.

Furthermore, it is intuitively clear that if full SH cannot be reached, it is preferable to have at least *partial SH*, i.e. the symbols "as homogeneous as possible". For each real symbol, the self-interference $\mathcal{N}_k$ of (4.48) is calculated. Symbols with the same self-interference belong to the same category. The partial SH criterion now requires that the number of such categories should be as small as possible. The best performance (SH) is reached if only one such category exists. This property may be proved by minimizing the union bound of pairwise errors.

Table 4.1: Characteristics of $N_t = 4, R_s = 1, T = 4$ examples

Scheme	Rank	$\widetilde{\mathcal{N}}$	$\frac{\mathcal{I}_2}{\mathcal{C}_2}$(dB)	XTRM	MSD	FO	SH	Rel $\mathcal{N}$
STTD-OTD	2	0	-0.79	yes	no	yes	yes	0:0
PSK-rank	4	$\frac{108}{121}$	-3.09	no	no	no	no	18:26:31:33
cyclic	2	3/4	-2.04	yes	yes	yes	yes	1:1:1:1
3+1	2	3/8	-1.14	no	yes	yes	no	1:3
2+2 ABBA	2	1/4	-0.79	yes	yes	yes	yes	1:1

Table 4.2: Operation point in E_b/N_0 at various target BERs for some $R_s = 1, N_t = 4$ schemes

Scheme	E_b/N_0 at target BER			
	10^{-1}	10^{-2}	10^{-3}	5×10^{-4}
PSK-rank	2.61	8.98	13.0	14.1
Cyclic	1.77	8.90	15.9	22.6
STTD-OTD	0.70	8.46	14.1	15.7
3+1	1.14	7.76	12.6	13.9
2+2 ABBA	0.60	7.26	12.1	13.5

4.7 PERFORMANCE OF EXAMPLES

Here, the rate $R_s = 1$ examples for 4 Tx antennas of section 4.2 will be discussed in terms of the design rules discussed in this chapter.

First it should be noted that STTD-OTD was designed to have explicit Tx diversity 2, and it has no self-interference. Of the non-orthogonal schemes, PSK-rank is not MSD, and its self-interference has a trace. All other are MSD and Frobenius orthogonal. Moreover, the 1+3 layered scheme and PSK-rank are non-SH (self-interference non-homogeneously distributed between symbols), whereas the cyclic scheme and 2+2 layered ABBA are SH.

These characteristics are gathered in Table 4.1. Numerical values for the self-interference per bit are in column $\widetilde{\mathcal{N}}$. These have been calculated with a normalization of the total transmit power to 1 per basis matrix. The ratios of the second-order mutual information and channel capacity $\mathcal{I}_2/\mathcal{C}_2$ are also reported. These are measured in dB. The column "XTRM" indicates whether the scheme fulfils the second-order information extremality equation (4.89). Finally, the column "rel $\mathcal{N}$" indicates the relative self-interference experienced by the layers in the matrix modulation. The schemes are named by their layering or possible nickname.

To assess the effects of various properties on performance, simulation results with these schemes can be found in Figure 4.1 and Table 4.2. All schemes have been simulated in i.i.d. Rayleigh block fading (block length 4), with one receive antenna, perfect channel estimates at Rx and maximum likelihood detection.

The following salient observations may be made from comparing the characteristics in Table 4.1 to performance results.

- Comparing the 1+1+1+1 layered "PSK-rank" and "cyclic" schemes, which have almost equal self-interference power $\widetilde{\mathcal{N}}$ (0.89 and 0.75, respectively), one may assess the effect of Frobenius orthogonality and MSD. "PSK-rank" has neither of these properties,

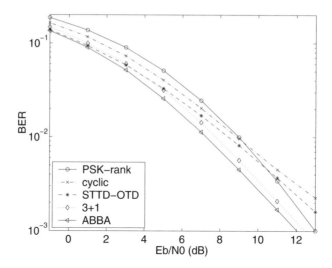

Fig. 4.1: BER of some rate 1 layered matrix modulators for 4 Tx antennas.

while "cyclic" has both. At BER 10^{-1}, "cyclic" performs 1 dB better. The difference shrinks around $E_b/N_0 =$ 5dB, where the full rank of "PSK-rank" starts to make a difference. The second-order information provided by these schemes reflects this; the information of "cyclic" is considerably larger.

- Comparing the Frobenius orthogonal schemes with MSD, the self-interference power $\tilde{\mathcal{N}}$ explains the performance differences at low E_b/N_0. For schemes with the same rank, the difference does not decrease for increasing E_b/N_0. The SH scheme "2+2 ABBA" uniformly outperforms (by 0.5 dB) the non-SH scheme "3+1", which again considerably outperforms the SH scheme "cyclic".

- A badly constructed non-orthogonal scheme with MSD ("cyclic") never outperforms a well designed scheme with only two-fold symbolwise diversity ("STTD-OTD").

- A well constructed non-orthogonal scheme with MSD ("2+2 ABBA") outperforms a scheme with only two-fold symbolwise diversity ("STTD-OTD") at all E_b/N_0, regardless of the schemes having the same true rank (=2). This is because a MSD scheme provides full diversity protection against error events with low Euclidean distance.

- Asymptotically, at low E_b/N_0, "2+2 ABBA" and "STTD-OTD" have the same performance. This may be understood from the fact that they provide the same mutual information. As we shall see, these schemes are related by a symmetry of information, which changes the diversity degree (and accordingly the self-interference and performance), but preserves information.

- Looking at the slopes of the performance curves at $E_b/N_0 =$ 12 dB, "STTD-OTD", "cyclic" and "PSK-rank" have reached their asymptotic regions with diversity 2, 2 and 4, respectively. The MSD schemes with low self-interference, "3+1" and "2+2 ABBA" still have a slope that is steeper than their asymptotic diversity of 2.

- Schemes that fulfil second-order information extremality (4.89) are FO and SH.

- The differences of the numeric values in dB of the loss of second-order information compared to capacity correlates to a high degree with performance differences at low SNR. This is a non-trivial fact which needs an analytical explanation. First it should be noted that equating the second-order information of two schemes a and b one gets the equation

$$\eta_a^2 \, \mathcal{I}_{2,\mathrm{a}} = \eta_b^2 \, \mathcal{I}_{2,\mathrm{b}} \,, \tag{4.99}$$

implying that the dB difference of η_a and η_b is exactly half the dB difference of the respective information coefficients. This assumes that all expansion coefficients are independent. This is not so, as there is only a limited set of unitary invariants for a given set of basis matrices. A rigorous analysis of the connection of performance and second-order information will be left to future work. Here we shall be content with the experimental observation quoted above.

4.8 SUMMARY

In this chapter, design rules for matrix modulation at low SNR have been considered. First, in (4.4), a linear modulation matrix $\mathbf{X}$ was expressed in terms of symbols and basis matrices $\mathbf{B}^{(k)}$. Using this structure, the squared modulation matrix $\mathbf{X}^\dagger \mathbf{X}$ was split into two parts in (4.8). The essential diversity properties are in the diversity matrix $\mathcal{D}$, whereas the self-interference induced by the matrix modulation is visible in the self-interference matrix $\mathcal{S}$.

Regarding performance at low SNR, the essential observations are:

- At low SNR (second-order) mutual information (4.82) explains differences in performance.

- More self-interference $\mapsto$ lower information.

- Performance at medium SNR may be improved by making the scheme maximally symbolwise diverse (MSD). This means that the transmission of each symbol is evenly divided over all antennas, using different symbol periods for different antennas.

- Design criteria for low SNR apply also at high SNR, in addition to the rank and determinant criteria of section 3.5.2.

A number of concrete and easy to fulfil design rules were found for matrix modulators to perform well and to provide much information at low SNR. The most important is Frobenius Orthogonality (FO):

- The self-interference matrix (4.10) should not have a trace. Equivalently, the basis matrices should be orthogonal with respect to the Frobenius norm (4.43).

Finally, the design rules discussed in this chapter were evaluated using a set of symbol rate $R_\mathrm{s} = 1$ matrix modulators for $N_\mathrm{t} = 4$ Tx antennas.

5

Increasing Symbol Rate: Quasi-orthogonal Layers

From the results of Chapter 4 it follows that for a linear matrix modulation

- it is information- and performance-optimal not to have any self-interference, and

- it is performance optimal to have maximal symbolwise diversity.

That is, complex orthogonal (unitary) designs are optimal, if their rate is sufficient to reach the channel capacity, which happens only for $N_t = 2, N_r = 1$ [37]. For higher numbers of antennas, more general matrix modulators are needed. In this chapter, complexity issues in designing linear matrix modulation are discussed. These include

- the choice of a suitable symbol rate, and

- the choice of the symbolwise diversity degree.

The symbol rate should be chosen so that the rank of the signal model equations is sufficient to solve the transmitted symbols by linear methods.

If the symbol rate and diversity degree of choice do not allow orthogonal designs, schemes with multiple "quasi-orthogonal" layers of mutually orthogonal symbols [51, 132, 142, 151, 156] are discussed. These allow simpler detection algorithms than generic non-orthogonal schemes. As examples, $N_t = 2$ and $N_t = 4$ schemes are discussed. In this chapter, as in the sequel of this book, only complex modulation signalling schemes will be considered. Thus the term "orthogonal design" will indicate a complex modulation orthogonal, i.e. a unitary design.

5.1 ORTHOGONAL DESIGNS

In this section, the performance optimality of orthogonal designs is discussed, along with the ensuing severe constraints on rates and information sub-optimality.

5.1.1 Performance Optimum for Linear Space–Time Codes

Combining the principles of minimal self-interference (4.45) and maximal symbolwise diversity (4.50) it follows immediately that no linear matrix modulation (4.4) with QPSK symbols can perform better than a QPSK orthogonal design with the same number of Tx and Rx antennas. An orthogonal design has no self-interference, and the diversity matrix (4.19) is optimal in the light of (4.50). This result is true irrespective of the rate, and provides a rigorous lower limit for the BER of a linear space–time code. Indeed, for a linear scheme with MSD, an orthogonal design employing the same modulation alphabet provides the matched filter bound for performance, i.e. the performance with genie interference cancellation.

In this section, the consequences of Radon–Hurwitz orthogonality will be investigated, and all square matrix orthogonal designs will be constructed.

5.1.2 Consequences of Unitarity and Linearity

As discussed in section 3.6, the defining characteristics of orthogonal designs are linearity (4.4) and (4.5) and orthogonality, which for unitary designs indicates unitarity, with the inner products of columns proportional to the sum of the squared amplitudes of the symbols:

$$\mathbf{X}^\dagger \mathbf{X} = \sum_k |x_k|^2 \, \mathbf{I}_{N_t} \, . \tag{5.1}$$

Pseudo-unitarity in Equation (5.1) is also required for non-square matrices with $T > N$. Combining linearity and unitarity, one gets the Radon–Hurwitz orthogonality equations

$$\mathcal{S}^{(kl)} = \mathbf{B}^{(k)\,\dagger}\mathbf{B}^{(l)} + \mathbf{B}^{(l)\,\dagger}\mathbf{B}^{(k)} = 2\delta_{lk}\,\mathbf{I}_{N_t} \, . \tag{5.2}$$

That is, for $k \neq l$, the basis matrices are RH-orthogonal, whereas each individual basis matrix is unitary.

Any solution of these equations defines a maximal diversity, rate Q/T unitary Space–Time Block Code (STBC). The real version of this equation was used in [18], to find orthogonal designs for real symbols. The complete set of square matrix solutions of (5.2) were found in [141].

As discussed in section 3.6, complex space–time block codes discussed in the literature belong to two categories. They are either rate-halving schemes, like (3.36), or square matrix schemes, like (3.28) and (3.37). From any matrix modulation, a scheme for lesser numbers of antennas can be constructed by deleting columns. Here, we concentrate on square matrix schemes. Thus, when proceeding with the analysis of Equation (5.2), we specialize to square matrices, i.e. $T = N_t$. Redefining

$$\mathbf{G}^{(k-1)} = \mathbf{B}^{(1)\,\dagger}\,\mathbf{B}^{(k)} \, , \quad k = 1, \ldots, 2Q \, , \tag{5.3}$$

we have $\mathbf{G}^{(0)} = \mathbf{I}_{N_t}$, and the form of the algebraic conditions (5.2) remains unchanged for the $\mathbf{G}$:s. From the relation between $\mathbf{B}^{(1)}$ and the other $\mathbf{B}^{(k)}$ we then see that the $\mathbf{G}^{(k)}$, $k \geq 1$ should be anti-Hermitian (a.k.a. skew Hermitian):

$$\mathbf{G}^{(k)\,\dagger} = -\mathbf{G}^{(k)} \, , \quad k = 1, \ldots, 2Q - 1 \, . \tag{5.4}$$

The algebraic relations of these remaining $\mathbf{G}$:s is

$$\mathbf{G}^{(k)}\,\mathbf{G}^{(j)} + \mathbf{G}^{(j)}\,\mathbf{G}^{(k)} = -2\delta_{jk}\mathbf{I}_{N_t}, \quad j, k = 1, \ldots, 2Q - 1 \, , \tag{5.5}$$

i.e. they anti-commute. This is the defining relation of generators of the Clifford algebra (see [163]). We thus get the following generic prescription for finding a space–time block code:

- Find a set of $2Q - 1$ anti-Hermitian $N_t \times N_t$ matrices $\mathbf{G}^{(k)}$, $k = 1, \ldots 2Q - 1$ that fulfil the Clifford algebra conditions (5.5).

- Take a unitary $N_t \times N_t$ matrix $\mathbf{B}^{(1)}$.

- Define $\mathbf{B}^{(k)} = \mathbf{B}^{(1)}\mathbf{G}^{(k-1)}$, $k = 2, \ldots 2Q$.

- Use the matrices $\mathbf{B}^{(k)}$, $k = 1, \ldots, 2Q$ to create a modulation matrix $\mathbf{X}(\mathbf{x})$ according to (4.4).

By construction, this prescription yields all possible space–time block codes with N_t antennas and symbol periods, full diversity, and rate Q/N_t.

In [18], strict constraints on the existence of rate $R_s = 1$ STBCs were found; they exist only for $N_t = 2$. Similarly, the theory of matrix representations of Clifford algebras gives very stringent conditions on the existence of block codes with arbitrary rate Q/N. These restrictions are derived in Appendix B. For any given number of symbols Q to be transmitted, there is a corresponding minimal dimension N_t for the STBC. The result is the following [141]:

For transmitting Q complex symbols in a unitary linear square modulation matrix, the number of antennas may be at most 2^{Q-1}.

Inverting this to yield results for the maximum allowable rate with a given N_t one gets Equation (3.38). Rates for some N_t may be found in Table 3.1.

5.1.3   Construction of Orthogonal Designs

For space–time block-coding purposes, real orthogonal designs are orthogonal $N_t \times N_t$ matrices, where all entries come from the set $\{\pm c_1, \pm c_2, \ldots, c_Q\}$ with $\mathbf{c}$ a Q-dimensional vector of real modulation symbols. Rate $R_s = 1$ designs have $Q = N_t$. For lower-rate designs, one should allow some of the matrix entries to take the value 0. For complex orthogonal (unitary) designs, the symbols may be complex, and the entries come from the set $\{0\} \cup \{\pm x_k, \pm j\, x_k\}_{k=1}^{Q}$.

The result on maximum number of antennas for a given number of symbols in Appendix B is constructive; orthogonal designs are easily constructed from the matrix representation (B.16) of the Clifford algebra with Q elements. We choose $\mathbf{B}^{(1)} = \mathbf{I}_{2^{Q-1}}$, $\mathbf{B}^{(k)} = \mathbf{G}^{(k-1)}$, $k = 1, \ldots, 2Q - 1$. Using (4.4) an orthogonal design may be written in terms of tensor (Kronecker) products of 2×2 matrices. Thus the $2^{Q-1} \times 2^{Q-1}$ matrix,

$$
\begin{aligned}
\mathbf{X}(\mathbf{x}) \;=\;& \operatorname{Re} x_1 \, \mathbf{I}_{2^{Q-1}} + \operatorname{Im} x_1 \; \otimes^{Q-1} \begin{bmatrix} 1 & 0 \\ 0 & -1 \end{bmatrix} \\
&+ \sum_{k=2}^{Q} \left(\otimes^{Q-k} \mathbf{I}_2 \right) \otimes \begin{bmatrix} 0 & x_k \\ x_k^* & 0 \end{bmatrix} \otimes \left(\otimes^{k-2} \begin{bmatrix} 1 & 0 \\ 0 & -1 \end{bmatrix} \right)
\end{aligned}
\tag{5.6}
$$

is a rate $Q/2^{Q-1}$ orthogonal design, which saturates the maximal rate of (3.38). The m:'th tensor power of a matrix A is defined as $\otimes^m A = \underbrace{A \otimes A \otimes \ldots \otimes A}_{m \text{ times}}$.

All possible $2^{Q-1} \times 2^{Q-1}$ orthogonal designs with rate $Q/2^{Q-1}$ can be created by a set of discrete operations including interchanging rows and/or columns, changing signs of symbols, permuting the set of real symbols, multiplying with unitary matrices with entries in $\{\pm 1, \pm j, 0\}$ etc.

To be explicit, for $Q = 2$, the orthogonal design (5.6) yields the 2×2 Alamouti code (3.28). For $Q = 3$, one gets the rate 3/4 orthogonal design (3.39). For $Q = 4$, one gets the rate 1/2 orthogonal design

$$
\mathbf{X}_{1/2} =
\begin{bmatrix}
x_1 & x_2 & x_3 & 0 & x_4 & 0 & 0 & 0 \\
-x_2^* & x_1^* & 0 & -x_3 & 0 & -x_4 & 0 & 0 \\
-x_3^* & 0 & x_1^* & x_2 & 0 & 0 & -x_4 & 0 \\
0 & x_3^* & -x_2^* & x_1 & 0 & 0 & 0 & x_4 \\
-x_4^* & 0 & 0 & 0 & x_1^* & x_2 & x_3 & 0 \\
0 & x_4^* & 0 & 0 & -x_2^* & x_1 & 0 & -x_3 \\
0 & 0 & x_4^* & 0 & -x_3^* & 0 & x_1 & x_2 \\
0 & 0 & 0 & -x_4^* & 0 & x_3^* & -x_2^* & x_1^*
\end{bmatrix}.
\tag{5.7}
$$

This has the 3/4 code (3.39) in the upper left and a complex conjugate inverted version in the lower right corner. Designs for $N_t = 5, 6, 7$ Tx antennas may be constructed by dropping columns from (5.7). It should be noted that the transmit power of a given antenna fluctuates heavily in time, leading to a high peak-to-average power ratio. This is disadvantageous from a power-amplifier point of view, as a wide linearity regime is required for the amplifier.

Unitary left and right symmetries: It follows directly from construction that equivalent STBCs may be constructed by acting on a STBC from the right and left with unitary matrices:

$$
\tilde{\mathbf{X}} = \mathbf{V} \mathbf{X} \mathbf{W},
\tag{5.8}
$$

where $\mathbf{V}$ is a $T \times T$ unitary matrix and $\mathbf{W}$ is a $N_t \times N_t$ unitary matrix. Here "equivalent" means that the performance and the ergodic mutual information in i.i.d. Rayleigh fading are the same. The existence of these symmetries is independent of orthogonality, they exist for any linear matrix modulator. The choice of $\mathbf{V}$ corresponds to the choice of $\mathbf{B}$ [1]. For an orthogonal design, the choice of $\mathbf{W}$ is a symmetry of the Radon–Hurwitz orthogonality equations (5.2). More generally, $\mathbf{W}$ may be thought of as a beam-forming matrix in the full signal model (1.1). In i.i.d. Rayleigh fading, these are symmetries of all performance and information optimization criteria for any linear scheme. As we have seen above, for i.i.d. Rayleigh, these criteria may be expressed in terms of unitary invariants of $\tilde{\mathbf{X}}^\dagger \tilde{\mathbf{X}} = \mathbf{W}^\dagger \mathbf{X}^\dagger \mathbf{X} \mathbf{W}$, and unitary invariants are by definition invariant under unitary transformations of this kind.

As an example of these symmetries, consider the two rate 3/4 schemes for $N_t = 4$, (3.37) and and (3.39). The unitary equivalence of these is given by the matrices

$$
\mathbf{V} =
\begin{bmatrix}
1 & 0 & 0 & 0 \\
0 & 1 & 0 & 0 \\
0 & 0 & -1 & 0 \\
0 & 0 & 0 & -1
\end{bmatrix}, \quad
\mathbf{W} =
\begin{bmatrix}
1 & 0 & 0 & 0 \\
0 & 1 & 0 & 0 \\
0 & 0 & \frac{1}{\sqrt{2}} & \frac{1}{\sqrt{2}} \\
0 & 0 & -\frac{1}{\sqrt{2}} & \frac{1}{\sqrt{2}}
\end{bmatrix}.
$$

Table 5.1: Loss in second-order mutual information from capacity for orthogonal designs with a different number of Rx antennas

N_t	R_s	$10\log_{10}(\mathcal{I}_2/\mathcal{C}_2)$			
		$N_r = 1$	$N_r = 2$	$N_r = 3$	$N_r = 4$
2	1	0	−0.97	−1.46	−1.76
3	3/4	−1.25	−2.71	−3.47	−3.94
4	3/4	−1.25	−3.01	−3.94	−4.52
5	1/2	−3.01	−4.97	−6.02	−6.69
6	1/2	−3.01	−5.12	−6.26	−6.99
7	1/2	−3.01	−5.23	−6.43	−7.22
8	1/2	−3.01	−5.31	−6.58	−7.40

5.1.4 Orthogonal Designs and Information

In [37, 133] it was observed that STTD (3.28) does not reach channel capacity if more than $N_r = 1$ receive antenna is used. Furthermore, in [37] it was proved that a rate $R_s = 1$ orthogonal design, if such an object existed, would reach channel capacity for any N_t, as long as $N_r = 1$. From the rate results of [18], quoted above, it follows that the only orthogonal design that reaches channel capacity is STTD (3.28), when received with $N_r = 1$ antenna.

As discussed in section 4.5.2, an expansion of information in SNR reveals this on the level of the second-order expansion coefficient. Table 5.1 shows the information loss of maximum rate complex orthogonal designs, for $N_r = 1, 2, 3, 4$ Rx antennas. The information loss is measured in decibels from the ratio of the second-order coefficients of mutual information and channel capacity. In section 4.7 it was noticed that these numbers indicate an order of magnitude in performance loss of schemes at low SNR.

To gain a further insight into this loss of capacity, comparisons of schemes with different rates concatenated with effective channel codes should be performed. Thus consider the rate $R_s = 3/4$ orthogonal design (3.39) for $N_t = 4$ Tx antennas, and compare it to the $R_s = 1$ ABBA scheme (4.35). These are concatenated with a turbo code, so that the total bandwidth efficiency is 1 bit/s/Hz, i.e. QPSK symbols are used, and the turbo code concatenated with the $R_s = 3/4$ orthogonal design is punctured to rate $R_c = 2/3$, whereas the turbo code concatenated with the $R_s = 1$ ABBA scheme has the rate $R_c = 1/2$. Figure 5.1 shows the performance of these schemes. There is one Rx antenna, fading is i.i.d. block Rayleigh with block length 4, and channel estimation is perfect. The turbo code is a rate 1/2 parallel concatenated convolutional code (PCCC) with constituent code generators 7_8 and 5_8, even puncturing. The frame of the turbo code is 100 bits, and it is decoded with five iterations. The turbo interleaver and the extra puncturing patterns are random. Between the channel code and the modulation matrix there is also a random BICM interleaver. The resulting code is a simple example of a space–time turbo code, based on BICM.

The difference at coded BER 10^{-3} is 0.6 dB. This should be compared to the second-order information, which for ABBA was 0.79 dB from capacity and for the orthogonal design 1.25 dB from capacity. The difference in second-order information, 0.46 dB, thus explains most of the coded modulation advantage of the $R_s = 1$ scheme.

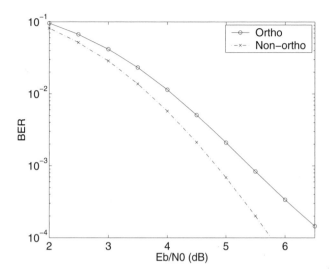

Fig. 5.1: Coded BER of a $R_s = 3/4$ orthogonal design and a $R_s = 1$ non-orthogonal matrix modulator, concatenated with punctured turbo codes to get the bandwidth efficiency 1 bps/Hz.

5.2 COMPLEXITY ISSUES: CHOOSING SYMBOL RATE AND TARGET TX DIVERSITY

In this section, complexity issues are discussed, which *a priori* should be taken into account when designing high-rate matrix modulators. First, an appropriate choice of symbol rate is discussed. The salient observation is that the role of the number of receive antennas N_r in matrix modulation design is not a straightforward one.

5.2.1 Reaching Capacity

Reaching channel capacity typically requires the number of "parallel streams" i.e. the symbol rate $R_s = Q/T$ to equal the number of Tx antennas N_t. That is, independent information streams should be transmitted from each transmit antenna, *irrespective of the number of receive antennas N_r*. For $N_r = 1$ it was shown in [96] that "unconstrained" vector modulation reaches channel capacity. For $N_r > 1$, the results in [16, 17] indicate the same. The only exceptions for this are for $N_r = 1$. Thus it was shown in [37, 133] that STTD (3.28) reaches channel capacity for $N_t = 2$, $N_r = 1$. Examples of schemes with $N_t > 2$ that reach channel capacity for $N_r = 1$ will be considered in Chapter 9. This feature can be proved on the second-order level in the expansion (4.80) of mutual information, as seen in section 4.5.2. The capacity results can be summarized to the statement that a capacity-reaching matrix modulation typically has

$$R_s = \frac{Q}{T} = N_t \tag{5.9}$$

parallel streams, except for a few examples for $N_r = 1$ receive antennas with lower symbol rate.

5.2.2 Linear Detection

Capacity and mutual information are often delusive measures when designing well performing realistic digital signalling schemes. The most important reason for this is that capacity is
an inherently non-digital concept, defined within a framework of continuous Gaussian inputs.
Also, reaching capacity requires using infinite block length codes; capacity is oblivious to
complexity. For this reason, the communication engineer designing a digital communication
system should take capacity argumentation with a pinch of salt.

However, as was seen in section 4.7, within a class of schemes with comparable complexity, (discrete input) mutual information is a valuable and easily accessible design tool.
Differences in mutual information directly translate to differences in performance at low
SNR. Here we shall examine to what extent this property holds as a function of the number
of parallel streams R_s and the rank of the channel matrix.

Consider the signal model (4.1): $\mathbf{Y} = \mathbf{XH} +$ noise, and the corresponding maximum
likelihood metric (4.64). In section 4.4.3 we saw that with Frobenius orthogonality, with $Q \leq$
$T \min(N_r, N_t)$, and a generic channel matrix away from singular points, one independent
measurement can be performed for each real symbol c_k which dominantly affects this symbol.
This corresponds to the fact that with a full rank channel matrix $\mathbf{H}$, one may use linear
algebraic methods to solve $T \min (N_t, N_r)$ complex numbers from the signal model (4.1),
knowing the received signals and the channel, except for some singular points depending on
the structure of $\mathbf{X}$. A priori, these singular points form a zero measure set.

Consider the case $N_r < N_t$, which is most relevant for the downlink direction of future
cellular communication systems. If $Q/T \leq N_r$, the linear term (4.65) invokes a Euclidean
metric in the matched filter projected signal space, which dominates the full metric. In
section 4.7, $R_s = N_r = 1 < N_t = 4$ schemes were investigated. The schemes do not reach
capacity, but mutual information was seen to be a reliable performance measure.

Conversely, if $N_t \geq R_s > N_r$, the mutual information provided is increased, but the
vestige of the underlying Euclidean structure is lost. Linear detection cannot be used. Benefits from the increased mutual information can only be enjoyed if sufficiently complicated
channel codes are concatenated with the matrix modulation, and if space–time detection is
performed jointly with decoding the concatenated code. This is the underlying philosophy
of space–time trellis codes and space–time turbo codes, where a capacity-reaching vector
modulation is used in MISO channels.

As an example of this phenomenon, consider one and two stream 4 bps/Hz schemes for
$N_t = 2$, received with one and two antennas. The two schemes may be taken as 16-QAM
STTD and unconstrained vector modulation (3.40) with QPSK symbols. These have $R_s = 1$
and $R_s = 2$ parallel streams, respectively. The second-order mutual information (compared
to channel capacity) for $N_r = 1, 2$ can be found in Table 5.2. Both schemes reach capacity
for $N_r = 1$, whereas only vector modulation reaches capacity for $N_r = 2$ [37, 133].

Performance of these schemes can be found in Figure 5.2. The channels are i.i.d. block
fading Rayleigh, block length 2. Perfect channel estimation is assumed, and ML detection
is used for both schemes. This, of course, is in contrast to the stated principle of linear
detection of this section. However, this numerical experiment indicates that for uncoded
BER, the essential feature is the vestige of a Euclidean structure in (4.65), not the linear
detection itself. The former is a prerequisite for having the latter. The question of how to
construct linear detectors for linear matrix modulators is addressed in Chapter 6.

With $N_r = 2$, the loss in information of STTD is visible as a slight performance penalty
at low E_b/N_0. The performance penalty suffered by the vector modulation when received

Table 5.2: Second-order mutual information for $N_t = 2$ schemes with 1 and 2 parallel streams.

Scheme	$\mathcal{I}_2/\mathcal{C}_2$ (dB)	
	$N_r = 1$	$N_r = 2$
16-QAM STTD	0	-0.97
QPSK vector	0	0

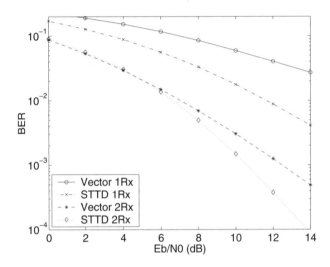

Fig. 5.2: BER of uncoded 4 bps/Hz schemes with $N_t = 2$ and one or two parallel streams, received with $N_r = 1$ and $N_r = 2$ Rx antennas. The performance penalty for a two-stream scheme is considerable, when receiving with one antenna.

with one Rx antenna is considerable. There is no trace of similarity in 16-QAM STTD and QPSK vector modulation performance at low E_b/N_0, notwithstanding that they provide the same (second-order) information. As the schemes provide the same information for $N_r = 1$, it should be possible to reach (near) error-free transmission for both schemes with the same rates below channel capacity. Intuitively it is clear that to bridge the discrepancy between matrix/vector modulation performance, a channel code concatenated with QPSK vector modulation should be much more complex than a channel code concatenated with 16-QAM STTD to reach comparable performance.

This leads to the conclusion that *for good performance of a matrix modulation, the signal model (4.1) should not be an underdetermined set of equations for the symbols in* **X** *for generic channels.* That is, the number of parallel streams should be

$$R_s = \frac{Q}{T} \leq \min(N_r, N_t) . \tag{5.10}$$

As a note for coming chapters, it should be noted that even for $N_r = 2$, 16-QAM STTD performs considerably better at high E_b/N_0 than QPSK vector modulation. The slope of the 16-QAM STTD performance curve corresponds to the combined Tx and Rx diversity degree of 4, whereas QPSK vector modulation performance has a combined diversity degree of 2 (Rx diversity only). With concatenated channel codes (e.g. D-BLAST [20]), some of the implicit Tx diversity in vector modulation may be turned into explicit temporal diversity. Another way is to add explicit Tx diversity to the matrix modulation itself. In section 9.2, a two-stream QPSK scheme will be constructed that performs uniformly better than 16-QAM STTD for $N_r = 2$.

5.2.3 Choosing the Symbol Rate

To keep complexity low, and the mutual information as high as possible, one should strike a balance between the requirements (5.9) and (5.10). Thus, for uncorrelated (i.i.d) channels, the symbol rate should be

$$R_s \approx \min\left(N_t, N_r\right) . \tag{5.11}$$

When $N_t > 1$ and $N_r = 1$, as in the immediate 3G evolution path, $R_s = 1$ should thus be used. Similarly, if two receive antennas are deployed at mobile stations, the rate $R_s = 2$ should be used, as long as $N_t \geq 2$. This means that constrained schemes with $R_s < N_t$ will be in great demand for future communication systems. In contrast, most MIMO research has concentrated on $N_r \geq N_t = R_s$.

In correlated channels, constrained schemes become even more relevant. The rank of the channel, or more exactly the practical rank number (Prank), which is the number of singular values of the channel matrix which are essentially non-zero, is not $\min(N_t, N_r)$, but

$$\text{Prank} \leq \min\left(N_t, N_r\right) . \tag{5.12}$$

Now the symbol rate should be chosen so that $R_s \leq \text{Prank}$.

Even with i.i.d. channels, the channel may be rank-deficient (5.12), or it may be near one of the singular points of the detection metric (4.64). The latter cases correspond to a rank-deficiency of the correlation matrix $\mathcal{R}$. With side information at the transmitter (e.g. feedback), the transmission scheme may be adaptively chosen so that $R_s \leq \text{Prank}$. An example of such an Adaptive Space–Time Modulation Arrangement (ASTMA) may be found in section 12.5.

5.2.4 Choosing Target Tx Diversity Degree

In section 4.7 the information and performance of the example schemes of Chapter 4 were analysed. It was observed that STTD-OTD and ABBA provided the same mutual information and had identical performance at low SNR. The symbolwise diversity degree of the schemes was 2 and 4, respectively. As a consequence, ABBA performed better at medium and high SNR. In Chapter 7 the performance difference at high SNR will be further increased, as it will be shown that it is always possible to increase the true asymptotic diversity degree to become equal to the symbolwise diversity degree, using diversity transforms.

The price of this improved performance is paid in complexity. "STTD-OTD" is Radon–Hurwitz orthogonal, and accordingly its ML metric (4.64) does not have the non-linear term (4.66); a matched filter detector reaches ML performance. With a non-orthogonal scheme such as ABBA the situation is different. Reaching ML performance typically requires a non-linear detector.

This performance/complexity tradeoff should be thoroughly analysed before using a matrix modulator. Depending on the operation point and the (set of) concatenated code(s), the complexity price for improved performance may or may not be worth paying. If the concatenated codes are strong, if the target FER is high (and accordingly the operation point low) or if there are other sources of diversity in the system, as AMC (Adaptive Modulation and Coding), ARQ (Automatic Retransmission reQuest), and/or power control, it is likely that it is not worth paying the complexity price. This is the case of WCDMA downlink with $N_t = 4$, $N_r = 1$. There, a simple matrix modulation like STTD-OTD, or STTD-PHOP (STTD + Phase Hopping, discussed in Chapter 8), is likely to be sufficient.

For different environments, schemes with different degrees of target Tx diversity may be designed. This means that the minimum rank of the diversity matrix $\mathcal{D}(\Delta)$ for non-vanishing one-symbol error events may be fixed. For a linear scheme this is equivalent to fixing the minimum rank of $\mathbf{B}^{(k)\dagger}\mathbf{B}^{(k)} = \frac{1}{2}\mathcal{S}^{(kk)}$. Applying the principle of unitary equivalence of $\mathcal{S}^{(kk)}$ (see (4.96)) then means that all $\mathcal{S}^{(kk)}$ should have the same rank.

For $N_t = 4$, $N_r = 1$, the principles of the previous section indicate that the symbol rate $R_s = 1$ should be chosen. If the target Tx diversity degree is chosen to be 2, STTD-OTD and STTD-PHOP fulfil all information optimization conditions, especially the unitary equivalence of $\mathcal{S}^{(kk)}$. From an information and performance point of view, in ergodic channels they are equivalent and optimal within the category of Tx diversity 2 schemes. In non-ergodic (typical pedestrian) channels, they show a performance difference favouring STTD-PHOP; see Chapter 8.

If the target Tx diversity degree 4 is pertinent, a MSD (Maximally Symbolwise Diverse) scheme should be chosen. This means that a truly non-orthogonal scheme should be used.

5.3 MULTIMODULATION SCHEMES

Before examining non-orthogonal space–time modulation in depth, a milder form of non-orthogonality may be investigated, with partly non-orthogonal bits in the symbols. In [164], the concept of multimodulation space–time block codes was discussed, relaxing the implicit assumption of much space–time block coding work that the symbols should be in the same modulation alphabet. This is especially suited for use with a $R_s < 1$ orthogonal scheme of the kind of Equation (3.37), where different symbols are scaled with different numbers. The symbols that acquire a lower power in a single-modulation scheme like that shown in (3.37) can be selected from another symbol alphabet with more bits, and transmitted with more power. For example, if symbols z_1, z_2 are taken in the QPSK alphabet, and z_3 is in 8-PSK, the code

$$C = \begin{bmatrix} z_1 & z_2 & z_3 & z_3 \\ -z_2^* & z_1^* & z_3 & -z_3 \\ z_3^* & z_3^* & -x_1 + \mathrm{j}\,y_2 & -x_2 + \mathrm{j}\,y_1 \\ z_3^* & -z_3^* & x_2 + \mathrm{j}\,y_1 & -x_1 - \mathrm{j}\,y_2 \end{bmatrix} \tag{5.13}$$

is a power-balanced version of (3.37) with bandwidth efficiency 7/4 bps/Hz. Similarly, if a 16-QAM symbol is used as z_3, one gets a rate 1 space–time block code. In [164], optimal power allocation between the multimodulation branches was investigated.

Performance of a QPSK/16-QAM multimodulation scheme with optimal power allocation is plotted in Figure 5.3. Performance is compared to a non-orthogonal matrix modulation with the same bandwidth efficiency, 2 bps/Hz, namely ABBA of (4.35). In low to medium SNR, ABBA performs ≈ 0.8 dB better. This should be contrasted to the difference in second-order information. When computing the information provided by a multimodulation

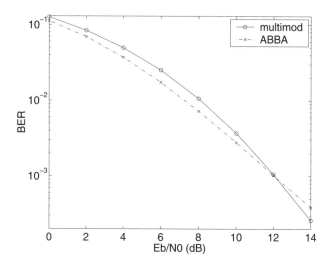

Fig. 5.3: BER performance of QPSK/16-QAM multimodulation space–time block code, compared to ABBA. Both schemes transmit 2 bps/Hz.

Table 5.3: Second-order mutual information for multimodulation (5.13) and ABBA

Scheme	$\mathcal{I}_2/\mathcal{C}_2$ (dB)	
	$N_r = 1$	$N_r = 2$
Multimodulation, $\lvert x_3 \rvert^2 = 2\lvert x_{1,2} \rvert^2$	-1.76	-3.52
ABBA	-0.79	-2.22

scheme, one should be careful with the normalization (as discussed in section 4.5). In the normalization calculation it was assumed that all symbols have the same average power. In a multimodulation scheme, symbols in a higher modulation are transmitted with a higher power. Thus the corresponding basis matrices should be scaled correspondingly. In the (5.13), the basis matrices for x_3 have been scaled by $\sqrt{2}$. This affects second-order information, as the scheme is not homogeneous anymore. The second-order mutual information for a multimodulation scheme with double power for x_3 is compared to ABBA in Table 5.3. The penalty in second-order information is almost 1 dB, which again predicts the order of magnitude of low SNR performance difference.

At high SNR, the full rank of the multimodulation scheme becomes visible, and there is a crossover after which it performs better than ABBA. However, with symbol rotations, as discussed in Chapter 7, ABBA can be optimized so that it maintains the 0.8 dB performance gain compared to the multimodulation scheme.

The conclusion of [164], supported by the simple mutual information argument above, is that it is better to relax space–time modulation orthogonality than modulation orthogonality.

5.4   MATRIX MODULATION WITH QUASI-ORTHOGONAL LAYERS

In this section, non-orthogonal matrix modulators will be constructed. To keep as many of the preferable properties of unitary designs as possible, non-unitary matrix modulation with Radon–Hurwitz orthogonal layers will be considered. To be specific, a layer is defined as a maximal collection of symbols that are RH-orthogonal among each other. Among the set of symbols in two different layers, at least one pair should be found that is not RH-orthogonal. Adopting the terminology of [142], this concept is called "quasi-orthogonality".

An intuitive reason for the quasi-orthogonal layering structure, i.e. to keep as much RH-orthogonality in the problem as possible, is to minimize the volume of non-generic channel realizations that affect the ML detection metric, see section 4.4.3.

The layering structure is described as a sum. A matrix modulation with Q complex symbols in L quasi-orthogonal layers, with layer l having Q_l complex symbols, is said to be $Q_1 + Q_2 + \ldots + Q_L$ layered. Sometimes there is a natural ordering of the layers, arising from a natural order of interference cancellation. If the layers have different numbers of symbols, the layer with the most symbols is the "lowest layer". The number of symbols in the largest quasi-orthogonal layer is often referred to as *the* quasi-orthogonality of a scheme. In Chapter 6 we shall see that this number is pertinent for comparing the detection complexity of different schemes.

The concept of layering considered here is related to the layering concept of BLAST [20] in that interference cancelling between layers may be performed.

5.4.1   Clifford Basis for Matrix Modulation

In section 5.1.3 and Appendix B it was found that the structure underlying RH-orthogonality is that of Clifford algebra. From this it follows that schemes with quasi-orthogonal layers should be defined using the Clifford algebra basis for complex matrices in the appropriate dimension, see section B.2 in Appendix B.

For the purpose of constructing layered schemes, a Clifford algebra of $2^L \times 2^L$ matrices may be defined as a vector space over the field of complex numbers with a basis of $2^L \times 2^L$ matrices which can be constructed e.g. from L-fold tensor (Kronecker) products of the 2×2 STTD basis (4.20)

$$\begin{aligned} \mathcal{B}_2 &= \left\{ \mathbf{I}_2, \begin{bmatrix} j & 0 \\ 0 & -j \end{bmatrix}, \begin{bmatrix} 0 & 1 \\ -1 & 0 \end{bmatrix}, \begin{bmatrix} 0 & j \\ j & 0 \end{bmatrix} \right\} \quad (5.14) \\ &\equiv \left\{ \mathbf{T}^{(1)}, \mathbf{T}^{(2)}, \mathbf{T}^{(3)}, \mathbf{T}^{(4)} \right\} . \end{aligned}$$

There are 4^L such tensor products and they constitute the 2^{2L} elements of a unitary Frobenius orthogonal basis for complex $2^L \times 2^L$ dimensional matrices, interpreted as an vector space over the complex numbers. In an alternative interpretation, more appropriate for designing matrix modulators, the space of complex $2^L \times 2^L$ matrices may be interpreted as a vector space over the real numbers. The corresponding 2^{2L+1}-dimensional Frobenius orthogonal basis consists of the 4^L tensor products of (5.14), and j times the same. This latter Clifford basis has a maximal anti-commuting anti-Hermitian subset of $2L+1$ matrices. Together with the identity matrix, this subset forms a basis for a maximal rate orthogonal design embedded in the Clifford algebra, see Appendix B. This design has complex symbol rate $(L+1)/2^L$.

A prime example of a Clifford basis is the STTD basis (5.14) itself. Any complex 2×2 matrix may be expanded in this basis, see (B.19). In the alternative interpretation, the basis is 8-dimensional, i.e. (5.14) and j times the same. The maximal anti-commuting subset has three matrices, and together with the identity the resulting orthogonal design is STTD (3.28).

Due to the tensor product structure, Clifford bases exist for any $N_t = 2^L$. For rectangular matrices, Clifford bases may be constructed analogously by stacking square bases. Thus a basis for $T \times N_t = 2 \times 4$ matrix modulation with transmit diversity 2 is given by

$$\mathcal{B}_{2 \times 4} = \begin{bmatrix} \mathcal{B}_2 & \mathbf{0}_2 \end{bmatrix} \cup \begin{bmatrix} \mathbf{0}_2 & \mathcal{B}_2 \end{bmatrix} . \tag{5.15}$$

When considering real symbols, one should take the basis $\mathcal{B}_{2 \times 4} \cup j \, \mathcal{B}_{2 \times 4}$.

For layered matrix modulation with $N_t = 4$, $T = 4$, a basis of 32 unitary Frobenius orthogonal matrices should be considered. It consists of the 16 matrices

$$\mathbf{B}^{(\mu \otimes \nu)} = \mathbf{T}^{(\mu)} \otimes \mathbf{T}^{(\nu)} , \quad \mu, \nu \in \{1, 2, 3, 4\} , \tag{5.16}$$

explicitly spelled out in (B.21), and j times the same. This basis was used in section 4.2 when discussing the structure of the quasi-orthogonal 4×4 schemes (4.32) and (4.35).

The maximal anti-commuting anti-Hermitian subset of (5.16) has five elements. One example is given by the matrices in (B.11). Together with $\mathbf{B}^{(1)} = \mathbf{I}_4$ (the four-dimensional identity matrix), and reordered as in (B.25), these basis matrices yield the unitary design (3.39), i.e.

$$\mathbf{X} = \begin{bmatrix} x_1 & x_2 & x_3 & 0 \\ -x_2^* & x_1^* & 0 & -x_3 \\ -x_3^* & 0 & x_1^* & x_2 \\ 0 & x_3^* & -x_2^* & x_1 \end{bmatrix} . \tag{5.17}$$

Here $\mathbf{B}^{(2j-1)}$ and $\mathbf{B}^{(2j)}$ in (B.25) are used to encode the real and imaginary part of x_j, respectively, as indicated in (4.4).

T-MSD: For a square matrix, a Clifford basis is maximally symbolwise diverse (MSD). The principle of MSD may be extended to a quasi-orthogonal layered scheme based on rectangular matrices, e.g. (5.15). In these cases, a Clifford basis is T-MSD, i.e. the basis matrices provide the maximal symbolwise diversity allowed by the temporal block length T. More exactly, the principle of T-MSD may be expressed as

$$\mathbf{B}^{(k)\dagger} \mathbf{B}^{(k)} \sim \mathbf{I}_{N_t}, \quad \text{if } T \geq N_t$$

$$\mathbf{B}^{(k)} \mathbf{B}^{(k)\dagger} \sim \mathbf{I}_{N_t}, \quad \text{if } T < N_t . \tag{5.18}$$

It can be proved that this is a prerequisite for performance optimality for a given T, N_t.

5.4.2   Idle Directions of a Modulation Matrix

With an appropriate choice of symbol rate discussed in section 5.2.3 and employing a Frobenius orthogonal unitary Clifford basis as discussed above, the path for constructing quasi-orthogonal layered schemes is clear.

As the starting point, a rate-deficient orthogonal design is chosen, and the rate should be increased. For this, the concepts of idle received signals and corresponding idle directions in a modulation matrix should be discussed. An idle received signal is a dimension in received signal space which carries *no information* about the transmitted signals. Such dimensions always exist if $Q < T \min(N_r, N_t)$. Corresponding to idle received signals, there are idle directions in the modulation matrix. These are matrix dimensions that are Frobenius-orthogonal to all employed basis matrices $\mathbf{B}^{(k)}$, and which at least partly project to idle received signals.

As an example, consider receiving STTD (3.28) with $N_r = 2$ Rx antennas. There are four real symbols in STTD. With $N_r = 2$, the received signal $\mathbf{Y}$ is a 2×2 matrix with eight real dimensions. Thus there is an idle received signal with four real dimensions. Now recall that the Clifford basis for complex 2×2 matrices has eight dimensions, $\mathcal{B}_2 \cup j\,\mathcal{B}_2$. STTD uses only four of these, namely $\mathcal{B}_2$. The idle directions of the modulation matrix are spanned by the remaining four basis elements, $j\,\mathcal{B}$, which all have non-trivial projections on the idle received signals.

As another example, consider receiving the rate 3/4 orthogonal design for $N_t = 4$ Tx antennas, (5.17), with one Rx antenna. There are six real symbols in (5.17). With $N_r = 1$, $\mathbf{Y}$ is a 1×4 vector with eight real dimensions. Thus there is an idle received signal with two real dimensions. To see what the idle directions in the modulation matrix are, recall that the 4×4 Clifford basis has 32 dimensions, and that six of these, namely (B.25), are used to construct (5.17). Now any of the $32 - 6 = 26$ remaining elements in the Clifford basis (5.16) has a non-trivial projection onto the idle received signal, and spans an idle dimension of the modulation matrix (5.17).

5.4.3 Layered Schemes for $N_t = 2$, $N_r = 2$

If $N_t = 2$, $N_r = 2$ and fading is i.i.d., the principles of section 5.2.3 tell us to design a symbol rate $R_s = 2$ scheme.

Vector Modulation: With target Tx diversity 1, one should simply use the well-known vector modulation

$$\mathbf{X} = \begin{bmatrix} x_1 & x_2 \end{bmatrix} . \tag{5.19}$$

Matrix Modulation: With target Tx diversity 2, it is easy to construct a 2+2 layered scheme based on the first example in the previous section. In addition to $\mathcal{B}_2$, one employs the idle directions $j\,\mathcal{B}_2$, or equivalently $\left\{ \begin{bmatrix} 1 & 0 \\ 0 & -1 \end{bmatrix} \mathbf{T}_\mu \right\}_{\mu=0}^{3}$.[1] The basis for a rate $R_s = 2$ matrix modulation is thus (with a suitably chosen ordering)

$$\mathbf{B}^{(1)} = \begin{bmatrix} 1 & 0 \\ 0 & 1 \end{bmatrix}; \ \mathbf{B}^{(2)} = \begin{bmatrix} j & 0 \\ 0 & -j \end{bmatrix}; \ \mathbf{B}^{(3)} = \begin{bmatrix} 0 & 1 \\ -1 & 0 \end{bmatrix}; \ \mathbf{B}^{(4)} = \begin{bmatrix} 0 & j \\ j & 0 \end{bmatrix};$$
$$\tag{5.20}$$
$$\mathbf{B}^{(5)} = \begin{bmatrix} 1 & 0 \\ 0 & -1 \end{bmatrix}; \ \mathbf{B}^{(6)} = \begin{bmatrix} j & 0 \\ 0 & j \end{bmatrix}; \ \mathbf{B}^{(7)} = \begin{bmatrix} 0 & 1 \\ 1 & 0 \end{bmatrix}; \ \mathbf{B}^{(8)} = \begin{bmatrix} 0 & j \\ -j & 0 \end{bmatrix},$$

---

[1] These two sets are equal up to a permutation.

and the corresponding modulation matrix can be written in terms of two overlapping STTD blocks as

$$
\begin{aligned}
\mathbf{X} &= \begin{bmatrix} x_1 + x_3 & x_2 + x_4 \\ -x_2^* + x_4^* & x_1^* - x_3^* \end{bmatrix} \\
&= \begin{bmatrix} x_1 & x_2 \\ -x_2^* & x_1^* \end{bmatrix} + \begin{bmatrix} 1 & 0 \\ 0 & -1 \end{bmatrix} \begin{bmatrix} x_3 & x_4 \\ -x_4^* & x_3^* \end{bmatrix} \\
&\equiv \mathbf{X}_A (x_3, x_4) + \begin{bmatrix} 1 & 0 \\ 0 & -1 \end{bmatrix} \mathbf{X}_B (x_3, x_4) .
\end{aligned}
\tag{5.21}
$$

The diversity matrix is

$$
\mathcal{D} = \sum_{k=1}^{4} |x_k|^2 \, \mathbf{I}_2 ,
\tag{5.22}
$$

and the self-interference is

$$
\mathcal{S} = 2 \begin{bmatrix} \mathrm{Re}\,[x_1 x_3^* - x_2 x_4^*] & x_1^* x_4 + x_2 x_3^* \\ x_1 x_4^* + x_2^* x_3 & -\mathrm{Re}\,[x_1 x_3^* - x_2 x_4^*] \end{bmatrix} .
\tag{5.23}
$$

All self-interference is between symbols in the quasi-orthogonal $\mathbf{X}_A$ and $\mathbf{X}_B$ blocks. As expected, this scheme is 2+2 layered. Also, the self-interference is traceless. As this scheme has $R_s > 1$, this is a non-trivial fact, and a direct consequence of using the Frobenius orthogonal Clifford basis for constructing the scheme.

At this stage, it is good to look at the differences between using complex and real symbols when constructing a modulation matrix. The form (5.21) was constructed from the basis (5.20) consisting of eight matrices. Each modulation matrix encoded one real symbol, and the real symbols were combined to complex symbols as $x_k = c_{2k-1} + \mathrm{j}\, c_{2k}$. The modulation matrix could equally well be constructed directly from the basis (5.14) consisting of four matrices, each basis matrix encoding a complex symbol. Absorbing the j's in $\mathbf{B}^{(2)}, \mathbf{B}^{(4)}$ to the corresponding symbols, the result is

$$
\mathbf{X} = \begin{bmatrix} x_1 + x_2 & x_3 + x_4 \\ -x_3 + x_4 & x_1 - x_2 \end{bmatrix} .
\tag{5.24}
$$

In this form, no complex conjugations are used. By construction, the forms (5.21) and (5.24) are equivalent, the only difference is in how the real symbols are combined to complex symbols. The form (5.21) has the advantage that the quasi-orthogonal layering is explicitly visible. This structure may be used to simplify detection, as discussed in Chapter 6. In Chapter 7 we shall see how this explicit layering structure may be used to construct optimal diversity transforms.

The scheme (5.21) was first discussed in [133], in the form (5.24). There the generic basis used was not the Clifford basis, but another basis (the "Weyl basis" discussed in Chapter 9) of unitary Frobenius orthogonal matrices. For 2×2 matrices, these two bases coincide. For larger matrices, the Weyl basis fails to capture the RH-orthogonal properties essential for constructing quasi-orthogonal layered schemes. In [143], the form (5.21) was considered as an example of "multi-stratum" space-time codes.

Performance of $R_s = 2$ vector and matrix modulation is reported in Figure 5.4. Simulation assumptions are i.i.d. block Rayleigh fading, block length two, perfect channel estimation, maximum likelihood detection. The matrix modulation performs uniformly better. However,

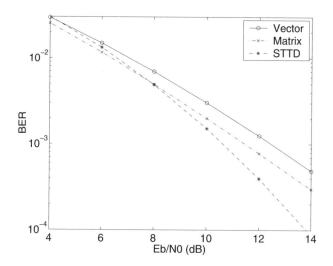

Fig. 5.4: BER of vector and matrix modulation with $N_t = 2$, $N_r = 2$ and two parallel streams, 4 bps/Hz.

the asymptotic diversity degree of the matrix and vector modulators are the same, i.e. the true Tx diversity of (5.21) is just one. Thus it performs worse than 16-QAM STTD at high SNR. This is due to pathological error events producing singular codeword difference matrices, and can be cured by applying suitable complex diversity transforms: see [51, 165] and Chapter 7 of this book.

5.4.4 Minimal Self-interference 3+1 Layered 4×4 Schemes

Now, the design rules found in Chapter 4 shall be put into use in designing layered matrix modulators for $N_t = 4$ Tx antennas that can be received with one Rx antenna. With target Tx diversity 2, the basis (5.15) may be used, so that a different half is used during a different pair of symbol periods, leading to STTD-OTD (4.21). Alternatively, different linear combinations of the two halves in (5.15) may be used during different pairs of symbol periods, leading to STTD-PHOP.

With target Tx diversity 4, layered matrix modulators should be constructed. The maximum rate orthogonal design with $T = N_t = 4$ has rate 3/4; an example is (5.17), employing six of the 32 Frobenius-orthogonal unitary basis elements of a Clifford basis. A symbol rate $Q/T = 1$ layered scheme with the maximal quasi-orthogonality of six real (or three complex) symbols can be constructed based on (5.17). A scheme with no idle received signals for $N_r = 1$ can be constructed by expanding the basis of (5.17) with any two Frobenius orthogonal matrices in the $32 - 6 = 26$ idle dimensions. The two extra real symbols will be denoted by c_7 and c_8, with the corresponding basis matrices $\mathbf{B}^{(7)}$ and $\mathbf{B}^{(8)}$.

This example will be considered in some detail, following [156], to show how the design principles of Chapter 4 work. As we shall see in the next section, the result is *not* an optimal $R_s = 1$ matrix modulator for $N_t = 4$ antennas. Optimal modulators are to be found in a different family of schemes.

A priori, the number of alternatives for of choosing two of 26 is huge; 650 combinations of matrices of the form (5.16) plus continuous rotations, corresponding to the Grassmannian of 2D planes embedded into 26 dimensions. Among these options, one particular optimal choice may be found, when the design principles are applied.

Minimal Self-Interference: Not all 650 combinations have the same self-interference (4.47). The minimal self-interference per real symbol is $\widetilde{\mathcal{N}} = 3/8$, and 100 combinations (plus continuous rotations) reach this value. It is interesting to note that these are exactly those of the 650 options for which the two additional symbols are Radon–Hurwitz orthogonal, i.e. with

$$\mathbf{B}^{(7)\,\dagger}\,\mathbf{B}^{(8)} + \mathbf{B}^{(8)\,\dagger}\,\mathbf{B}^{(7)} = 0 \,. \tag{5.25}$$

Of all rate 1 layered 4×4 schemes built upon (5.17), the minimal self-interference schemes are thus $3+1$ layered. Generic schemes may be $3 + \frac{1}{2} + \frac{1}{2}$ layered, in terms of the complex symbol layering count used here. The "3" layer consisting of (5.17) is the "lower layer", and the "1" layer consisting of c_7, c_8 is the upper layer.

Partial SH: From the layering count it is clear that it is impossible to have symbol homogeneity (SH) with a 3+1 scheme. The total self-interference experienced by the complex symbol in the "1" layer equals the total self-interference experienced by the three symbols in the "3" layer. Also, none of these 3+1 schemes fulfil the information extremality equations (4.89).

In section 4.6.2, the principle of partial SH was discussed, meaning that it is preferable to have as few categories with different symbolwise self-interference $\mathcal{N}_k$ as possible. For the 3+1 schemes this means that the complex symbols in the "3" layer should all experience the same $\mathcal{N}_k$. That is, the symbols in the "3" layer should be SH among each other. As a consequence of the complex symbol description, the two real symbols in the "1" layer already experience the same $\mathcal{N}_k$. Requiring partial SH narrows the choice of the "1" layer to 10 pairs (plus continuous rotations). For each of these, the basis matrices for the fourth complex symbol fulfil

$$\mathbf{B}^{(8)} = \pm\mathrm{j}\,\mathbf{B}^{(7)} \,. \tag{5.26}$$

The real symbols in the "1" layer thus combine to a complex symbol $x_4 = c_7 + \mathrm{j}\,c_8$, and this symbol is encoded without complex conjugation.

One example of the 10 partially symbol homogeneous 3+1 schemes is

$$\mathbf{X}_{3+1} = \begin{bmatrix} x_1 & x_2 & x_3 & x_4 \\ -x_2^* & x_1^* & x_4 & -x_3 \\ -x_3^* & -x_4 & x_1^* & x_2 \\ -x_4 & x_3^* & -x_2^* & x_1 \end{bmatrix}, \tag{5.27}$$

the example scheme (4.32) in Chapter 4. For this scheme, the relative self-interference of the eight real symbols is

$$\mathcal{N}_1 : \mathcal{N}_2 : \mathcal{N}_3 : \mathcal{N}_4 : \mathcal{N}_5 : \mathcal{N}_6 : \mathcal{N}_7 : \mathcal{N}_8 \; = \; 1:1:1:1:1:1:3:3 \,.$$

That is, there are exactly two categories of symbols with different self-interference, which are the "3" layer and the "1" layer.

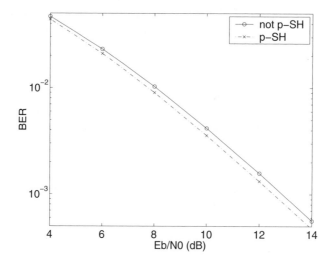

Fig. 5.5: BER of two 3+1 layered schemes, one with partial SH, one without.

An alternative 3+1 scheme which has the same average self-interference, but is not even partially SH, is

$$\mathbf{X}_{\widetilde{3+1}} = \begin{bmatrix} x_1 & x_2 & x_3 & x_4 \\ -x_2^* & x_1^* & x_4 & -x_3 \\ -x_3^* & x_4^* & x_1^* & x_2 \\ x_4^* & x_3^* & -x_2^* & x_1 \end{bmatrix}. \tag{5.28}$$

Here, the symbol x_4 is partly encoded with a complex conjugate. For this scheme,

$$\mathcal{N}_1 : \mathcal{N}_2 : \mathcal{N}_3 : \mathcal{N}_4 : \mathcal{N}_5 : \mathcal{N}_6 : \mathcal{N}_7 : \mathcal{N}_8 = 2 : 0 : 2 : 0 : 1 : 1 : 3 : 3.$$

The imaginary parts of x_1 and x_2 do not interfere with x_4 at all. Instead, the real parts of x_1 and x_2 experience double self-interference. There are four categories of symbols with different self-interference.

Simulated performance results of the partially SH scheme (5.27) and the not even partially SH scheme (5.28) are reported in Figure 5.5. There is a constant gain of ≈ 0.25 dB from being partially SH. This is not much, but it comes for free.

Power Fluctuations: Of the 10 partially SH alternatives, six encode x_4 in positions overlapping with the symbols in (5.17), whereas four use the off-diagonal zeros. From the point of view of minimizing the peak-to-average fluctuations, the latter alternatives are preferable. With PSK modulation, the corresponding schemes are power-balanced.

Complete Interference Cancellation for Upper Layer: The choice may still be narrowed down to a single scheme, if a specific detection scheme is kept in mind. To find this direction, a Euclidean metric in the received signal space, based on matched filter projections, should be found. For a non-orthogonal scheme, or a scheme with $Q < TN_r$, a relation of received signal space Euclidean metric, and the matrix modulation basis of the form (4.67) is not possible. As observed, the rate 3/4 scheme (5.17), received with one antenna, has two idle receive dimensions. The question arises, what kind of projections of which idle direction(s) of the modulation matrix span the remaining complex dimension in received signal space? The solution for this can be found from the following.

Of the 10 partially SH alternatives with (5.26), there is one that fulfils

$$\mathbf{H}^T \, \mathbf{B}^{(7)\dagger} \, \mathbf{X}_{3/4} \, \mathbf{H} \; = \; 0 \,, \tag{5.29}$$

namely

$$\mathbf{B}^{(7)} = -\mathrm{j} \, \mathbf{B}^{(3 \otimes 4)} = \begin{bmatrix} 0 & 0 & 0 & 1 \\ 0 & 0 & 1 & 0 \\ 0 & -1 & 0 & 0 \\ -1 & 0 & 0 & 0 \end{bmatrix} \tag{5.30}$$

or some complex number times the same. Inspired by (5.29), one may construct the "twisted complex projection"

$$\tilde{z}_4 = \tilde{z}^{(\mathbb{R})}{}_7 + \mathrm{j} \, \tilde{z}^{(\mathbb{R})}{}_8 = \mathrm{Tr}\left(\mathbf{H}^T \, \mathbf{B}^{(7)\dagger} \, \mathbf{Y}\right) . \tag{5.31}$$

Note that here, as opposed to the matched filter projections in (4.59), $\mathbf{H}$ is transposed, not Hermitian-conjugated. Equation (5.29) means that $\tilde{z}_4$ yields an estimate of the "upper layer" x_4 where *all interference from the "lower layer"* x_1, x_2, x_3 *is cancelled.*

It is important to note that $\tilde{z}_4$ is *not* the matched filter projector corresponding to x_4 appearing in the maximum likelihood metric (4.64). The matched filter projector z_4 should be constructed according to (4.59) and (4.61), and suffers from interference from the lower layer. However, $\tilde{z}_4$ may be used as a decorrelating pre-stage in an iterative IC (Interference Cancelation) receiver. This shall be discussed in Chapter 6.

To continue with finding an Euclidean metric in received signal space, the twisted projection operators $\tilde{\mathcal{P}}_{7,8}$ corresponding to the twisted matched filter (5.31) should be constructed. These are constructed as in (4.60), except that the elements in the equivalent channel matrix $\mathcal{H}$ in (4.60) are replaced by expressions calculated from $\mathbf{B}^{(7)}\mathbf{H}^*$, not from $\mathbf{B}^{(7)}\mathbf{H}$ as in (4.57).

The real matched filter projection operators $\mathcal{P}_k$, $k = 1, \ldots, 6$ corresponding to the orthogonal design (5.17), together with the twisted projection operators $\tilde{\mathcal{P}}_{7,8}$ fulfil the completeness relation

$$\sum_{k=1}^{6} \mathcal{P}_k + \sum_{k=7}^{8} \tilde{\mathcal{P}}_k \; = \; \mathbf{I}_4 \,. \tag{5.32}$$

From this it follows that

$$\mathrm{Tr}(\mathbf{H}^\dagger\mathbf{H}) \; \mathrm{Tr}(\mathbf{Y}^\dagger\mathbf{Y}) = \sum_{k=1}^{3} |z_k|^2 \, + \, |\tilde{z}_4|^2 \,, \tag{5.33}$$

which provides the sought-for Euclidean metric in received signal space. In particular, the existence of this metric indicates that the noise experienced by $\tilde{z}_4$ is independent from the noise experienced by z_k, $k = 1, 2, 3$. A decorrelating pre-stage in an iterative IC receiver based on $\tilde{z}_4$ thus has *unbiased noise.*

Choosing a 3+1 scheme with minimal self-interference, partial SH, and an unbiased IC pre-stage thus singles out the choice of the idle direction in the modulation matrix to just one. The corresponding 3+1 layered matrix modulation is (5.27).

5.4.5   Minimal Self-interference 2+2 Layered 4×4 Schemes

The 3+1 matrix modulation discussed in the previous section does not fulfil the information extremality equation (4.89). From this it follows that a matrix modulation providing more information should exist.

In [132], the self-interference of $T \times N_t = 4 \times 4$ linear matrix modulators with rate $Q/T = 1$ was minimized with no boundary conditions besides maximal symbolwise diversity (4.50). The result was a family of 2+2 layered quasi-orthogonal schemes with self-interference $\widetilde{N} = 1/4$ per symbol. These fulfil the information extremality equations (4.89). The simplest is ABBA (4.35), i.e.

$$\mathbf{X}_{\mathrm{ABBA}} = \left[\begin{array}{cc} \mathbf{X}_A(x_1, x_2) & \mathbf{X}_B(x_3, x_4) \\ \mathbf{X}_B(x_3, x_4) & \mathbf{X}_A(x_1, x_2) \end{array} \right], \tag{5.34}$$

discussed as an example in Chapter 4. In ABBA, one quasi-orthogonal layer consists of the A part, the other of the B part. There is no natural ordering of the layers.

The basis matrices of ABBA were written out in (4.36). From the point of view adopted in this chapter it is worth noticing that they are elements in the 4×4 Clifford basis (5.16).

Schemes belonging to the same family (with the same non-orthogonality structure and same self-interference) were also discussed in [133, 142, 151]. The schemes reported in [142, 151] involve negations and complex conjugations of the A and B blocks, and can be considered as selecting a different set of eight basis matrices from the Clifford basis (5.16). These discrete transformations do not change performance nor information. The self-interference is still given by terms of the type (4.37).

It is not as straightforward to see that ABBA (restricted to three Tx antennas), and the orthogonal but non-orthonormal $N_t = 3$, $T = 4$, $Q = 4$ linear dispersion code reported in [133, Eq. (34)] are equivalent. Demonstrating this provides a good warm-up for the concepts to be discussed in Chapter 7.

The scheme in [133] was presented in terms of real and imaginary parts of four symbols, α_j, β_j. If these are mapped to new complex symbols as $x_1 = \alpha_1 + j(\beta_2 + \beta_3)/\sqrt{2}$, $x_2 = -\alpha_2 - j\beta_1$, $x_3 = \alpha_3 + j\beta_4$, $x_4 = \alpha_4 + j(\beta_3 - \beta_2)/\sqrt{2}$, the modulation matrix may be written in the form

$$\mathbf{X} = \left[\begin{array}{cccc} x_1 + x_3 & \frac{-x_2^* - x_4^*}{\sqrt{2}} & 0 \\ \frac{x_2 + x_4}{\sqrt{2}} & x_1^* & \frac{x_2^* - x_4^*}{\sqrt{2}} \\ 0 & \frac{-x_2 + x_4}{\sqrt{2}} & x_1 - x_3 \\ \frac{-x_2 - x_4}{\sqrt{2}} & -x_3^* & \frac{x_2^* - x_4^*}{\sqrt{2}} \end{array} \right] \tag{5.35}$$

The mapping of the symbols is partially a discrete symmetry of the performance and partially a specific realization of a continuous performance-changing symmetry. These concepts will be discussed in more detail in Chapter 7.

For (5.35), the diversity matrix is of the form (4.30), and the self-interference is diagonal,

$$S = \mathrm{Re}\left[x_1 x_3^* + x_2 x_4^*\right] \left[\begin{array}{ccc} 1 & 0 & 0 \\ 0 & 0 & 0 \\ 0 & 0 & -1 \end{array} \right]. \tag{5.36}$$

Thus the columns in (5.35) are indeed orthogonal but not orthonormal. Also, for this scheme it is a non-trivial fact that S is traceless.

Now a right unitary transformation of the kind (5.8) may be used. For i.i.d. Rayleigh fading, these transformations preserve information and performance. Choosing

$$\mathbf{W} = \frac{1}{\sqrt{2}} \begin{bmatrix} 1 & 0 & -1 \\ 0 & \sqrt{2} & 0 \\ 1 & 0 & 1 \end{bmatrix}, \tag{5.37}$$

and mapping the modulation matrix to $\mathbf{XW}$, the self-interference (5.36) becomes

$$\mathbf{W}^\dagger \, \mathcal{S} \, \mathbf{W} = \mathrm{Re}\,[x_1 x_3^* + x_2 x_4^*] \begin{bmatrix} 0 & 0 & 1 \\ 0 & 0 & 0 \\ 1 & 0 & 0 \end{bmatrix} \tag{5.38}$$

which is exactly the self-interference (4.37) of ABBA restricted to $N_t = 3$.

The scheme (5.35) was found in [133] by computer search, where the mutual information was maximized. The fact that schemes in the same family are found by maximizing mutual information and by minimizing self-interference, and that the schemes fulfil the information extremality condition (4.89), can be seen as a confirmation of the considerations in section 4.5. It should be noted that in [133] it was found that (5.35) was only a local information maximum for $N_t = 3$ Tx antennas. A matrix modulator providing more information was found by concatenating STTD (3.28) with antenna shuffling. This can be seen as a confirmation of the heuristic design rules of Chapter 4, as a modulator in the ABBA family, restricted to $N_t = 3$, is not symbol homogeneous. In contrast, STTD with antenna shuffling is SH. For $N_t = 4$, we shall see in Chapter 8 that STTD with antenna shuffling (i.e. the STTD-OTD scheme (4.21)) and ABBA provide the same information, and can be mapped to each other with symbol rotations.

5.4.6 Quasi-orthogonal Schemes for $N_t > 4$, $N_r = 1$

Here extending symbol rate $R_s = 1$ MISO schemes to more than 4 Tx antennas are considered. The natural generalization of ABBA to 5–8 antennas is the 2+2+2+2 layered ABBA [2] scheme [132],

$$\mathbf{X}_{\mathrm{ABBA}^2} = \begin{bmatrix} \mathbf{X}_A & \mathbf{X}_B & \mathbf{X}_C & \mathbf{X}_D \\ \mathbf{X}_B & \mathbf{X}_A & \mathbf{X}_D & \mathbf{X}_C \\ \mathbf{X}_C & \mathbf{X}_D & \mathbf{X}_A & \mathbf{X}_B \\ \mathbf{X}_D & \mathbf{X}_C & \mathbf{X}_B & \mathbf{X}_A \end{bmatrix}. \tag{5.39}$$

Here $\mathbf{X}_A, \mathbf{X}_B, \mathbf{X}_C, \mathbf{X}_D$ are STTD blocks (3.28) depending on the symbols $(x_1, x_2), (x_3, x_4)$, $(x_5, x_6), (x_7, x_8)$, respectively.

The self-interference matrix for this code is scarce, similarly to (4.37), so it can easily be inverted to produce a reliable LMMSE detection scheme, see Chapter 6.

ABBA[2] should be contrasted with other 8×8 rate 1 schemes. A 4+4 layered power-balanced scheme may be constructed by overlaying two 8×8 rate 1/2 orthogonal designs

Table 5.4: Characteristics of $N_t = 8$, $R_s = 1$, $T = 8$ schemes.

Scheme	Layering	$\widetilde{N}$	$\frac{\mathcal{I}_2}{\mathcal{C}_2}$(dB)	XTRM	MSD	FO	SH
ABBA^2	2+2+2+2	3/8	−1.25	yes	yes	yes	yes
OD+OD	4 + 4	5/8	−1.92	yes	yes	yes	yes
Cyclic	1+1+1+1+1+1+1+1	7/8	−2.50	yes	yes	yes	yes

(5.7):

$$
\mathbf{X}_{\text{OD+OD}} =
\begin{bmatrix}
x_1 & x_2 & x_3 & x_8 & x_4 & x_7 & x_6 & x_5 \\
-x_2^* & x_1^* & -x_8 & -x_3 & -x_7 & -x_4 & x_5^* & -x_6^* \\
-x_3^* & -x_8 & x_1^* & x_2 & x_6 & x_5^* & -x_4 & -x_7^* \\
x_8 & x_3^* & -x_2^* & x_1 & x_5 & -x_6^* & x_7^* & x_4 \\
-x_4^* & x_7 & x_6 & x_5^* & x_1^* & x_2 & x_3 & -x_8^* \\
-x_7 & x_4^* & x_5 & -x_6^* & -x_2^* & x_1 & x_8^* & -x_3 \\
x_6 & x_5 & x_4^* & -x_7^* & -x_3^* & x_8^* & x_1 & x_2 \\
x_5^* & -x_6^* & x_7^* & -x_4^* & -x_8^* & x_3^* & -x_2^* & x_1^*
\end{bmatrix}
\tag{5.40}
$$

$$
= \; \mathbf{X}_{1/2}(x_1, x_2, x_3, x_4) \; + \; \text{Reverse}\left[\mathbf{X}_{1/2}(x_5, x_6, x_7, x_8)\right]
$$

Here "Reverse" means that the order of columns in the matrix is reversed. A multi-stratum code for eight Tx antennas with two strata, constructed according to [143], would provide the same information as (5.40), but would not be power-balanced.

Furthermore, a cyclic scheme, generalizing the 4×4 cyclic (4.29) may easily be constructed. This scheme is 1+1+1+1+1+1+1+1 layered.

Of these schemes, ABBA^2 and OD+OD are quasi-orthogonal. Their 16 basis matrices are a subset of the Clifford basis for 8×8 matrices. The cyclic scheme is not quasi-orthogonal; each complex symbol interferes with all other symbols. Its basis matrices are not a simple subset of a Clifford basis. Other characteristics of (5.39), (5.40) and a cyclic scheme are collected in Table 5.4.

All schemes satisfy the qualitative design criteria discussed in Chapter 4: they are MSD, SH, and Frobenius orthogonal (FO). All schemes even fulfil the second-order information extremality equation (4.89) (XTRM). The difference is in the self-interference per bit, $\widetilde{N}$, and correspondingly, in the provided second-order information. The self-interference of ABBA^2 is considerably smaller, and thus it is superior to OD+OD, which is superior to the cyclic scheme.

The information extremality should now be interpreted so that ABBA^2 is a minimum of information, at least locally in matrix modulation space, whereas the cyclic scheme is likely to be at least a local maximum. OD+OD may be a local maximum, minimum or an inflection point in the family of Frobenius orthogonal schemes with MSD. The information extrema are SH schemes.

This leads to the conjecture that for any number of Tx antennas, a delay optimal rate $R_s = 1$ matrix modulation with minimal self-interference may be constructed by iterating the ABBA scheme.

5.5 SUMMARY

In this chapter, complexity issues in matrix modulation design were addressed. It was argued that a balance between information and detection complexity may be found by choosing

$$R_\mathrm{s} \approx \mathrm{Prank} \leq \min(N_\mathrm{t}, N_\mathrm{r}) . \qquad (5.41)$$

Here Prank is the practical rank of the channel, i.e. the number of essentially non-zero eigenvalues of $\mathbf{H}^\dagger\mathbf{H}$. For correlated channels Prank may be considerably less than $\min(N_\mathrm{t}, N_\mathrm{r})$, see section 2.4. Furthermore, it was argued that a suitable choice of the symbolwise diversity degree depends on the operational point.

In uncorrelated channels the rate of choice is $\min(N_\mathrm{t}, N_\mathrm{r})$. Regarding linear matrix modulation reaching this rate, the results of this chapter can be summarized as follows.

- Complex orthogonal designs work only for $N_\mathrm{t} = 2, \ N_\mathrm{r} = 1$.

- For more antennas, quasi-orthogonal layered schemes, or schemes with suboptimal symbolwise diversity, should be chosen.

The results of Chapter 4 indicate that quasi-orthogonal layered schemes provide more information and better performance at low SNR than generic non-orthogonal schemes. As will be seen in Chapter 6, quasi-orthogonality may be used to simplify detection as well. In Chapter 7 we shall use unitary diversity transforms to increase the (asymptotic) diversity degree, essential for high SNR performance, to equal the symbolwise diversity degree of a matrix modulation.

6

Receiver Algorithms

All performance results and comparisons in previous chapters were based on Maximum Likelihood (ML) detection. While ML or Maximum-*a-Posteriori* (MAP) detection is often required to capture the full diversity benefit and performance gain with space–time matrix modulators, sub-optimal low-complexity detection may need to be used in practice. Reduced complexity detection strategies are extremely relevant e.g. with high symbol rate MIMO modems, where the number of signal states is often enormous. For example, a matrix modulator involving eight QPSK symbols could have as many as $2^{16} = 65\ 536$ states. In these cases exhaustive search over all possible bit combinations is not feasible. Luckily, in many cases such a search it is not necessary, since the same performance can be achieved with a search over a small subset of all possible states, or by using efficient iterative detection strategies.

In this chapter we provide a non-exhaustive set of low complexity receiver algorithms. Most of these are directly applicable for the non-orthogonal MIMO and MISO systems developed elsewhere in this book.

6.1 CHANNEL ESTIMATION ISSUES

The receivers for linear matrix modulators, or for more primitive linear MIMO transmitters, are intimately related to those developed for multiuser detection or channel equalization problems [166]. Essentially, the spreading code matrix applied in CDMA is analogous to the equivalent channel matrix. Similarly, the CDMA code correlation matrix is analogous to the equivalent channel correlation matrix (4.57).

The elements of the equivalent channel matrix are obtained via channel estimation, and the structure of the matrix depends on the selected matrix modulation. It is therefore useful to briefly summarize how the equivalent channel matrix may be obtained.

The equivalent channel matrix can be constructed from the channel estimates. In WCDMA systems, channel estimation can be done using common and dedicated pilot channels. Typically the power of common pilot channels is sufficiently high for channel estimation purposes. While the sole use of common pilot channels has been found to be sufficient with STTD when $N_t = 2$, $N_r = 1$ many matrix modulators employ e.g. $N_t = 4$ transmit antennas and $N_r = 2$ receive antennas. Hence, the number of channel coefficients is increased, and alternative channel estimation paradigms may be needed, provided that the number of additional common pilot channels cannot always be increased when N_t increases. A similar case appears when considering space-time block codes in conjunction with GSM/EDGE [167, 168]. In these studies it was shown that in practice the diversity or performance gain from a two-antenna space–time block code (when compared to delay diversity of phase hopping) is compromised due to significant degradation in channel estimation accuracy. In these cases the performance gain may be recovered if more complex channel estimation techniques are applied. Iterative channel estimation, by using decision feedback channel estimation using soft information provided by a channel decoder has been found to be effective [169, 170]. Alternatively, or in addition, non-iterative semiblind subspace methods based on second-order statistics may come in handy.

Recent results show that a pilot-based channel estimate can be combined with a blind channel estimate [171] to further improve the quality of the channel estimate in the presence of space–time coding. In blind subspace methods, the observation space is first partitioned into signal subspace and noise subspace. Then MIMO channel parameters are estimated based on orthogonality between signal and noise subspaces. This paradigm allows one to identify MIMO channel up to a right multiplication of an invertible matrix. To resolve the ambiguity with respect to matrix multiplication a small number of transmitted symbols (e.g. pilot symbols at a dedicated channel) should be known at the receiver. More generally, any prior information on the MIMO channel (e.g. obtained from pilot signals) or the symbol alphabet should be used to improve the accuracy and robustness of blind estimation [172]. Such prior information may be designed into the frame. Indeed, symbols may be precoded in an appropriate fashion [173], to resolve the ambiguity problem.

In the following sections, we assume that the channel estimates have been obtained prior to detection.

6.2 MAXIMUM LIKELIHOOD DETECTION

Detection of linear matrix modulation is based on the received signal

$$\mathbf{y} = \mathcal{H}\mathbf{x} + \mathbf{n} \tag{6.1}$$

where $\mathcal{H}$ is the equivalent channel matrix (4.57) depending on the selected space–time modulation matrices. Channel noise is conventionally assumed to be zero-mean Gaussian. As discussed in section 4.4, $\mathcal{H}$ serves as the space–time matched filter matrix for the matrix modulation. The space–time matched filter has the same structure as the original code, and inherits the orthogonality or non-orthogonality properties of the space–time matrix modulation.

Assuming that the receiver has perfect channel state information, the receiver can construct the equivalent channel matrix and compute sufficient statistics for symbol detection,

$$\mathbf{z} = \mathcal{H}^\dagger \mathcal{H}\mathbf{x} + \text{noise} \tag{6.2}$$

where the noise is in general coloured complex Gaussian noise. With these assumptions a parametric Maximum Likelihood (ML) detector solves

$$\hat{\mathbf{x}}_{\mathrm{ml}} = \arg \max_{\mathbf{x} \in \mathbb{A}} \Omega(\mathbf{x}) \tag{6.3}$$

where $\mathbb{A}$ is the modulation alphabet and Ω is the likelihood function (4.64), which after dropping constant terms reads

$$\Omega(\mathbf{x}) = 2 \operatorname{Re}(\mathbf{x}^\dagger \mathbf{z}) - \mathbf{x}^\dagger \mathcal{H}^\dagger \mathcal{H} \mathbf{x}. \tag{6.4}$$

This problem may be rewritten as

$$\hat{\mathbf{x}}_{\mathrm{ml}} = \arg \min_{\mathbf{x}} \|\mathbf{z} - \mathcal{H}^\dagger \mathcal{H} \mathbf{x}\|_{\mathbf{Q}}^2, \tag{6.5}$$

where

$$\|\mathbf{v}\|_{\mathbf{Q}}^2 = \mathbf{v}^\dagger \mathbf{Q} \mathbf{v}$$

and $\mathbf{Q} = \mathcal{R}^{-1}$ is the weighting matrix where the equivalent channel correlation matrix is $\mathcal{R} = \mathcal{H}^\dagger \mathcal{H}$; see (4.58).

As is well known, the complexity of the ML problem is in general exponential in the model dimension. Here the problem dimension is essentially dictated by the dimensions of the modulation matrix. With orthogonal designs the off-diagonal elements of the equivalent correlation matrix vanish, and optimal ML detection is performed for each symbol independently of each other. However, the equivalent correlation matrix for non-orthogonal schemes entertains non-zero off-diagonal elements, and this leads to a combinatorial optimization problem. In space–time coding problems the ML solution is feasible if the dimensionality of $\mathcal{R}$ is sufficiently low, or it has some special structure that can be exploited. Unfortunately, a low model dimension tends to require either a low symbol rate or a small number of antennas.

Soft Output Detection: Soft decisions contain reliability information which is useful in a concatenated coding scheme where the matrix modulation is concatenated with an outer channel code, e.g. a turbo code or a convolutional code as in WCDMA. We consider below the signal model $\bar{\mathbf{y}} = \bar{\mathcal{H}} \bar{\mathbf{x}} + \bar{\mathbf{n}}$ of Equation (4.54), involving real input-real output matched filtering with BPSK coordinate constellations. Note that the results in this section may be extended to the generic real valued signal model (4.51), as the central results may be expressed in terms of the real symbol correlation matrix $\bar{\mathcal{R}}$.

The desired input to the channel decoder is the *a posteriori* log-likelihood ratio (LLR) of a transmitted bit

$$\Lambda(\bar{x}_j) = \log \frac{P(\bar{x}_j = 1 \mid \bar{\mathbf{y}})}{P(\bar{x}_j = -1 \mid \bar{\mathbf{y}})} \tag{6.6}$$

This can be expanded, using Bayes' rule, as

$$\Lambda(\bar{x}_j) = \log \frac{P(\bar{\mathbf{y}} \mid \bar{x}_j = 1)}{P(\bar{\mathbf{y}} \mid \bar{x}_j = -1)} + \log \frac{P(\bar{x}_j = 1)}{P(\bar{x}_j = -1)}, \tag{6.7}$$

where $\log(P(\bar{x}_j = 1)/P(\bar{x}_j = -1))$ represents the *a priori* LLR for bit $\bar{x}_j$. This can be obtained from some other receiver stage, e.g. from the channel decoder. Initially, it is typically assumed that the *a priori* distribution is symmetric, in which case $\log(P(\bar{x}_j = $

$1)/P(\bar{x}_j = -1)) = 0$. For a linear model with independent bits, corresponding to one space–time block, the LLR for bit j is given by

$$\Lambda_j = \log \frac{\sum_{\bar{\mathbf{x}} \in \mathbb{S}_j^+} \exp(-\|\bar{\mathbf{z}} - \bar{\mathcal{R}}\bar{\mathbf{x}}\|^2_{(\sigma^2\bar{\mathcal{R}})^{-1}}) \prod_{l \neq j} P(\bar{x}_l)}{\sum_{\bar{\mathbf{x}} \in \mathbb{S}_j^-} \exp(-\|\bar{\mathbf{z}} - \bar{\mathcal{R}}\bar{\mathbf{x}}\|^2_{(\sigma^2\bar{\mathcal{R}})^{-1}}) \prod_{l \neq j} P(\bar{x}_l)}, \tag{6.8}$$

where $\mathbb{S}_j^+$ is the set of those transmitted symbol vectors that correspond to transmitted bit sequences in which j has value 1 and $\mathbb{S}_j^-$ is the set of those transmitted vectors that corresponds to transmitted bit sequences in which j has value -1 [174]. Soft output detection can be carried out equivalently before matched filtering,

$$\Lambda_j = \log \frac{\sum_{\bar{\mathbf{x}} \in \mathbb{S}_j^+} \exp(-\|\bar{\mathbf{y}} - \bar{\mathcal{H}}\bar{\mathbf{x}}\|^2_{\sigma^{-2}}) \prod_{l \neq j} P(\bar{x}_l)}{\sum_{\bar{\mathbf{x}} \in \mathbb{S}_j^-} \exp(-\|\bar{\mathbf{y}} - \bar{\mathcal{H}}\bar{\mathbf{x}}\|^2_{\sigma^{-2}}) \prod_{l \neq j} P(\bar{x}_l)}. \tag{6.9}$$

It is clear that computing reliability information or LLR increases the receiver complexity. Euclidean distances need to be computed to all signal states, and the noise covariance needs to be known or estimated. Computationally simpler solutions for computing soft decisions typically operate with a reduced set of signal states. Indeed, max-log approximation yields

$$\Lambda_j \approx -\max_{\bar{\mathbf{x}} \in \mathbb{S}_j^+} \|\bar{\mathbf{y}} - \bar{\mathcal{H}}\bar{\mathbf{x}}\|^2/\sigma^2 + \max_{\bar{\mathbf{x}} \in \mathbb{S}_j^-} \|\bar{\mathbf{y}} - \bar{\mathcal{H}}\bar{\mathbf{x}}\|^2/\sigma^2. \tag{6.10}$$

This simplifies detection if the set of nearest neighbours can be restricted by other methods. Alternative centralized multiuser receivers can be browsed, e.g. from [166, 175–177].

Sphere Decoder: The complexity of the ML detector can be reduced by reduced search ML algorithms. These provide near-ML performance with significantly reduced complexity. Another option is to apply recently developed lattice decoding algorithms. Lattice decoding algorithms, e.g. the sphere decoder [178, 179], have gained popularity recently in both multiuser detection and MIMO detection problems [180–182]. Sphere decoding is often significantly less complex than the full ML search, provided that the signal model can be posed to follow the lattice representation [183]. The standard lattice representation utilized by the sphere decoder follows, when the signal model is decoupled into real and imaginary parts, and the modulation alphabet comprises integers. In space–time coding applications, decoupling can be applied to the (complex symbol) equivalent channel matrix, and the columns of matrix (see Equation (4.54))

$$\bar{\mathcal{H}} = \begin{bmatrix} \mathrm{Re}\,\mathcal{H} & -\mathrm{Im}\,\mathcal{H} \\ \mathrm{Im}\,\mathcal{H} & \mathrm{Re}\,\mathcal{H} \end{bmatrix} \tag{6.11}$$

are the basis vectors of the lattice, and the real and imaginary parts of elements of $\mathbf{x}$ represent the coordinates. Solutions to such lattice problems can be solved efficiently via local search in a sphere around the received signal sample with complexity between $O((2Q)^3)$ and $O((2Q)^6)$, where $2Q$ is the dimension of the signal model. The complexity depends on how one determines the radius of the sphere and various other implementation issues. However, detector complexity is essentially unchanged for all QAM alphabets. Hence, the decoder is particularly attractive when the data rate is increased via high-order modulation.

6.3 QUASI-ORTHOGONALITY ASSISTED MAXIMUM LIKELIHOOD DETECTION

Maximum likelihood detection involves a full configuration search over all symbols. For schemes with quasi-orthogonal layers, a simplified ML detection scheme can be devised with equivalent performance. This detection scheme shall be called Quasi-Orthogonality assisted ML (QOML). The idea of QOML is that after the interference of all quasi-orthogonal layers but one is cancelled, the last may be detected with the linear space–time block code metric (4.65). Here, the terminology "upper layer" will be used for the layers that are detected first, and "lower layer" for the layer that is detected last. If the quasi-orthogonal layers in the scheme have different number of symbols, one of the layers with most symbols should be chosen to be the lower layer.

The (real symbol) equivalent channel correlation matrix (4.58) may be written in block form as as

$$\bar{\mathcal{R}} = \mathcal{D}_{\mathbf{H}} + \mathcal{S}_{\mathbf{H}} \tag{6.12}$$
$$\mathcal{D}_{\mathcal{H}} = \mathbf{h}^{\dagger}\mathbf{h}\ \mathbf{I}_{2Q}$$
$$\mathcal{S}_{\mathcal{H}} = \begin{bmatrix} 0 & \mathcal{R}_{\mathrm{u}\times\mathrm{l}}^{\mathrm{T}} \\ \mathcal{R}_{\mathrm{u}\times\mathrm{l}} & \mathcal{S}_{\mathrm{u}\times\mathrm{u}} \end{bmatrix}, \tag{6.13}$$

where $\mathbf{I}_{2Q}$ is a $2Q \times 2Q$ dimensional identity matrix, $\mathcal{R}_{\mathrm{u}\times\mathrm{l}}$ is the inter-layer correlation and $\mathcal{S}_{\mathrm{u}\times\mathrm{u}}$ is the self-interference of the "upper layer". Note that this need not be vanishing, if the "upper layer" consists of several quasi-orthogonal layers. For clarity, the symbol vector may be split into upper and lower layer parts $\bar{\mathbf{x}}^{(\mathrm{u})}$ and $\bar{\mathbf{x}}^{(\mathrm{l})}$, which are considered as vectors of real symbols, i.e. real and imaginary parts of the complex symbols. Correspondingly, the matched filter outputs are split into $\bar{\mathbf{z}}^{(\mathrm{u})}, \bar{\mathbf{z}}^{(\mathrm{l})}$. In terms of these, the matched filter outputs in (4.55), expressed in terms of the transmitted symbols are

$$\bar{\mathbf{z}}^{(\mathrm{l})} = \mathbf{h}^{\dagger}\mathbf{h}\ \bar{\mathbf{x}}^{(\mathrm{l})} + \mathcal{R}_{\mathrm{u}\times\mathrm{l}}^{\mathrm{T}}\ \bar{\mathbf{x}}^{(\mathrm{u})}$$
$$\bar{\mathbf{z}}^{(\mathrm{u})} = \left(\mathbf{h}^{\dagger}\mathbf{h}\ \mathbf{I}_{\mathrm{u}} + \mathcal{S}_{\mathrm{u}\times\mathrm{u}}\right) \bar{\mathbf{x}}^{(\mathrm{u})} + \mathcal{R}_{\mathrm{u}\times\mathrm{l}}\ \bar{\mathbf{x}}^{(\mathrm{l})}.$$

Thus for any given value of $\bar{\mathbf{x}}^{(\mathrm{u})}$, the metric (4.64) is linear. QOML considers the possible values of $\bar{\mathbf{x}}^{(\mathrm{u})}$. With a given value $\hat{\bar{\mathbf{x}}}^{(\mathrm{u})}$, the lower layer matched filter outputs $\bar{\mathbf{z}}^{(\mathrm{l})}$ with upper layer interference cancelled are

$$\bar{\mathbf{z}}^{(\mathrm{l,IC})}\left(\hat{\bar{\mathbf{x}}}^{(\mathrm{u})}\right) = \bar{\mathbf{z}}^{(\mathrm{l})} - \mathcal{R}_{\mathrm{u}\times\mathrm{l}}^{\mathrm{T}}\ \hat{\bar{\mathbf{x}}}^{(\mathrm{u})} \tag{6.14}$$

For simplicity, we shall assume that the lower-layer complex symbols are QPSK, so the corresponding real symbols are BPSK. The lower layer BPSK symbols corresponding to $\hat{\bar{\mathbf{x}}}^{(\mathrm{u})}$ are

$$\hat{\bar{\mathbf{x}}}^{(\mathrm{l})}\left(\hat{\bar{\mathbf{x}}}^{(\mathrm{u})}\right) = \mathrm{sign}\left(\bar{\mathbf{z}}^{(\mathrm{l,IC})}\left(\hat{\bar{\mathbf{x}}}^{(\mathrm{u})}\right)\right) \tag{6.15}$$

In (4.64), the normalization (4.65) may be omitted, as well as the part in (4.66) depending on the diversity matrix (and correspondingly of the squared symbol powers, which are constant for QPSK). All diagonal elements of the correlation matrix are the same, $\mathbf{h}^{\dagger}\mathbf{h}$, so this factor may also be removed from the metric. For each $\hat{\bar{\mathbf{x}}}^{(\mathrm{u})}$ the metric calculated is thus

$$\Omega\left(\hat{\bar{\mathbf{x}}}^{(\mathrm{u})}\right) = \left\|\bar{\mathbf{z}}^{(\mathrm{u})} - \left(\mathbf{h}^{\dagger}\mathbf{h}\ \mathbf{I}_{\mathrm{u}} + \mathcal{S}_{\mathrm{u}\times\mathrm{u}}\right)\ \hat{\bar{\mathbf{x}}}^{(\mathrm{u})}\right\|^2 + \left\|\bar{\mathbf{z}}^{(\mathrm{l,IC})} - \mathbf{h}^{\dagger}\mathbf{h}\ \hat{\bar{\mathbf{x}}}^{(\mathrm{l})}\left(\hat{\bar{\mathbf{x}}}^{(\mathrm{u})}\right)\right\|^2 \tag{6.16}$$

The vector $\hat{\hat{\mathbf{x}}}^{(u)}$ with the smallest $\Omega\left(\hat{\hat{\mathbf{x}}}^{(u)}\right)$ is chosen, together with the corresponding $\hat{\hat{\mathbf{x}}}^{(l)}\left(\hat{\hat{\mathbf{x}}}^{(u)}\right)$. This algorithm leads to considerable complexity reduction, as ML detection complexity grows exponentially in the number of symbols searched over. For a two-layer scheme with equal numbers of symbols in the layer, QOML takes a square root of detection complexity. With small constellation sizes, QOML is of the same order of complexity as IC detection, and it often performs better. For non-symbol-homogeneous schemes the reduction is even larger. Thus if QPSK symbols are used on the upper layer of (5.27), QOML detection would involve going through four upper layer states, whereas full ML would consider 256 states. This result extends to more complicated non-orthogonal schemes, discussed below.

6.4 LINEAR RECEIVERS

The linear MMSE (LMMSE) detector solves [176]

$$\mathbf{L}_k = \arg\min_{\mathbf{L}_k} E(\|x_k - \mathbf{L}_k^\dagger \mathbf{y}\|^2).$$

The well-known solution is

$$\mathbf{L} = (\mathcal{H}^\dagger \mathcal{H} + \sigma^2 \mathbf{I})^{-1}\mathcal{H}, \tag{6.17}$$

where $\mathcal{H}$ is the equivalent channel matrix and σ^2 is the noise power. After filtering, the symbol decision is based on

$$\mathbf{z}_k = \mathbf{L}_k^\dagger \mathbf{y}, \tag{6.18}$$

where $\mathbf{y} = \mathcal{H}\mathbf{x} + \mathbf{n}$. The formulation above assumes that the receiver has estimated the matrix $\mathcal{H}$ and channel noise power σ^2 with sufficient accuracy.

We consider again the real-valued signal model with BPSK coordinate constellations. In this model, the error probability of linear receivers can be expressed in closed form,

$$P_b = \frac{2}{K|\{\bar{\mathbf{x}}\}|}\sum_{k=1}^{K}\sum_{\bar{\mathbf{x}};\bar{x}_k=1} Q\left(\frac{a_k(\bar{\mathbf{L}}^\dagger\bar{\mathcal{R}})_{k,k}(1+\sum_{j\neq k}a_j(\bar{\mathbf{L}}^\dagger\bar{\mathcal{R}})_{k,j}\bar{x}_j)}{\sigma\sqrt{(\bar{\mathbf{L}}^\dagger\bar{\mathcal{R}}\bar{\mathbf{L}})_{k,k}}}\right), \tag{6.19}$$

where $\bar{\mathbf{L}}$ contains the coefficients of linear detector using the real model and Q denotes the complementary error function. The transmit amplitude of stream k is denoted by a_k. Clearly, the number of terms in the expression increases exponentially in the number of parallel streams K. A simpler performance estimate can be obtained by invoking the Gaussian approximation [166, 184]. using coefficients

$$\gamma_{k,j} = a_j(\bar{\mathbf{L}}^\dagger\bar{\mathcal{R}})_{k,j},$$

$$\beta_k = \frac{a_k(\bar{\mathbf{L}}^\dagger\bar{\mathcal{R}})_{k,k}}{\sigma\sqrt{(\bar{\mathbf{L}}^\dagger\bar{\mathcal{R}}\bar{\mathbf{L}})_{k,k}}},$$

and

$$\lambda_k^2 = \frac{\beta_k^2\sum_{j\neq k}\gamma_{k,j}^2}{\gamma_{k,k}}.$$

Using these notations, a computationally attractive and accurate approximation to error probability is given by

$$P_b = \frac{1}{K} \sum_{k=1}^{K} Q(\frac{\beta_k}{\sqrt{1 + \lambda_k^2}}). \tag{6.20}$$

The fraction $\gamma_{k,j}/\gamma_{k,k}$ quantifies interference leakage between the kth and jth stream. This vanishes for the decorrelating detector, $\lambda_k^2 = 0, \forall k$. The derivation of this result and a discussion on the validity of the approximation can be found in [166, 184].

In practice, multiple access interference (intra- or inter-cell) and inter-path interference affects the decision statistics and the signal model is partly unknown. In these cases the LMMSE detector can be implemented in a decentralized, blind or adaptive fashion. A blind multiuser detector, which operates without a training sequence, is given in [185, 186] with the introduction of a Minimum Output Energy (MOE) detector. In addition, [187] proposed a subspace-based formulation for the blind MMSE receiver, which is summarized below for the space–time coded channel.

Define the autocorrelation matrix of the received symbols by $\mathbf{R} = \mathrm{E} \langle \mathbf{yy}^\dagger \rangle$. Performing the eigenvalue decomposition of $\mathbf{R}$ we obtain

$$\mathbf{R} = [\mathbf{U}_s \ \mathbf{U}_n] \begin{bmatrix} \mathbf{\Lambda}_s & \\ & \mathbf{\Lambda}_n \end{bmatrix} [\mathbf{U}_s \ \mathbf{U}_n]^\dagger, \tag{6.21}$$

where $\mathbf{U}_s$ and $\mathbf{\Lambda}_s$ denote the eigenvectors and eigenvalues matrix of the signal subspace, and $\mathbf{U}_n$ and $\mathbf{\Lambda}_n$ denote the eigenvectors and values of the noise subspace, respectively. Matrix $\mathbf{\Lambda}_s = \mathrm{diag}(\lambda_1, ..., \lambda_d)$ contains the d largest eigenvalues in descending order of magnitude. Matrix $\mathbf{U}_n$ contains the eigenvectors corresponding to eigenvalue σ^2. The blind Linear MMSE (LMMSE) filter for the arises as

$$\mathbf{L} = \mathbf{U}_s \mathbf{\Lambda}_s^{-1} \mathbf{U}_s^\dagger \mathcal{H}. \tag{6.22}$$

In practice, the eigenvalue decomposition can be applied to a sample estimate of the autocorrelation matrix, $\hat{\mathbf{R}} = 1/P \sum_{i=1}^{P} \mathbf{y}^{(i)} \mathbf{y}^{(i)^\dagger}$. When the number of signals is known, the dimension of the signal subspace d is easily determined. This is the case with matrix modulators. A caveat with the "blind" approach, as described above, is the fact that $\mathcal{H}$ is assumed to be known. A fully blind approach would first estimate $\mathcal{H}$ blindly using similar subspace-based signal processing, and then utilize blind detection. This approach would yield linear decisions, short of any possible ambiguity problems due to blind channel estimation. However, even if $\mathcal{H}$ is assumed to be known, the detector has some attractive characteristics. Namely, the same detector can be used in conjunction with orthogonal or non-orthogonal matrix modulation, and also in cases where the model is only partly identified.

6.5 ITERATIVE RECEIVERS

Iterative (interference cancellation) receivers are popular in multiuser detection community. One of the first iterative solutions, the multistage detector [188], performs iterative updates in the vicinity of the prevailing tentative decision. The Expectation-Maximization (EM) algorithm [189] provides yet another appealing framework for the detection of non-orthogonal schemes. Multiuser detection problems with different EM decompositions were considered in [190].

Multistage Detection: Multistage interference cancellation provides a sub-optimum iterative solution to the ML detection problem. The multistage receiver iterates the following equation

$$\hat{\bar{\mathbf{x}}}^t = \text{sign}[\bar{\mathbf{z}} - \bar{\mathcal{R}}\hat{\bar{\mathbf{x}}}^{t-1} + \text{diag}(\bar{\mathcal{R}})\hat{\bar{\mathbf{x}}}^{t-1}], \tag{6.23}$$

where, in analogy with (6.11), we have decoupled the correlation matrix into real and imaginary parts:

$$\bar{\mathcal{R}} = \begin{bmatrix} \text{Re}\,\mathcal{R} & -\text{Im}\,\mathcal{R} \\ \text{Im}\,\mathcal{R} & \text{Re}\,\mathcal{R}. \end{bmatrix} \tag{6.24}$$

The decoupled model allows us to express QPSK modulation and demodulation algorithms in real space using simpler notation, and is used below to simplify notation. The decision function sign(.) should be optimally replaced by $\tanh(.)$, which delivers soft decisions leading to algorithms described in [190]. The iterations in (6.23) are typically initialized with a linear prestage $\mathbf{Lz}$, although other initial values are also possible.

Gradient Guided Search: Multistage detection is only one possible iterative detection strategy. Hu and Blum [191] proposed an alternative iterative (gradient guided) fixed-radius sphere search algorithm which sequentially updates a current estimate $\bar{\mathbf{x}}^{t-1}$ according to

$$\bar{\mathbf{x}}^t = \arg \max_{\|\bar{\mathbf{x}} - \bar{\mathbf{x}}^{t-1}\|_H < r} \Omega(\mathbf{x}), \tag{6.25}$$

where $\| \cdot \|_H$ denotes Hamming weight metric. With a sufficiently large r, the algorithm is identical to maximum likelihood. A more interesting and computationally feasible special case occurs when $r = 1$. Assume below, for simplicity, that each element in $\mathbf{x}$ belongs to the QPSK alphabet, i.e. each element of $\bar{\mathbf{x}}$ belongs to the BPSK alphabet, and further that $r = 1$. Then, it can be shown that each component is updated according to

$$\hat{\bar{x}}_k^t = \begin{cases} -\hat{\bar{x}}_k^{t-1} & \text{if } k = \arg\max_j \omega_j^{t-1} \text{ and } \omega_k^{t-1} > 0 \\ \hat{\bar{x}}_k^{t-1} & \text{otherwise,} \end{cases} \tag{6.26}$$

where

$$\omega^{t-1} = -\hat{\bar{\mathbf{x}}}^{t-1} \odot (\bar{\mathbf{z}} - (\bar{\mathcal{R}} - \text{diag}(\bar{\mathcal{R}}))\hat{\bar{\mathbf{x}}}^{(t-1)}), \tag{6.27}$$

where operator $\text{diag}(\cdot)$ nulls the off-diagonal values of the argument matrix and $\odot$ denotes the Hadamard product. It is apparent that this is a form of iterative interference cancellation, where the cancellation order is conjectured from residual signal statistics. Essentially, the element of the bit vector that most increases the likelihood is updated in each iteration round. This is different from what is typically used in successive interference cancellation (SIC) where channel powers determine the cancellation order. With full diversity matrix modulation SIC would lead to an arbitrary cancellation order since the diagonal elements of the equivalent channel matrix are all identical. The initial estimate for the fixed-radius algorithm can be obtained from some linear receive pre-stage. Methods based on group detection and iterative algorithms [192, 193] have recently been proposed, mainly to improve the error performance of the SIC algorithm.

BLAST: The algorithms described above were initially developed for the multiuser detection problem. In contrast, the V-BLAST detector was originally developed for the BLAST MIMO testbed [21]. The V-BLAST detector performs ordered successive interference cancellation (OSIC), with a dynamic cancellation order in each iteration epoch. The individual signal dimensions are detected sequentially, and then cancelled from the received signal to form a residual signal. Linear detection is then applied to the residual signal using a signal model that contains only the undetected signals. This process is repeated until all signal components are detected. In addition to the analogy to multiuser detection receivers, the V-BLAST detector has been shown to be equivalent to Generalized Decision Feedback Equalization in [194].

6.6 JOINT DECODING AND DETECTION

A heuristic approach for hard decision feedback using the channel decoder output was proposed in [195] in conjunction with very low rate convolutional codes. In this solution the decoder re-encodes the data (to obtain the redundant bits) by using hard decisions on the decoded information bits. Recoded data is used to reconstruct the interference signal, which is cancelled from the received signal. More refined receivers, where information contained in channel code is utilized in a more efficient manner, were later proposed.

A computationally challenging solution that uses a complete trellis description over multiple streams was proposed in [196]. Considerably simpler, yet efficient, receiver structures were proposed in [174, 197–205]. Many of these solutions compute the soft decision metrics for each coded bit. The soft-weighted bits are used when reconstructing the interference signal. These solutions are also closely related to turbo equalization [206], an approach developed for inter-symbol interference (ISI) channels. Based on the analogy above, it is not surprising that similar joint decoding-detection solutions are also applicable in the presence of non-orthogonal matrix modulation.

A natural drawback with joint decoding–detection solutions is increased latency. Therefore, such efficient strategies are applicable only for selected services. Joint decoding and detection using iterative structures where soft information (e.g. *a posteriori* probabilities) provides perhaps the simplest and modular enhancement to the aforementioned iterative detection concepts. With these iterative receivers, the soft cancellation values for the channel coded bits/symbols are provided (at least in part) by the channel decoder. In particular, the *a posteriori mean* (APM) estimate of the transmitted symbols (assuming BPSK symbols again for simplicity) is determined as follows:

$$\hat{x}_k^i = P(x_k^i = 1|\mathbf{y}; \mathbb{X}) - P(x_k^i = -1|\mathbf{y}; \mathbb{X}) \ \forall k, i, \tag{6.28}$$

where $\mathbb{X}$ is the set of codewords. The notation above stresses the fact that the codeword set is used in obtaining the marginal probabilities for each channel bit. This can be done by augmenting the MAP or ML decoder appropriately [207, 208]. A straightforward variant is obtained with the use of symbol-by-symbol MAP decisions

$$\hat{x}_k^i = \text{sign}[Pr(x_k^i = 1|\mathbf{y}; \mathbb{X}) - Pr(x_k^i = -1|\mathbf{y}; \mathbb{X})] \ \forall k, i. \tag{6.29}$$

Note that the probabilities in (6.28) and (6.29) can be computed within the turbo decoder with negligible additional complexity. In fact, the soft output for the systematic bit is readily available in turbo decoder output and in addition to this we only need to compute the probabilities for the coded bits.

Once the probabilities are computed one has to decide the order in which signals are cancelled from each other. However, with multiuser or multichannel decoding, the received signal power may not be the desired criterion for determining the cancellation order. More refined criteria is based on signal quality at the output of individual decoders. Thus, performance measures related to channel decoding metrics are more appropriate. In addition, error-detection codes may improve the speed of convergence of the iterative algorithms. The signals that have passed detection can be assumed to be reliable and the *a posteriori* probabilities for each channel bit can be quantized to 1.

6.7 EXAMPLE: LINEAR DETECTION FOR ABBA

Consider a particular example involving the ABBA scheme (5.34). For detection, the interference caused by symbols on each other, related to the non-orthonormality, has to be cancelled. Detection of schemes in the ABBA family was discussed in [132, 151], with a parallel interference cancellation (PIC) method. In [142], maximum likelihood detection (ML) detection was used. The simple structure of the non-orthonormality matrix results in an easily implementable LMMSE detector, which may be used as a pre-stage for PIC.

Here, only detection in a single-path channel is considered. The channel coefficients are $\mathbf{h} = [h_1 \ h_2 \ h_3 \ h_4]^\mathrm{T}$, and the received signal vector is

$$\mathbf{y} = \mathbf{X}_{ABBA}\, \mathbf{h} + \text{noise} .$$

If the second and fourth elements are conjugated, the received signals can be written in the form

$$\tilde{\mathbf{y}} = \begin{bmatrix} y_1 \\ y_2^* \\ y_3 \\ y_4^* \end{bmatrix} = \mathcal{H} \begin{bmatrix} x_1 \\ x_2 \\ x_3 \\ x_4 \end{bmatrix} + \text{noise}, \tag{6.30}$$

where the complex symbol equivalent channel matrix is

$$\mathcal{H} = \begin{bmatrix} h_1 & h_2 & h_3 & h_4 \\ h_2^* & -h_1^* & h_4^* & -h_3^* \\ h_3 & h_4 & h_1 & h_2 \\ h_4^* & -h_3^* & h_2^* & -h_1^* \end{bmatrix} . \tag{6.31}$$

Linear processing with the Hermitian conjugate of the estimated equivalent channel matrix gives the MF outputs

$$\mathbf{z} = \mathcal{H}^\dagger\, \tilde{\mathbf{y}} = \left(\sum_{i=1}^{4} |h_i|^2\, \mathbf{I} + \mathcal{S}_\mathbf{H} \right) \mathbf{x} + \text{noise} . \tag{6.32}$$

where the self-interference part of the channel correlation matrix reflects the non-orthogonality of the scheme,

$$\mathcal{S}_\mathbf{H} = 2\, \mathrm{Re}\,[h_1 h_3^* + h_2 h_4^*] \begin{bmatrix} 0 & 0 & 1 & 0 \\ 0 & 0 & 0 & 1 \\ 1 & 0 & 0 & 0 \\ 0 & 1 & 0 & 0 \end{bmatrix} . \tag{6.33}$$

The dependence on the symbols x_i in (6.32) can be easily inverted, leading to a LMMSE detector. The linear detector assumes the form

$$\mathbf{L} = \frac{1}{1-d^2} \begin{bmatrix} 1 & 0 & d & 0 \\ 0 & 1 & 0 & d \\ d & 0 & 1 & 0 \\ 0 & d & 0 & 1 \end{bmatrix}, \tag{6.34}$$

$$d = \frac{2\,\mathrm{Re}\,[h_1 h_3^* + h_2 h_4^*]}{\mathbf{h}^\dagger \mathbf{h} + e} \tag{6.35}$$

With different values of e, different estimates for the symbols can be constructed. Decorrelation estimates are found by making $e = 0$. Indeed,

$$\mathbf{L}_{|e=0} \ \mathbf{z} = \sum_{i=1}^{4} |h_i|^2 \mathbf{x} + \text{noise} . \tag{6.36}$$

Similarly, when e equals an estimate of the sum of SNRs over the four channels, we obtain an LMMSE estimate of the transmitted symbols.

This linear detector may be used as a pre-stage for iterative parallel interference cancellation. The interference due to estimates obtained by the pre-stage is cancelled from the matched filter output, and the symbols are iteratively detected.

The worst case occurs when the off-diagonal elements almost equal the diagonal elements in the linear estimate (6.32). This happens if all channels are close to each other. In this case, the interfering symbols x_1, x_3 and x_2, x_4 cannot be distinguished, and detection breaks down.

6.8 PERFORMANCE

Figure 6.1 shows the performance of four antenna ABBA using different detectors. The performance degradation when applying linear detection strategies is significant but still comparable with limited diversity schemes, such as STTD-OTD. Maximum likelihood detector and the gradient guided detector with Hamming sphere $r = 1$ and LMMSE initial decisions achieve identical performance.

Increasing the symbol rate from one (ABBA) to two highlights the need to use a sophisticated receiver. Double STTD (DSTTD) [209,210] transmits two copies of STTD (Alamouti code) in parallel from four transmit antennas using modulation matrix

$$\mathbf{X}(x_1, ..., x_4) = \begin{bmatrix} \mathbf{X}(x_1, x_2) & \mathbf{X}(x_3, x_4) \end{bmatrix} . \tag{6.37}$$

Two receive antennas are used to detect the DSTTD signal. We assume the channel coefficients are i.i.d. Rayleigh. Four symbols are transmitted in two symbol intervals. Thus, the scheme has symbol rate two. If the QPSK modulation alphabet is used the spectral efficiency is 4 bps/Hz.

To motivate the need for advanced receivers we quantify the loss imposed by suboptimal receivers in Figure 6.2. Indeed, Figure 6.2 depicts the simulated performance for DSTTD assuming four transmit antennas and two receive antennas. Here, linear detection methods are clearly inferior to Maximum Likelihood and sub-optimal iterative detectors. As an example, the LMMSE receiver requires 3.5 dB more power to achieve BER = 10^{-2}, when compared

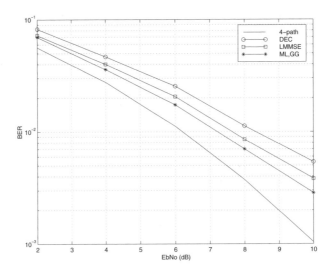

Fig. 6.1: Performance of ABBA in flat Rayleigh fading 4 Tx- 1 Rx channel using decorrelating detector ($- \circ -$), LMMSE detector ($-\square-$), Maximum Likelihood (ML) and gradient guided detector (GG) ($- * -$). Theoretical error probability assuming orthogonal four-path diversity is shown for reference ($-$) (2 bps/Hz).

to the Maximum Likelihood detector. Figure 6.2 also highlights the fact that a receiver which attains ML performance for one particular transmission scheme may not be able to repeat the achievement for another transmission scheme. In particular, gradient guided search is unable to match ML performance with DSTTD. Nevertheless, GG detector outperforms linear detectors by several decibels, especially when the signal-to-noise ratio is high. The lower (matched filter) bound on performance with diversity order four and eight is shown for reference. It is seen that DSTTD performance follows the order four performance. This is not satisfactory since in the considered 4 Tx-2 Rx architecture we would like to approach order eight performance. We will return to more efficient (and complex) high symbol rate schemes in Chapter 9.

6.9 SUMMARY

In this chapter we summarized the main principles behind Maximum Likelihood detection and several linear and non-linear low-complexity detectors. These were addressed in the context of linear space–time modulation concepts that use the equivalent channel matrix in detection. The structure of the equivalent channel correlation matrix largely determines the number of states the receiver needs to consider for optimal (ML) detection. With non-orthogonal high-rate matrix modulation the number of states can be rather large. Therefore, with high symbol rate it is likely that sub-optimal receiver algorithms need to be used in practice. Many of these sub-optimal algorithms can follow ML performance sufficiently well to warrant their use. Linear receivers, on the other hand, often do not perform satisfactorily, in particular when used in conjunction with a highly parallel MIMO system. Linear receivers

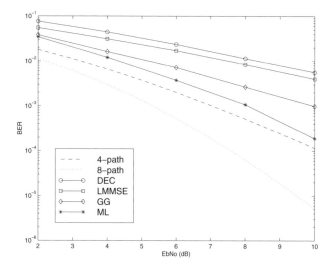

Fig. 6.2: Performance of DSTTD in a flat Rayleigh fading 4 Tx- 2 Rx channel using decorrelating detector $(-\circ-)$, LMMSE detector $(-\square-)$, gradient guided (GG) detector
$(-\diamond-)$, and Maximum Likelihood (ML) detector $(-*-)$. Theoretical error
probabilities assuming orthogonal four-path and eight-path diversity are shown for
reference by $(--)$ and $(\cdots)$ (4 bps/Hz).

also tend to waste some diversity. Thus, to get a desired diversity order, efficient baseband
detection algorithms need to be adopted. More advanced detectors may also mitigate the
need to employ an additional receiver antenna to obtain a given diversity order.

7
Matrix Modulation: High SNR Aspects

Constructing linear matrix modulators that perform well at low SNR was considered in Chapter 4. The essence was to maximize the mutual information [133], or equivalently, to minimize the total self-interference [132].

After maximizing the mutual information, some degrees of freedom are left unspecified in choosing the matrix modulation. In [135] it was argued that for given symbol constellations, these variables may be chosen so that a suitable performance measure is optimized. Thus the well-known rank [129] and determinant [19] criteria for space–time code design may be analysed, once mutual information is maximized. Generically, the degrees of freedom remaining after maximizing the mutual information are (possibly matrix-valued) constellation rotations. Such rotations have been discussed in [52, 86, 164, 211, 212] for rate $R_s = 1$ MISO systems, and in [51, 86, 213] for MIMO systems. Such constellation rotations may be considered as diversity transforms (sometimes referred to as precoding) of the kind discussed in section 3.4, where the diversity transform fights against the self-interference created by the matrix modulation.

A complementary approach to improve matrix modulation performance is to extend the scheme in the temporal direction, discussed in [132, 133, 156].

In this chapter, these methods to optimize linear matrix modulation at high SNR are discussed. They are applied after using the design principles for low SNR, discussed in Chapter 4, and the complexity principles discussed in Chapter 5. The steps in the design procedure of Chapters 4 and 5 may be summarized as

1. Choose an appropriate symbol rate, typically $\sim \min(N_t, N_r)$.

2. Choose a target degree of transmit diversity.

3. Maximize the provided mutual information (=optimize performance at low SNR).

In this chapter, a fourth step is added:

4. Optimize performance at high SNR within a set of transformations leaving the mutual information invariant.

The three first steps operate on the level of the modulation matrix. (i.e. symbols, not symbol differences). They determine performance at low to medium SNR. Thus Maximal Symbolwise Diversity (4.50) is a minimal version of diversity-optimality. It gives full diversity protection against one-symbol error events. This can be understood in terms of the trace criterion [130], which states that the trace of the distance matrix acts as a Euclidean distance in codeword difference matrix space. For error events with multiple symbol errors, the Euclidean distance is big. At medium SNR, diversity protection (product distance) is less important than Euclidean distance, and single-symbol error events (with small Euclidean distance) dominate performance. In this SNR regime it is thus most important to have full diversity for single-symbol error events, i.e. MSD.

At high SNR, performance is determined by the properties of the codeword difference matrix. Due to non-orthogonality, some multiple symbol error events may have non-optimal diversity protection, despite MSD. As an example, consider ABBA (4.35),

$$
\mathbf{X}_{\mathrm{ABBA}}(x_1, x_2, x_3, x_4) = \left[\begin{array}{cccc} x_1 & x_2 & x_3 & x_4 \\ -x_2^* & x_1^* & -x_4^* & x_3^* \\ x_3 & x_4 & x_1 & x_2 \\ -x_4^* & x_3^* & -x_2^* & x_1^* \end{array} \right] \equiv \left[\begin{array}{cc} \mathbf{X}_A & \mathbf{X}_B \\ \mathbf{X}_B & \mathbf{X}_A \end{array} \right]. \tag{7.1}
$$

It is easy to see that modulators of this form are performance sub-optimal. Consider an error event where the same error Δ is made in detecting x_1 and x_3, and no errors are made in detecting x_2 and x_4. Then the codeword difference matrix takes the form

$$
\mathbf{D}^{(\mathrm{ce})} = \left[\begin{array}{cccc} \Delta & 0 & \Delta & 0 \\ 0 & \Delta^* & 0 & \Delta^* \\ \Delta & 0 & \Delta & 0 \\ 0 & \Delta^* & 0 & \Delta^* \end{array} \right], \tag{7.2}
$$

and is clearly singular. The distance matrix $\mathbf{D}^{(\mathrm{ce})\dagger}\mathbf{D}^{(\mathrm{ce})}$ has only rank 2, which is the diversity protection against error events of this type. At high SNR, these error events dominate performance, and asymptotically ABBA has diversity 2.

This is a generic property of non-orthogonal matrix modulation. Typically, if all symbols are taken in the same modulation constellation, pathologic error events give rise to singular distance matrices and thus have a reduced diversity protection. To design a well performing scheme, one may use performance-changing symbol rotations to rotate the constellations away from these singular points. This leads to an investigation of symmetries of mutual information.

Symmetries and Groups of Transformations: Following standard mathematical terminology [214, 215], a symmetry of a quantity is a set of transformations leaving the quantity invariant. Such transformations typically form a group. Thus a *symmetry group* is a set of transformations leaving a quantity invariant, with a group structure. For more discussion on this, see Appendix A.

As discussed in [133], a matrix modulation providing a given mutual information has a large symmetry group, which (at least partly) is not a symmetry of performance. Such performance-changing symmetries of information are symbol rotations removing pathological error events.

Performance Measures: With the term "performance", we mean bit, symbol or block error rate. Of these, bit error rate is the most pertinent. It should be stressed that in this Chapter, all results are in i.i.d. Rayleigh channels. When generalizing to other channel statistics, some new features may appear. Such are discussed in Chapter 8.

The fourth design step discussed above is analysed on the level of the codeword difference matrix. Performance may be optimized by selecting the proper transformation from the symmetries of information. In i.i.d Rayleigh fading, various space–time code design criteria aim at optimizing performance at high SNR. Thus the rank and determinant criteria may be used, or some suitable refined criteria. In this chapter, the following criteria are considered:

- *The rank criterion* [129], discussed in section 3.5.2.

- *The determinant criterion (MAX-MIN-DET)* [19], discussed in section 3.5.2.

- *The bit-error and multiplicity enhanced determinant criterion (MAX-MIN-DET/Errs/ Mult):* The MAX-MIN-DET criterion minimizes the pairwise probability of code-word errors, i.e. the block error rate (BLER). Here we are interested in space–time modulators, and a more relevant measure to minimize is the bit error rate (BER). Different error events have different multiplicity; an error event with higher multiplicity gives rise to higher BLER and BER. For this, the diversity protection given by the determinant should be divided by the number of bit-errors induced by a given pairwise error event, and the multiplicity of the error event.

- *The union bound of bit-error probabilities (Min-UB):* This may be accessed using simulations, and has been used in [134]. In some cases, it is possible to compute the union bound exactly.

It should be noted that in continuously parameterized situations, where the distance matrix is singular at some discrete points, the determinant may be analytically continued to the singular points. In such cases the rank criterion is a part of the determinant criterion. All cases discussed in this chapter are of this type.

Diversity transforms may be considered in continuous [109, 114] sets or in discrete sets of algebraic rotations [111, 112]. Here we follow the former approach, although simplifying to algebraic rotations may be done in a straight forward manner. Considering diversity transforms, the main problem addressed in this chapter is to choose a subset of all possible rotations to consider, without loosing generality, following [216]. The pertinent low-dimensional rotations in the high dimensional set of all possible symbol rotations are identified. Within this limited set of degrees of freedom, MAX-MIN-DET optimal constellation rotations may sometimes be solved in closed form. The solutions are algebraic rotations.

As toy models of the methods discussed in this chapter, both symbol rotation and extending block, non-orthogonal matrix modulators for $N_t = 4$ transmit antennas and symbol rate 1 are considered. In order to benefit fully from the inherent transmit diversity, the schemes extend over $T \geq N_t$ symbol periods. We shall see that for these toy models, the design problem becomes analytically tractable, and we shall find optimal schemes for some symbol constellations. These schemes are mainly based on ABBA (7.1), or the 3+1 layered quasi-orthogonal scheme (5.27).

7.1 SYMMETRIES OF INFORMATION AND PERFORMANCE

The mutual information of a linear matrix modulator was discussed in section 4.5, and it reads

$$
\begin{aligned}
\mathcal{I} &= \frac{1}{T} \, \mathrm{E}\Big\langle \log \det \Big(\mathbf{I}_{TN_r} + \eta \, \widetilde{\mathcal{H}}\widetilde{\mathbf{Q}}\widetilde{\mathcal{H}}^\dagger\Big) \Big\rangle \\
&= \frac{1}{T} \, \mathrm{E}\Big\langle \log \det \Big(\mathbf{I}_{2Q} + \eta \, \widetilde{\mathbf{Q}}\widetilde{\mathcal{H}}^\dagger\widetilde{\mathcal{H}}\Big) \Big\rangle,
\end{aligned}
\tag{7.3}
$$

expressed in terms of the (real symbol) equivalent channel matrix $\widetilde{\mathcal{H}}$ of (4.57). Here $E\langle\rangle$ is the expectation value over channel realizations $\mathbf{H}$, $\mathbf{I}_M$ is the $M \times M$ identity matrix, η is the signal-to-noise ratio per transmit antenna, and $\widetilde{\mathbf{Q}}$ is the real symbol covariance matrix, normalized by the total transmit power. The matrix elements of the correlation matrix in (7.3) can be found in (4.58).

Ergodic and Non-ergodic: Here, we shall refine the analysis of information with respect to averaging over channel realizations. The expression for a fixed channel realization,

$$
\mathcal{I}(\mathbf{H}) = \frac{1}{T} \, \log \det \Big(\mathbf{I}_{TN_r} + \eta \, \widetilde{\mathcal{H}}\widetilde{\mathbf{Q}}\widetilde{\mathcal{H}}^\dagger\Big)
\tag{7.4}
$$

is called the *non-ergodic* mutual information. The expression (7.3), where $\mathbf{H}$ is averaged over, is the *ergodic* mutual information.

The difference in these two is a practical one. In fast fading, it is likely that during a code block, all possible channel states, or at least a sufficiently rich choice of such states, are sampled, and the ergodic information is relevant. In slow fading, it is likely that the channel is almost constant during a code block (of a concatenated code), and the non-ergodic information provided by the scheme is pertinent. This may lead to *outage*, i.e. the channel realization is not capable of supporting the scheme being transmitted, and a frame error ensues. Robust methods to overcome outages induced by pathologic channel states incommensurate with the structure of a particular modulation matrix will be discussed in Chapter 8.

7.1.1 Orthogonal Real Symbol Symmetry

Based on the expression (7.3) for the mutual information, the following symmetry of the mutual information was found in [133]:

$$
\tilde{\mathbf{c}} = \mathbf{O}\,\mathbf{c}\,, \quad \mathbf{O} \in O(2Q)\,.
\tag{7.5}
$$

This symmetry leaves the covariance $\widetilde{\mathbf{Q}}$ invariant. The symmetry group is $O(2Q)$, meaning that any orthogonal combination of the real symbols c_k leads to the same mutual information. From (7.3) it is clear that $O(2Q)$ is a symmetry of the non-ergodic mutual information, i.e. it is a symmetry already *before* integrating over the channel realizations. This orthogonal real symbol symmetry is large; it has $2Q^2 - Q$ dimensions. These may be characterized as $2Q^2 - Q$ one-parameter "Givens rotations" $\mathbf{O}^{(kl)}$ mixing any pair of real symbols c_k, c_l,

$$
\mathbf{O}^{(kl)}(\phi) \left[\begin{array}{c} c_k \\ c_l \end{array} \right] = \left[\begin{array}{c} \cos\phi\, c_k + \sin\phi\, c_l \\ \cos\phi\, c_l - \sin\phi\, c_k \end{array} \right].
\tag{7.6}
$$

Any orthogonal rotation may be expressed in terms of a sequence of Givens rotations acting with different parameters ϕ on different symbol pairs.

Complex Symbol Rotations: In most cases, we are only interested in the part of $O(2Q)$ that preserves the complex symbols. This is the group $U(Q)$ of unitary $Q \times Q$ matrices rotating the complex symbols x_k and mixing them with each other. The reason for concentrating on the complex symbol preserving orthogonal rotations is symbol homogeneity, discussed in section 4.3; the unitary symbol rotations preserve homogeneity between real and imaginary parts of complex symbols. Generic orthogonal rotations treat real and imaginary parts of symbols in a different way.

7.1.2 Unitary Left and Right Symmetries

Other symmetries in the system are the left and right unitary symmetries (5.8) of *performance* (in i.i.d. Rayleigh fading channels) discussed in [133, 141]. These act by multiplying the modulation matrix from left and/or right with fixed unitary matrices:

$$\tilde{\mathbf{X}} = \mathbf{V} \mathbf{X} \mathbf{W}, \quad \mathbf{V} \in U(T), \quad \mathbf{W} \in U(N_t). \tag{7.7}$$

The right and left transformations are symmetries of the performance for the very reason that the properties of the codeword difference and distance matrices that determine performance are unitarily invariant properties of the matrices, like the rank, the determinant and the trace.

More exactly, consider the maximum likelihood detection metric (4.64) in i.i.d. Rayleigh fading. As discussed in Appendix A, the bit error probability (BEP) for a given channel realization is a function of the metric (4.64). The average BEP is an integral over channel realizations. As the i.i.d. Rayleigh pdf (A.3) is invariant under unitary rotations, only unitarily invariant properties of the maximum likelihood metric (4.64) are relevant. The metric (4.64) may be expressed in terms of the equivalent channel correlation matrix (4.58), with elements

$$\left(\operatorname{Re} \widetilde{\mathcal{H}}^{\dagger} \widetilde{\mathcal{H}} \right)_{kl} = \operatorname{Tr} \left[\mathcal{S}^{(kl)} \mathbf{H} \mathbf{H}^{\dagger} \right]$$

in terms of the Radon–Hurwitz matrices $\mathcal{S}^{(kl)}$ of Equation (4.11). As all ergodic quantities may be expressed up to a unitary symmetry acting on $\mathbf{H}$, *the bit error probability for any SNR may be expressed in terms of unitarily invariant properties of* $\mathcal{S}^{(kl)}$. For a discussion of unitary invariants, see Appendix A. The design criteria discussed above and in the literature (rank, determinant, trace) concentrate on one specific unitary invariant, with the argument that the determinant of the codeword difference matrix, for example, is the most relevant for performance at high SNR.

Symbolwise Diversity Properties and Matched Filter Bound: The matched filter bound characterizes performance when all self-interference has been mitigated. As such, it provides a lower bound for BEP. This means that the self-interference given by $\mathcal{S}^{(kl)}$, $k \neq l$ is omitted. Thus the MF bound in i.i.d. Rayleigh fading is completely characterized by the unitarily invariant properties of the matrices $\mathcal{S}^{(kk)}$. These properties will be collectively called *symbolwise diversity properties*. The symbolwise diversity degree (4.12) is the most

Table 7.1: Symmetries of performance and mutual information in i.i.d. Rayleigh fading

Symmetry group	O(2Q)	U(Q)	U(T)	U(N_t)
Type	real symbol	complex symbol	left	right
Dimension	$2Q^2 - Q$	Q^2	T^2	N_t^2
Symmetry of				
ergodic information	yes	yes	yes	yes
non-ergodic information	yes	yes	yes	possibly/partly
performance	possibly/partly		yes	yes

important of these properties. These invariants may be described as a collection of trace invariants (4.7) of the diversity matrix (4.9),

$$d_n(\mathbf{X}) = \mathrm{Tr}\,(\mathcal{D})^n\,,\ n = 1,\ldots,\ \min(N_t, T)\,. \tag{7.8}$$

The functions d_n are homogeneous multinomials in the real symbol powers c_k^2. A difference in any of the d_n indicates an essential difference in the distribution of symbol power between antennas.

Part of the orthogonal symbol rotations may change performance by changing symbolwise diversity properties. The true diversity properties are given by the invariants of the full $\mathbf{X}^\dagger\mathbf{X}$ matrix.

Left and Right Unitary Symmetries of Mutual Information: From the structure of the correlation matrix matrix (4.58) and mutual information (7.3) it is clear that both the left and right unitary transformations are symmetries of the ergodic mutual information, if the channels are i.i.d. Rayleigh. This is a direct consequence of the invariance of the channel pdf under unitary transformations, see Appendix A.

The left-symmetry U(T) is a symmetry of the *non-ergodic* mutual information, irrespective of the channel realization. The right-symmetry U(N_t) is only partly or not at all a symmetry of the non-ergodic mutual information.

It is important to know whether the right symmetry, i.e. the beam-forming matrix $\mathbf{W}$, is a symmetry of performance and information before or after averaging over the channel states. If it is a non-ergodic symmetry, i.e. a symmetry before averaging, it has no effect at all. If it is a symmetry only after averaging, it may be used to ergodize (randomize) the scheme, which improves coded performance in slow fading. This aspect will be discussed in Chapter 8.

The properties of the symmetries discussed are collected in Table 7.1.

7.1.3 Examples

Some orthogonal symbol rotations may change performance. As examples, the orthogonal symbol symmetries of mutual information of the best known schemes in the literature are analysed.

Vector Modulation: We have $T = 1$, $Q = N_t$, the diversity matrix (4.16) and the self-interference matrix (4.17). The unitary trace invariants (4.7) characterizing performance are easiest to investigate in terms of the temporal square matrix, the 1×1 matrix $\mathbf{X}\,\mathbf{X}^\dagger$ (4.18).

The invariants are

$$\mathrm{Tr}\left(\mathbf{X}\,\mathbf{X}^\dagger\right)^n = \left(\sum_{k=1}^Q |x_k|^2\right)^n . \tag{7.9}$$

Any orthogonal real symbol rotation is a symmetry of all these invariants, and thus of performance.

Vector Modulation Extended in Time: In [133], rotations changing performance and symbolwise diversity degree were constructed for vector modulation extended in time. Thus consider T consecutive vectors (4.15). We have $Q = TN_t$, and the modulation matrix is simply a collection of T row vectors of length N_t, with the symbol x_{tn} as the t, nth element. The diversity matrix part of the $T \times T$ matrix $\mathbf{X}\,\mathbf{X}^\dagger$ is

$$\mathcal{D}_T = \begin{bmatrix} \sum_{n=1}^{N_t} |x_{1n}|^2 & 0 & \cdots & 0 \\ 0 & \sum_{n=1}^{N_t} |x_{2n}|^2 & \cdots & 0 \\ \vdots & \vdots & \ddots & \vdots \\ 0 & 0 & \cdots & \sum_{n=1}^{N_t} |x_{Tn}|^2 \end{bmatrix} \tag{7.10}$$

whereas the self-interference part is

$$\mathcal{S}_T = \sum_{n=1}^{N_t} \begin{bmatrix} 0 & x_{1n}^* x_{2n} & \cdots & x_{1n}^* x_{Tn} \\ x_{2n}^* x_{1n} & 0 & \cdots & x_{2n}^* x_{Tn} \\ \vdots & \vdots & \ddots & \vdots \\ x_{Tn}^* x_{1n} & x_{Tn}^* x_{2n} & \cdots & 0 \end{bmatrix} . \tag{7.11}$$

This shows how the temporal self-interference $\mathcal{S}_T$ should be taken with a pinch of salt. As the transmission of a symbol does not extend over time, and different time instances are considered orthogonal, there is no temporal self-interference. The elements in (7.11) are just artifacts of analysing performance in terms of $\mathbf{X}\,\mathbf{X}^\dagger$ instead of $\mathbf{X}^\dagger\mathbf{X}$. From the temporal diversity matrix, however, it is clear that the diversity properties (7.8) immediately change if a symbol rotation is applied which mixes symbols transmitted at different times. The rotations considered in [133] are exactly of this type.

Orthogonal Designs: For an orthogonal design, we have $\mathbf{X}^\dagger\mathbf{X} = \sum_{k=1}^{2Q} c_k^2\,\mathbf{I}_{N_t}$. All real symbol rotations preserve the quadratic form $\sum_{k=1}^{2Q} c_k^2$, and are thus symmetries of performance.

7.2   OPTIMIZING PERFORMANCE WITH ORTHOGONAL SYMBOL ROTATIONS

In [133] it was observed that part of $O(2Q)$ relates e.g. the V-BLAST modulation to a $N_t \times N_t$ non-orthogonal matrix modulation. This part of $O(2Q)$ completely changes the Tx diversity properties of the scheme by changing the symbolwise diversity degree. This changing of the diversity degree is independent of the modulation alphabet in question, in line with the underlying information-theoretical ideology of striving for capacity-achieving continuous Gaussian modulation. Here we shall investigate how to use the part of this symmetry that *preserves* the symbolwise diversity degree, but changes performance.

The motivation for this is the design method discussed above—the target diversity degree is fixed by a complexity constraint from the beginning. Using methods in Chapters 4 and 5, an information maximizing matrix modulation with this target symbolwise diversity degree may be constructed. Then the task is to optimize the performance of this scheme.

Specifically, we shall be interested in the part of $O(2Q)$ which preserves the terms in the the diversity matrix but changes the values of the matrix elements of the self-interference. That is, we are interested in symbol rotations that preserve the MF bound but changes the details of self-interference to yield better performance.

The bit error rate of a class of non-orthogonal matrix modulators with a chosen degree of Tx diversity may be optimized within this subset of orthogonal symbol rotations. The details of optimal symbol rotations are specific for the finite modulation alphabets employed, but slightly sub-optimal rotations may be shared by a set of modulations (QPSK, M-QAM).

7.2.1 Symbol Rotations that Preserve Performance

First we shall identify the symbol rotations preserving performance for i.i.d Rayleigh fading, following [86,216]. As discussed above, performance is determined by the unitarily invariant properties of the distance matrix $\mathbf{D}^{(ce)\dagger}\mathbf{D}^{(ce)}$; its rank, trace, determinant or, in general, its eigenvalues.

From linearity it follows directly that the part of $O(2Q)$ preserving performance consists of the symmetries of the $\min(N_t, T)$ functions

$$f_n(\mathbf{c}) = \mathrm{Tr}\,(\mathbf{X}^\dagger \mathbf{X})^n\,,\ n = 1,\ldots,\ \min(N_t, T)\,. \tag{7.12}$$

A symmetry of $f_n(\mathbf{c})$ is a symmetry of $f_n(\boldsymbol{\Delta})$, and thus of all trace invariants of the distance matrix $\mathbf{D}^{(ce)\dagger}\mathbf{D}^{(ce)}$, and of performance.

The unitary invariants d_n of the diversity matrix, defined in (7.8), are parts of the invariants f_n. It should be noted that f_1 is the power constraint, which is trivially invariant under orthogonal symbol rotations. For schemes with traceless self-interference, $f_1 = d_1$. Also, it is worth noting that typically it is enough to investigate the invariances of the matrix elements of $\mathbf{X}^\dagger \mathbf{X}$ (or $\mathbf{X}\mathbf{X}^\dagger$, if $T < N_t$). Transformations that change the matrix elements but preserve the unitary invariants typically do not exist.

The part of $O(2Q)$ that changes performance is a quotient space of $O(2Q)$ with respect to the symmetries of performance.

Example: Unitary Invariants of ABBA.   As an example which a comparatively simple structure, we investigate the space of symbol rotations acting on ABBA (7.1) in detail. To analyse the performance properties of ABBA, the unitary invariants of the scheme should be constructed. Due to the 2×2 block structure of the Hermitian square of ABBA (4.30) and (4.37), it can be written in the form

$$\mathbf{X}_{\mathrm{ABBA}}^\dagger \mathbf{X}_{\mathrm{ABBA}} = \begin{bmatrix} a & b \\ b & a \end{bmatrix} \otimes \mathbf{I}_2\,, \tag{7.13}$$

where

$$a = \sum_{k=1}^{4} |x_k|^2 \tag{7.14}$$

$$b = 2\,\mathrm{Re}\,[x_1^* x_3 + x_2^* x_4]\,. \tag{7.15}$$

The trace invariants are

$$
\begin{aligned}
\mathrm{Tr}\, \mathbf{X}^\dagger \mathbf{X} &= 4\,a \\
\mathrm{Tr}\left(\mathbf{X}^\dagger \mathbf{X}\right)^2 &= 4\left(a^2 + b^2\right) \\
\mathrm{Tr}\left(\mathbf{X}^\dagger \mathbf{X}\right)^3 &= 4\,a\left(a^2 + 3\,b^2\right) \\
\mathrm{Tr}\left(\mathbf{X}^\dagger \mathbf{X}\right)^4 &= 4\left(a^4 + 6a^2 b^2 + b^4\right) \\
\det[\mathbf{X}^\dagger \mathbf{X}] &= \left(a^2 - b^2\right)^2 ,
\end{aligned}
\tag{7.16}
$$

The last expression is a direct consequence of (A.10) in Appendix A.

The invariants of the distance matrix can be constructed by replacing the symbols x_k in a and b with the symbol differences Δ_k. From (7.13), and the form of b, it follows that it is sufficient to maximize the determinant for the symbol difference pairs Δ_1, Δ_3 and Δ_2, Δ_4 separately, to avoid the pathological error events discussed above. With $\Delta_2 = \Delta_4 = 0$, we have

$$
\det[\mathbf{D}^{(ce)\dagger}\mathbf{D}^{(ce)}] = |\Delta_1^2 - \Delta_3^2|^4 .
\tag{7.17}
$$

This is a remarkably simple form of the determinant, which is amenable to complete analysis of the problem.

First, the rank criterion now dictates the choice of symbol constellations so that the determinant is never zero. To avoid all pathological error events, the modulation alphabets of x_1, x_2 and x_3, x_4 should be designed so that they have no overlapping points. Non-overlapping constellations may be constructed by constellation rotations. These may be scalar- or matrix-valued. Performance may further be optimized by judiciously choosing the modulation according to e.g. a determinant criterion. The optimal rotation may be different for different modulation alphabets.

In order to search systematically for an optimal scheme, the relevant directions in the symmetry group of the mutual information should be identified.

Example: Symbol Rotations Preserving Performance of ABBA. The real symbol symmetry is O(8) and it is 28-dimensional. The unitary complex symbol rotations form the group U(4) which is 16-dimensional. We shall concentrate on complex symbol rotations. From (7.16) it follows that for ABBA, all orthogonal rotations that are symmetries of performance leave $\mathbf{X}^\dagger \mathbf{X}$, i.e. a and b invariant.

By definition, all orthogonal rotations leave the squared symbol sum a invariant. To investigate symmetries of b it may be written in terms of the complex symbols as

$$
b = \mathbf{x}^\dagger \begin{bmatrix} \mathbf{0}_2 & \mathbf{I}_2 \\ \mathbf{I}_2 & \mathbf{0}_2 \end{bmatrix} \mathbf{x} = \mathbf{x}^\dagger \mathbf{W}^\dagger \begin{bmatrix} \mathbf{I}_2 & \mathbf{0}_2 \\ \mathbf{0}_2 & -\mathbf{I}_2 \end{bmatrix} \mathbf{W} \mathbf{x} ,
\tag{7.18}
$$

where $\mathbf{x}$ is the vector of four complex symbols, and $\mathbf{W} = \begin{bmatrix} \mathbf{I}_2 & \mathbf{I}_2 \\ \mathbf{I}_2 & -\mathbf{I}_2 \end{bmatrix} / \sqrt{2}$. Thus U(4) rotations $\mathbf{U}$ that leave b invariant are such that $\mathbf{U}$ commutes with

$$
\mathbf{W} \begin{bmatrix} \mathbf{I}_2 & \mathbf{0}_2 \\ \mathbf{0}_2 & -\mathbf{I}_2 \end{bmatrix} \mathbf{W} .
\tag{7.19}
$$

This means that

$$
\mathbf{W}\mathbf{U}\mathbf{W} = \begin{bmatrix} \mathbf{U}_1 & \mathbf{0}_2 \\ \mathbf{0}_2 & \mathbf{U}_2 \end{bmatrix}
$$

$$\Leftrightarrow \mathbf{U} = \tfrac{1}{2} \begin{bmatrix} \mathbf{U}_1 + \mathbf{U}_2 & \mathbf{U}_1 - \mathbf{U}_2 \\ \mathbf{U}_1 - \mathbf{U}_2 & \mathbf{U}_1 + \mathbf{U}_2 \end{bmatrix}. \tag{7.20}$$

where $\mathbf{U}_1$ and $\mathbf{U}_2$ are 2×2 unitary matrices, and $\mathbf{0}_2$ is the 2×2 zero matrix. Unitary 4×4 matrices of the form (7.20) are parameterized by eight parameters. Thus exactly half of the complex symbol rotating symmetry of information preserves performance, and half of it changes it. It should be noted that with $\mathbf{U}_1 = \mathbf{U}_2$ the performance preserving symbol rotations are of the form

$$\mathbf{U} = \begin{bmatrix} \mathbf{U}_1 & \mathbf{0}_2 \\ \mathbf{0}_2 & \mathbf{U}_1 \end{bmatrix}, \tag{7.21}$$

i.e. the same rotation is operating within the two quasi-orthogonal layers x_1, x_2 and x_3, x_4.

The quotient space of performance-changing unitary rotations may be parameterized in terms of an exponential map of a set of anti-Hermitian matrices as

$$\mathbf{U} = \exp \begin{bmatrix} u_3 & -\mathrm{j}\, u_4 \\ \mathrm{j}\, u_4 & -u_3 \end{bmatrix}. \tag{7.22}$$

Here u_3, u_4 are anti-Hermitian 2×2 matrices [216]. Recall that the exponent of an anti-Hermitian matrix is unitary.

7.2.2 Symbol Rotations that Change Diversity Degree

Part of the symbol rotations that change performance may change the MF bound, i.e. symbolwise diversity degree as well. These are true diversity transforms in the spirit of section 3.4.

The splitting of $\mathbf{X}^\dagger \mathbf{X}$ to $\mathcal{D}$ and $\mathcal{S}$ in (4.9) is not invariant under orthogonal rotations After an orthogonal symbol rotation, $\mathcal{D}$ has to be recalculated according to (4.9), and the properties of this matrix may have changed. To find the orthogonal rotations that change the MF bound, it is sufficient to consider Givens rotations that change the diversity properties.

In [86, 216] it was found that Givens rotations changing the symbolwise diversity properties of $\mathbf{X}$ are such that they mix real symbols c_k, c_l with

$$\mathcal{S}^{(k,l)} \neq 0 \quad \text{and/or} \quad \mathcal{S}^{(k,k)} \neq \mathcal{S}^{(l,l)}. \tag{7.23}$$

That is, symbols with different diversity, or which interfere with each other, should be mixed to change the diversity properties of $\mathbf{X}$.

This can be proved in a straightforward manner using properties of orthogonal rotations. Consider one-parameter families of Givens rotations (7.6) mixing symbols c_k and c_l. These leave all $c_i, i \neq k, l$ invariant, and map

$$\begin{bmatrix} c_k \\ c_l \end{bmatrix} \mapsto \begin{bmatrix} \tilde{c}_k \\ \tilde{c}_l \end{bmatrix} = \begin{bmatrix} \cos \phi \; c_k + \sin \phi \; c_l \\ \cos \phi \; c_l - \sin \phi \; c_k \end{bmatrix}. \tag{7.24}$$

Equivalently, the symbol rotation may be considered as a "rotation" in the space of basis matrices. Such an equivalent transformation leaves all $\mathbf{B}^{(i)}, i \neq k, l$ invariant, and maps

$$\begin{bmatrix} \mathbf{B}^{(k)} \\ \mathbf{B}^{(l)} \end{bmatrix} \mapsto \begin{bmatrix} \tilde{\mathbf{B}}^{(k)} \\ \tilde{\mathbf{B}}^{(l)} \end{bmatrix} = \begin{bmatrix} \cos \phi \; \mathbf{B}^{(k)} - \sin \phi \; \mathbf{B}^{(l)} \\ \cos \phi \; \mathbf{B}^{(l)} + \sin \phi \; \mathbf{B}^{(k)} \end{bmatrix}. \tag{7.25}$$

After the transformation, the diversity matrix is given by the Hermitian squares of these transformed basis matrices:

$$\tilde{\mathcal{D}} = \tfrac{1}{2} \sum_{i=1}^{2Q} c_i^2 \, \tilde{S}^{(i,i)} , \tag{7.26}$$

where c_i are the non-transformed symbols, and the transformed squares of basis matrices, expressed in terms of the non-transformed ones, are

$$
\begin{array}{rcl}
\tilde{S}^{(k,k)} & = & \cos^2 \phi \; S^{(k,k)} + \sin^2 \phi \; S^{(l,l)} - \sin 2\phi \; S^{(k,l)} \\
\tilde{S}^{(l,l)} & = & \cos^2 \phi \; S^{(l,l)} + \sin^2 \phi \; S^{(k,k)} + \sin 2\phi \; S^{(k,l)} \\
\tilde{S}^{(i,i)} & = & S^{(i,i)} , i \neq k, l
\end{array}
$$

The variation of the diversity matrix is

$$
\begin{array}{rcl}
2\Delta\mathcal{D} & = & 2\tilde{\mathcal{D}} - 2\mathcal{D} \\
& = & c_k^2 \left((\cos^2 \phi - 1) \; S^{(k,k)} + \sin^2 \phi \; S^{(l,l)} - \sin 2\phi \; S^{(k,l)} \right) \\
& & + c_l^2 \left((\cos^2 \phi - 1) \; S^{(l,l)} + \sin^2 \phi \; S^{(k,k)} + \sin 2\phi \; S^{(k,l)} \right) .
\end{array} \tag{7.27}
$$

For generic ϕ, c_k and c_l, the variation $\Delta\mathcal{D}$ vanishes if and only if the symbols c_l and c_k have the same diversity; $S^{(k,k)} = S^{(l,l)}$, and they are orthogonally encoded; $S^{(k,l)} = 0$. It should be noted that changing $\mathcal{D}$ is a necessary but not sufficient condition for a transformation to change the MF bound. The bound is changed if the unitary invariants (7.8) of $\mathcal{D}$ are changed. Typically it is sufficient for $\mathcal{D}$ to change.

Example: MSD Changing Symbol Rotations for ABBA.   ABBA (7.1) was designed to have maximal symbolwise diversity (MSD). When optimizing performance, it is of interest to preserve this property.

From (7.15) we see that the non-orthogonality of ABBA is within the pairs x_1, x_3 and x_2, x_4, and that all symbols have the same $S^{(kk)}$. Thus, according to (7.23), the symbol rotations that change the MF bound mix x_1 with x_3 and/or x_2 with x_4. These constitute the diagonal part of u_4 in (7.22). Thus unitary rotations of the form

$$\mathbf{U} = \exp \begin{bmatrix} \mathbf{0}_2 & -\mathrm{j}\, d_4 \\ \mathrm{j}\, d_4 & \mathbf{0}_2 \end{bmatrix} \tag{7.28}$$

with $d_4 = \begin{bmatrix} \mathrm{j}\, r & 0 \\ 0 & \mathrm{j}\, p \end{bmatrix}$, $r, p \in \mathbb{R}$ change the diversity degree.

7.2.3 Performance-changing, Diversity-preserving Symbol Rotations

When following the design method adopted in this chapter, one is interested in symbol rotations that do not change the symbolwise diversity properties. These may be characterized based on the analysis above. Givens rotations mixing RH-orthogonally encoded symbols c_k, c_l with the same diversity matrix,

$$S^{(k,l)} = 0 \text{ and } S^{(k,k)} = S^{(l,l)} . \tag{7.29}$$

do not change the symbolwise diversity properties. It should be stressed that this is not necessarily true if transformations mixing different symbol pairs are concatenated. Intuitively this is clear. Consider three symbols, c_k, c_l, c_m, and suppose that the pairs (k, l) and (l, m) fulfil (7.29), but the pair (k, m) does not. Now if a transformation mixing c_k, c_l is followed by a transformation mixing c_l, c_m, the concatenated transformation has a part where c_l is mixed with c_m. According to (7.23), this part of the total transformation changes the diversity properties of $\mathbf{X}$.

This leads to the requirement that performance-changing, symbolwise diversity preserving symbol rotations are such that they mix only sets of symbols that are all RH-orthogonally encoded, and that all have the same diversity matrix.

Only symbols within a quasi-orthogonal layer should be mixed by a diversity preserving symbol rotation.

Example: Performance Changing, MSD Preserving Rotations of ABBA.   The symbolwise diversity changing part of U(16) found above is two-dimensional. As the performance preserving part was eight-dimensional, this leaves *a priori* six dimensions for the part that changes performance but preserves MSD. However, only symbol rotations within quasi-orthogonal layers should be considered. With the 2+2 layers x_1, x_2 and x_3, x_4 in (7.1), such rotations are of the form

$$\mathbf{U} = \begin{bmatrix} \mathbf{U}_1 & \mathbf{0}_2 \\ \mathbf{0}_2 & \mathbf{U}_2 \end{bmatrix} \tag{7.30}$$

with $\mathbf{U}_1, \mathbf{U}_2$ unitary 2×2 matrices with eight dimensions together. Of these, rotations that operate in the same way in both layers are symmetries of performance, see (7.21). These have four dimensions. This leaves four dimensions to complex symbol rotations that change performance but preserve MSD. These may be parameterized as

$$\mathbf{U} = \begin{bmatrix} \mathbf{I}_2 & \mathbf{0}_2 \\ \mathbf{0}_2 & \mathbf{U}_2 \end{bmatrix}, \tag{7.31}$$

i.e. only unitary rotations within the second layer are considered, which rotate the symbols x_3, x_4 with the matrix $\mathbf{U}_2$.

The action of this transformation can be expressed on the level of the modulation matrix as

$$\mathbf{X}_{ABBA}(\mathbf{x}) = \begin{bmatrix} \mathbf{X}_A & \mathbf{X}_B \\ \mathbf{X}_B & \mathbf{X}_A \end{bmatrix} \mapsto \mathbf{X}_{ABBA}(\mathbf{Ux}) = \begin{bmatrix} \mathbf{X}_A & \mathbf{V}\,\mathbf{X}_B\,\tilde{\mathbf{U}} \\ \mathbf{V}\,\mathbf{X}_B\,\tilde{\mathbf{U}} & \mathbf{X}_A \end{bmatrix}, \tag{7.32}$$

where $\mathbf{V} = \begin{bmatrix} \det \mathbf{U}_2 & 0 \\ 0 & (\det \mathbf{U}_2)^* \end{bmatrix}$ represents the overall phase of $\mathbf{U}_2$, and $\tilde{\mathbf{U}}$ is the part of $\mathbf{U}_2$ with unit determinant.

Performance of ABBA may now be optimized within this four-dimensional space of complex symbol rotations. Compared to the 28 dimensions of the group of all orthogonal real symbol rotations, this is a considerable simplification.

7.3 EXPLICIT PERFORMANCE OPTIMA FOR ABBA

In this section, the MAX-MIN-DET criterion and refined versions of it are considered when choosing optimal rotations within the four-dimensional set of MSD-preserving, performance-changing complex symbol rotations for ABBA.

Table 7.2: Critical error events for the one-parameter rotations (7.33)

$\det\left(\mathbf{D}^{(ce)\dagger}\mathbf{D}^{(ce)}\right)/256$	Ratio of all error events	Bit errors	Example errors$/\sqrt{2}$	
			Δ_1	Δ_3
$\sin^4\phi$	1.6%	2	1	1
$\left(\frac{5}{4}-\sin 2\phi\right)^2$	1.6%	3	1	1+j

7.3.1 Scalar Constellation Rotations

Most of the performance gain for ABBA may be captured by a simple scalar symbol rotation. For equal power constellations, the resulting schemes are power-balanced. These are formed from one-parameter subgroups of performance-changing rotations. All such are equivalent to the form (7.32), with

$$V = \begin{bmatrix} e^{j\phi} & 0 \\ 0 & e^{-j\phi} \end{bmatrix}, \quad \tilde{U} = I_2, \tag{7.33}$$

or vice versa. These yield schemes of the form (7.1), where the constellations of x_3, x_4 are simply rotated by ϕ. (The latter choice of matrices V and $\tilde{U}$ gives a rotation of $-\phi$ to the x_4 constellation. As the factor b in (7.15) depends only on the real part of products of symbols, this difference is irrelevant from a performance point of view.)

The rotation should be chosen by a suitable MAX-MIN criterion and the optimum choice depends on the modulation alphabet.

QPSK modulation: The QPSK modulation alphabet is $\{\pm 1 \pm j\}/\sqrt{2}$. The possible values for Δ are $\sqrt{2}\{0, \pm 1, \pm j, \pm 1 \pm j\}$. These have 0, 1 and 2 bit errors, respectively. The squared symbol differences that show up in the determinant distance (7.17) are 0 with no bit errors, ± 2 with one bit-error, and $\pm 2j$ with two bit errors. These points are plotted in Figure (7.1), for non-rotated and rotated constellations.

By symmetry considerations, the phase rotation ϕ can be restricted to the interval $[0, \pi/4]$.

First, consider MAX-MIN-DET, following [211, 216]. With scalar rotated symbols, the self-interference (7.15) only couples x_1 with x_3 and x_2 with x_4. Thus it is sufficient to consider these pairs separately. The minimum value of the determinant with $\Delta_2 = \Delta_4 = 0$, in Equation (7.17), should thus be maximized, leading to an optimum rotation angle. The same rotation angle should then be chosen between x_2 and x_4. Note that maximizing (7.17) *maximizes the Euclidean distance between squared symbol difference constellations* $\{\Delta_1^2\}$ *and* $\{\Delta_3^2\}$.

A *critical error event* is such that MIN-DET is given by this error event at least in part of the parameter interval. In the interval $\phi \in [0, \pi/4]$, the critical error events belong to two classes. One class consists of events where the same one-bit error is made in x_1 and x_3 (or in x_2 and x_4), and no other errors are made. The other class consists of events where a one-bit error is made in e.g. x_1, and a two-bit error is made in x_3. The determinants and multiplicities of these error classes can be found in Table 7.2.

At the MAX-MIN-DET optimum phase rotation, the determinants of these error events are equal. A sufficient condition for this is

$$4\sin(2\phi) - 2\cos(2\phi) = 3, \tag{7.34}$$

which has the solution

$$\phi = \tfrac{1}{2} \arccos\left[(2\sqrt{11} - 3)/10\right] \approx 0.191\,\pi \approx 34.4° \tag{7.35}$$

in the interval of interest. This point can be directly understood in terms of the squared error constellations, and the form (7.17) of the determinant of the distance matrix. At this value of ϕ, the minimum distance between points in the Δ^2 constellations for x_1 and x_3 is maximized, see first plot in Figure 7.1. The other plot in Figure 7.1 shows the squared symbol difference constellations Δ^2 for $\phi = \pi/4$, with marginally worse distance properties.

The same error events are critical if the multiplicities and numbers of induced bit errors are taken into account. The MAX-MIN-DET/Errs/Mult optimum phase rotation can be found by equating the determinants divided by the respective multiplicities and numbers of errors. This leads to

$$4\sin(2\phi) - \sqrt{6}\cos(2\phi) = 5 - \sqrt{6}\,, \tag{7.36}$$

which has the solution

$$\begin{aligned}
\phi &= \tfrac{1}{2} \arccos\left[\left(6 - 5\sqrt{6} + 4\sqrt{10\sqrt{6} - 9}\right)/22\right] \\
&\approx 0.179\,\pi \approx 32.2°
\end{aligned} \tag{7.37}$$

in the interval of interest.

Exact analytical union bounds of bit error probabilities of pairwise error events can also be calculated. For $E_b/N_0 = 10$ dB, the Min-UB is achieved at $\phi \approx 0.22\,\pi$.

As an example of the behaviour of MIN-DETs, the QPSK MIN-DETs are plotted as a function of ϕ in Figure 7.2. The plot is characterized by the fact that for QPSK, the determinant of 2-symbol, 2- and 4-bit error events, and some 4-symbol, 6- and 8-bit error events vanish at $\phi = 0, \pi/2$. For these values, the true diversity degree of QPSK ABBA is only two. The determinant of 2-symbol, 3-bit error events never vanishes, but in the vicinity of $\phi = \pi/4$ these errors are critical.

Essence of Symbol Rotations: All MAX-MIN-DETs and Min-UBs are well away from these singular values of ϕ. This is the essence of optimizing non-orthogonal matrix modulation at high SNR: the constellations should be rotated away from singular points of the

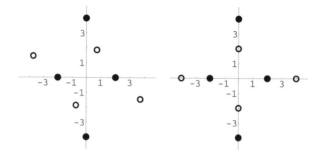

Fig. 7.1: QPSK modulation: values for Δ^2 (disk) and $(e^{j\phi}\Delta)^2$ (circle). Left plot: $\phi = 0.191\,\pi$ (MAX-MIN-DET), right plot: $\phi = \pi/4$ (set-partitioned 8-PSK).

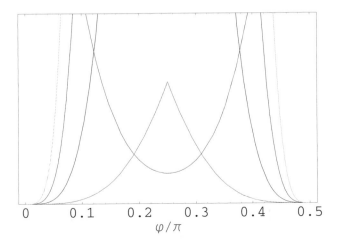

Fig. 7.2: MIN-DET's for events with 8, 6, 4, 2, 3, bit errors in 4, 4, 2, 2, 2, symbols respectively, listed from the edges towards the center. The critical error events for $\phi \neq n\pi/2$ have 2 and 3 bit errors. MAX-MIN-DET at $\phi = 0.19\pi$ and $\phi = 0.31\pi$.

distance matrix which correspond to pathological error events. The exact rotation is not too important, as long as the constellation is well away from the singular points. From the form of the invariants (7.16) it is clear that for QPSK, any scheme with $\phi \neq n\pi/2, n \in \mathbb{Z}$ in (7.32) automatically fulfils the main criterion that singularities in the codeword difference matrix are avoided.

The analytic results quoted above may be confirmed by simulations. The differences between performance in the region between $\phi = \pi/6$ and $\phi = \pi/4$ are negligible, of the order of 0.1 dB. The true optimum is very close to the values given by various MIN-DET criteria.

The sub-optimal solution with $\phi = \pi/4$ has the advantage that if 8-PSK symbols are implemented in the system, the $\phi = \pi/4$ rotated ABBA is easily realized as a set partitioning of 8-PSK; symbols x_1, x_2 could be transmitted from a 8-PSK subset $e^{j\pi/2\,m+\pi/4}, m = 0, \ldots, 3$, whereas symbols x_3, x_4 would be in the subset $e^{j\pi/2\,m}, m = 0, \ldots, 3$.

16-QAM Modulation: As an alternative, consider a 16-QAM constellation with power two,

$$z = \sqrt{\tfrac{4}{5}} \left(m - \tfrac{3}{2} + (n - \tfrac{3}{2})j\right), \quad m, n = 0, \ldots, 3 \,. \tag{7.38}$$

The possible values for Δ are $2/\sqrt{5}(m - m' + (n - n')j)$ for $m, m', n, n' = 0, \ldots, 3$, i.e. $2/\sqrt{5}(m'' + n''j)$ for $m'', n'' = -3, \ldots, 3$. Depending on the bit-encoding, different symbol errors carry different numbers of bit errors. If the four bits are Gray-encoded, the parameters may be chosen as

$$m = 2b_1 b_3 + b_1 - b_3 + 1 \,, \quad n = 2b_2 b_4 + b_2 - b_4 + 1 \,, \tag{7.39}$$

where the bits are $b_i = \{0, 1\}$. The lattice of symbol differences with corresponding numbers of bit-errors can be found in Figure 7.3.

```
2   3   2   1   2   3   2

3   4   3   2   3   4   3

2   3   2   1   2   3   2

1   2   1   ·   1   2   1

2   3   2   1   2   3   2

3   4   3   2   3   4   3

2   3   2   1   2   3   2
```

Fig. 7.3: Lattice of possible values for symbol differences Δ for Gray-encoded 16-QAM modulation, with the corresponding number of bit-errors indicated.

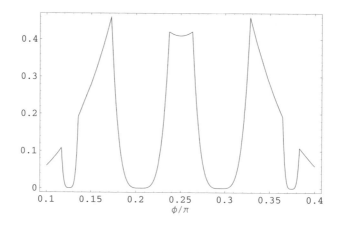

Fig. 7.4: MIN-DET as a function of rotation angle ϕ, ABBA with 16-QAM modulation.

The MIN-DET as a function of ϕ is plotted in Figure 7.4. Compared to the corresponding MIN-DET for QPSK in Figure 7.2, it is notable that now the singular points are not restricted to $\phi = n\,\pi/2$, $n \in \mathbb{Z}$. Additional singularities have appeared for smaller rotations, corresponding to additional commensurate distances in the error lattice in Figure 7.3. The region of "good" rotation angles has been infringed upon by these additional singularities.

In the vicinity of MAX-MIN-DET, the critical error events are two-bit and six-bit symbol errors, with properties summarized in Table 7.3. It is notable that due to the proliferation of different error events, the ratio of critical event to all events has been considerably reduced from the QPSK case. Equating these determinants yields the MAX-MIN-DET rotation angle

$$\phi = \tfrac{1}{2} \arccos\left(\frac{8}{17}\right) \approx 0.172\,\pi \approx 31° \,, \tag{7.40}$$

with the corresponding value of the minimal determinant being ≈ 0.46. It is remarkable how the minimum determinant is much smaller than the one for QPSK modulation.

Table 7.3: Critical error events for one-parameter rotations (7.33), 16-QAM symbols

$\det\left(\mathbf{D}^{(ce)\dagger}\mathbf{D}^{(ce)}\right) \times (5/8)^4$	Ratio of all error events	Bit errors	Example errors$\times \sqrt{5}/2$	
			Δ_1	Δ_3
$\sin^4 \phi$	0.013%	2	1	1
$(3\cos\phi - 4\sin\phi)^4$	0.0034%	6	$2-j$	$1-2j$

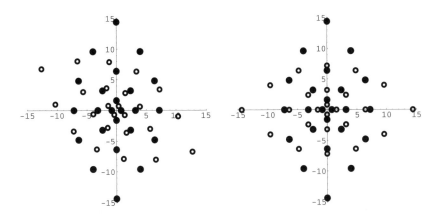

Fig. 7.5: 16QAM modulation: values for Δ^2 (disk) and $(e^{j\phi}\Delta)^2$ (circle). Left plot: $\phi = 0.172\,\pi$ (MAX-MIN-DET), right plot: $\phi = \pi/4$.

In Figure 7.5, the squared 16-QAM error constellations and rotated versions of it are plotted. For MAX-MIN-DET, the distance between overlapping square constellation points should be minimized. Again, $\phi = \pi/4$ is a sub-optimal solution, which gives almost as good performance as MAX-MIN-DET. Indeed, from the lattice structure of a regular M-QAM modulation it follows that $\phi = \pi/4$ is always in a regime away from singularities.

The essential observations of this 16-QAM exercise are

- With increasing modulation size, more singularities appear and the proportion of parameter space providing essentially non-singular rotations is narrowed. This makes resulting determinant distances smaller, and infringes on gains obtained by constellation rotations.

- However, except for the singularity with no rotations, the other singularities are mild in that only very few determinants become critical.

- The achievable minimal determinants are much reduced due to increased constellation size.

- Due to these facts, the performance gains due to rotations reaching full diversity are pushed towards higher SNR.

- If a rotation is chosen, the simplest $\phi = \pi/4$ is recommended.

7.3.2 Matrix Rotations

If the unitary right rotation $\widetilde{\mathbf{U}}$ in (7.32) is chosen to be a non-diagonal matrix, the resulting scheme is power-unbalanced even for equal power constellations. A generic 2×2 unitary matrix with unit determinant may be parameterized as

$$\widetilde{\mathbf{U}} = \left[\begin{array}{cc} \mu & \nu \\ -\nu^* & \mu^* \end{array} \right] ; \; \mu, \nu \in \mathbb{C}, \; |\mu|^2 + |\nu|^2 = 1 \qquad (7.41)$$

In addition, we have

$$\mathbf{V} = \left[\begin{array}{cc} e^{j\rho} & 0 \\ 0 & e^{-j\rho} \end{array} \right] . \qquad (7.42)$$

Acting with these on the symbols x_3, x_4 in the B blocks yields the rotated complex symbols

$$\begin{array}{rl} \tilde{x}_3 = & e^{j\rho}(\mu\, x_3 - \nu\, x_4^*) \\ \tilde{x}_4 = & e^{j\rho}(\mu\, x_4 + \nu\, x_3^*) . \end{array} \qquad (7.43)$$

From a heuristic argument it is clear that the power-unbalanced schemes created by allowing $\nu \neq 0$ may have better performance than the power-balanced ones with $\nu = 0$; the intersymbol interference is spread among more symbols and is thus more Gaussian.

A more rigorous symmetry argument may also be formulated, based on convexity of error functions. This symmetry principle would be similar to the principle of symbol homogeneity, but elevated to act on symbol pairs, triplets etc. Thus it would be desirable that the protection against a number of bit errors in a set of symbols would not depend on the selection of symbols. As we shall see, such a symmetry principle has to be taken with a pinch of salt; full symmetry cannot be reached. Experimenting with the example under consideration seems to indicate an "as much symmetry as possible" principle.

Symmetries between error events are constructed by equating the corresponding determinants. In the previous section, two critical error events were equated to find an optimum for the single parameter ϕ. Here, we have 4 parameters. To fix these, the determinants of five critical error events may be equated.

MAX-MIN-DET may be found by equating the determinants of as many critical two-bit error events as possible, without sacrificing the determinant distance of events with more bit-errors [216]. The reason why two-bit error events should be concentrated on is diagonal dominance, see e.g. (4.40). Error events with multiple errors have larger Euclidean distance and thus have a stronger response to a constellation rotation—a small rotation is enough to make the determinant substantially larger than the determinant of any two-bit error event. Also, arranging as much symmetry as possible in two-bit error events automatically yields a high symmetry for events with more bit errors as well.

Intra-layer error events are not plagued by self-interference. All possible inter-layer two-bit errors for ABBA with the four-parameter constellation rotation (7.43) are shown in Table 7.4. The determinants are functions of the variables μ, ν, ρ of the form

$$\det \left(\mathbf{D}^{(ce)\dagger} \mathbf{D}^{(ce)} \right) = (a^2 - b^2)^2 , \qquad (7.44)$$

with a and b given by (7.14) and (7.15) with symbols replaced by symbol differences. For all two-bit error events, $a = 4$. The values of the non-orthogonality b can be found in Table 7.4. The possible one-bit errors for QPSK modulation are $\Delta^{(1)} \in \{\pm 1, \pm j\}\sqrt{2}$. Note that for a given one-bit error in a complex symbol $\Delta^{(1)}$, all possible one-bit errors in another complex symbol are $\pm\Delta^{(1)}$ or $\pm j\Delta^{(1)}$.

Now the four parameters of the constellation rotations may be chosen so that all error events in Table 7.4 are symmetric. This is achieved by taking

$$\rho = 0, \quad \mu = \nu = \frac{1}{\sqrt{2}}e^{j\pi/4}. \tag{7.45}$$

With this choice, the determinant of all inter-layer two-bit error events are equal, with the value 144. There is a caveat, though: some events with more bit errors are singular with the parameters (7.45). Consider e.g. the four- or eight-bit, four-symbol error events $\Delta_1 = \Delta_2 = \pm j\Delta_3 = \pm\Delta_4 = \Delta$. For these, the determinant vanishes. With a comparatively low operation point (in E_b/N_0) the higher Euclidean protection of these errors may be enough. For high E_b/N_0, this choice of parameters is not acceptable, and the requirement for complete symmetry has to be relaxed. At least one real parameter in (7.43) has to be liberated to change the performance of events with more bit errors.

Relaxing one parameter, six of the eight types of inter-layer two-bit errors may be made symmetric. First, it is clear from the non-orthogonalities in Table 7.4 that choosing

$$\rho = 0 \tag{7.46}$$

is essential to have a high degree of symmetry between error events. With $\rho = 0$, we have four classes of inter-layer two-bit errors. By fixing two real parameters, an arbitrary choice of three of these may be made equal. First, equate the determinants of the two first error events in Table 7.4

$$\text{Re}\,[\mu] = \text{Im}\,[\mu]. \tag{7.47}$$

Next, equate these two with the fifth:

$$\text{Re}\,[\mu] = \text{Re}\,[\nu]. \tag{7.48}$$

As the complex numbers μ, ν are constrained by $|\mu|^2 + |\nu|^2 = 1$, there is still one real degree of freedom to specify. This should be fixed so that the chosen performance measure

Table 7.4: The non-orthogonality b for all types of inter-layer two-bit error events: an one-bit error in the A layer, another in the B layer. The four-parameter constellation rotation (7.43) is applied on the B layer.

$\|b\|/4$	Errors			
	Δ_1	Δ_2	Δ_3	Δ_4
$\text{Re}\,[e^{j\rho}\mu]$	$\Delta^{(1)}$	0	$\pm\Delta^{(1)}$	0
$\text{Im}\,[e^{j\rho}\mu]$	$\Delta^{(1)}$	0	$\pm j\Delta^{(1)}$	0
$\text{Re}\,[e^{-j\rho}\mu]$	0	$\Delta^{(1)}$	0	$\pm\Delta^{(1)}$
$\text{Im}\,[e^{-j\rho}\mu]$	0	$\Delta^{(1)}$	0	$\pm j\Delta^{(1)}$
$\text{Re}\,[e^{j\rho}\nu]$	$\Delta^{(1)}$	0	0	$\pm\Delta^{(1)}$
$\text{Im}\,[e^{j\rho}\nu]$	$\Delta^{(1)}$	0	0	$\pm j\Delta^{(1)}$
$\text{Re}\,[e^{-j\rho}\nu]$	0	$\Delta^{(1)}$	$\pm\Delta^{(1)}$	0
$\text{Im}\,[e^{-j\rho}\nu]$	0	$\Delta^{(1)}$	$\pm j\Delta^{(1)}$	0

Table 7.5: Properties of spectra of determinants for various choices of parameters in (7.43). All have $\rho = 0$

Scheme	μ, ν	$|\{\text{dets}\}|$	MIN-DET	Inter-layer 2-bit errors MIN-DET	fraction
ABBA	1, 0	15	0	0	25%
				256	75%
ϕ-ABBA	$e^{j\phi_{\text{OPT}}}, 0$	47	26	26	25%
				119	25%
				256	50%
OPT-ABBA	$\sqrt{\frac{2}{3}}e^{j\pi/4}, \sqrt{\frac{1}{3}}$	22	114	114	75%
				256	25%
$\mu = \nu$-ABBA	$\frac{1}{\sqrt{2}}e^{j\pi/4}, \frac{1}{\sqrt{2}}e^{j\pi/4}$	22	0	144	100%

is maximized. The MAX-MIN-DET is in a region where in addition to the two-bit error events, a number of six-bit, four-symbol error events are critical. Thus the error event $\Delta_1 = -\Delta_4 = (1+j)\sqrt{2}$, $\Delta_2 = -\Delta_3 = \sqrt{2}$, for example, has Euclidean distance $a = 12$, and gives rise to the non-orthogonality

$$b = 4\,\text{Re}\left[e^{j\rho}\left(2\,\text{Re}\left[(1-j)\mu\right] - \nu^* + 2\nu\right)\right] . \tag{7.49}$$

Applying (7.46)–(7.48), this non-orthogonality reads $b = 20\,\text{Re}\left[\mu\right]$. Equating the determinant of these six-bit, four-symbol error events with those of the critical two-bit error events above leads to the equation

$$\left(144 - 400\,\text{Re}\left[\mu\right]^2\right)^2 = \left(16 - 16\,\text{Re}\left[\mu\right]^2\right)^2 \tag{7.50}$$

One solution of this is $\text{Re}\left[\mu\right] = 1/\sqrt{3}$, leading to

$$\mu = \sqrt{\frac{2}{3}}e^{j\pi/4} , \quad \nu = \sqrt{\frac{1}{3}} , \quad \rho = 0 . \tag{7.51}$$

Computer searches of the full four-dimensional parameter range indicate that this is one of a discrete set of optimal choices of parameters, i.e. MAX-MIN-DET.

The multiplicity and bit error enhanced system is difficult to analyse, as the number of critical error events is not big enough to fix all four parameters. The minimum of UB of pairwise bit errors may be calculated. At $E_b/N_0 = 10$ dB, the optimum is indistinguishable from the MAX-MIN-DET solution (7.51).

Some characteristics of the spectra of determinants for various choices of parameters are shown in Table 7.5. There one may see how the degree of symmetry between inter-layer two-bit error events increases. With non-rotated ABBA, all inter-layer error events are either critical (DET = 0), or orthogonal (DET = 256). With increasing symmetry, an increasing part of the orthogonal error events are sacrificed to increase the DET of the interfering error events. The cardinality of the set of different DET values is denoted by $|\{\text{dets}\}|$.

The cumulative distribution of product distances (fourth roots of determinants) is displayed in Figure 7.6. There this phenomenon may be observed in a wider perspective: the

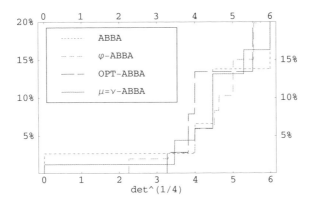

Fig. 7.6: Cumulative distribution of percentage of bit errors that have at least a given product distance $\det^{1/4}$ for the QPSK ABBA variants of Table 7.5.

price for increasing MIN-DET is paid by decreasing the DET of other error events. OPT-ABBA has almost 13% of all error events with nearly minimal determinant.

The guiding principle for optimizing non-orthogonal matrix modulators within the orthogonal rotations considered in this chapter would thus be to maximize the symmetry between inter-layer two-bit error events, as long as other error events do not become critical. If full symmetry is not possible, the remaining inter-layer two-bit error events should be kept orthogonal.

7.3.3 Performance

The simulated performances of ABBA and its optimized versions can be found in Figure 7.7. The simulation parameters are uncorrelated Rayleigh block fading, with block length 4, perfect channel state information at Rx, none at Tx. Maximum Likelihood (ML) detection is used for all schemes. For QPSK modulation, the complexity of a ML detection is comparable to the complexity of a iterative interference canceller. For higher order modulation, linear detection schemes would be preferable. Simple interference cancellers are not capable of taking advantage of the performance benefits allowed by constellation rotations. One has to rely on restricted ML type of detectors, e.g. the sphere decoder.

One-parameter optimized ABBA performs within 0.7 dB of the matched filter bound of a fourth-order Tx-diversity QPSK scheme with no self-interference. The four-parameter optimized scheme performs within 0.5 dB from the bound.

It is notable that at least in the regime of SNRs plotted, the benefit from full four-parameter optimization is negligible compared to the gain from one-parameter optimization. The fine-tuning of the spectrum of determinants is necessary only at high SNR. At medium SNR, it is enough just to make the error events with the smallest Euclidean distance non-singular. At low SNR, the benefits from diversity are negligible, and it would be preferable to sacrifice MSD in favour of orthogonality.

Finally, in Figure 7.8, some eight-antenna extensions are considered. It is straightforward to construct full diversity schemes for more than four antennas with higher rates than those of orthogonal designs, simply by using higher dimensional orthogonal designs as the A and

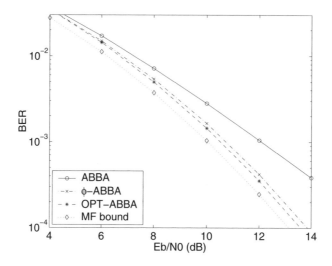

Fig. 7.7: The simulated performance of QPSK ABBA (7.1), a one-parameter optimized version and a four-parameter optimized version, compared to the matched filter bound of orthogonal four-fold Tx diversity.

B blocks in ABBA. These may be transformed to have full diversity by simple constellation rotations in the B block, just as for 4 Tx ABBA. Thus using the rate $R_s = 3/4$ orthogonal design (3.39) as a constituent in ABBA, one gets a $R_s = 3/4$ matrix modulation for $N_t = 8$ antennas. The performance of this may be compared to a generalization of STTD-OTD (4.21) to STBC-OTD, with higher dimensional orthogonal designs as constituents. With the rate 3/4 orthogonal design (3.39), the resulting $R_s = 3/4$ scheme would be orthogonal, with diversity 4. Also, $R_s = 1$ STTD-OTD may be trivially extended to $R_s = 1$ $N_t = 8$ STTD-OTD, simply by doubling the dimensionality of the OTD part. Also, any non-orthogonal matrix modulation may be concatenated with OTD to double the number of transmission antennas. ABBA-OTD would thus be a rate 1 scheme for $N_t = 8$ antennas.

It is seen from Figure 7.8 that 8 Tx ($R_s = 3/4$) ABBA with a $\phi = \pi/4$ constellation rotation trails optimal eight-fold diversity Rayleigh performance (the eight-antenna matched filter bound). Recall that an $N_t = 8$ orthogonal design has $R_s = 1/2$. The loss in performance due to the rate increase from 1/2 to 3/4 is remarkably small. 8 Tx STBC-OTD ($R_s = 3/4$) performance is identical to that of the four-antenna orthogonal design. For comparison, the performance of 8 Tx ($R_s = 1$) ABBA-OTD is plotted, with and without $\phi = \pi/4$ constellation rotation, as well as that of 8 Tx ($R_s = 1$) STTD-OTD. At BER $= 10^{-3}$ rotated 8 Tx ABBA outperforms STBC-OTD by about 2 dB.

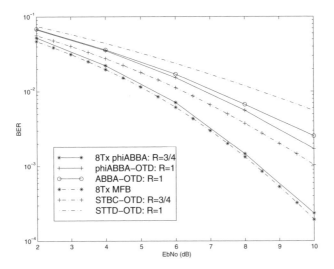

Fig. 7.8: Performance of 8 Tx matrix modulators with and without rotated constellations and full diversity. STBC-OTD performance is identical to theoretical performance of diversity order corresponding to the STBC part.

7.4 IMPROVED PERFORMANCE BY EXTENDING BLOCK

In this section, a complementary approach to improve performance at high SNR is considered, namely extending the block length to be longer than the desired diversity degree. This approach has been discussed in [132, 133, 156]. In [133], $N_t = 3$ was considered. STTD (3.28) was concatenated with antenna shuffling, and diversity transforms were used to improve performance. Here, specific examples of extended block space–time modulation for $N_t = 4, N_r = 1$ with $T = 8$ and $T = 16$ shall be considered, with heuristic designs building on ABBA and (5.27). We shall see that in well performing cases, extending block length indeed falls within the paradigm of exploiting symmetries of information—for successful schemes, the schemes with extended block provide the same information as the shorter block scheme, but can be designed to have full diversity.

7.4.1 Extending ABBA

A 4+4 layered 8×4 scheme with rate $Q/T = 1$ can be constructed by extending ABBA to $AB^2C^2D^2A$:

$$\mathbf{X}_{AB^2C^2D^2A} = \begin{bmatrix} \mathbf{X}_A & \mathbf{X}_B \\ \mathbf{X}_B & \mathbf{X}_C \\ \mathbf{X}_C & \mathbf{X}_D \\ \mathbf{X}_D & \mathbf{X}_A \end{bmatrix}. \tag{7.52}$$

Here $\mathbf{X}_A, \mathbf{X}_B, \mathbf{X}_C, \mathbf{X}_D$ are 2×2 STTD blocks (3.28) with independent symbols. Blocks $\mathbf{X}_A$ and $\mathbf{X}_C$ are in the same layer, as are $\mathbf{X}_B$ and $\mathbf{X}_D$.

Similarly, a 8+8 layered 16×4 scheme may be constructed as $AB^2C^2D^2E^2F^2G^2H^2A$. These schemes can be interpreted as tail-biting concatenations of delay diversity (with delay 2) and STTD (3.28).

This construction, however, does not improve the minimum rank of the scheme. If the same errors are made in detecting $\mathbf{X}_A$ and $\mathbf{X}_C$, as well as in $\mathbf{X}_B$ and $\mathbf{X}_D$, the codeword difference matrix of (7.52) reduces to one of ABBA, which may be singular. When the modulation matrix becomes longer, singular error events become less frequent, however, and their detrimental effect on performance becomes visible at ever higher SNR.

7.4.2 Extending 3+1 Layered

To improve the linear estimate of the data encoded on the idle direction (see section 5.4.2) of the rate 3/4 orthogonal design (5.17) one may employ additional coding [156]. A rate-effective way for this is to use space–time block coding on the idle direction. Rows of a modulation matrix (the higher layer matrix) may be inserted into the idle directions of a number of lower-layer matrices, thus generalizing the 3+1 layered scheme (5.27).

A 6+2 layered 8×4 scheme with rate $Q/T = 1$ can be constructed by transmitting two complex symbols on the idle directions of two consecutive rate 3/4 4×4 blocks (3.39). The lower layer consists of six symbols. The two upper-layer symbols can be transmitted e.g. with a repeated STTD (3.28) on the upper layer, resulting in the modulation matrix

$$
\mathbf{X}_{6+2} =
\left[
\begin{array}{cccc}
x_1 & x_2 & x_3 & 0 \\
-x_2^* & x_1^* & 0 & -x_3 \\
-x_3^* & 0 & x_1^* & x_2 \\
0 & x_3^* & -x_2^* & x_1 \\
\hline
x_4 & x_5 & x_6 & 0 \\
-x_5^* & x_4^* & 0 & -x_6 \\
-x_6^* & 0 & x_4^* & x_5 \\
0 & x_6^* & -x_5^* & x_4
\end{array}
\right]
+
\left[
\begin{array}{cccc}
0 & 0 & 0 & x_8 \\
0 & 0 & x_8 & 0 \\
0 & -x_7 & 0 & 0 \\
-x_7 & 0 & 0 & 0 \\
\hline
0 & 0 & 0 & x_7^* \\
0 & 0 & x_7^* & 0 \\
0 & x_8^* & 0 & 0 \\
x_8^* & 0 & 0 & 0
\end{array}
\right] .
\tag{7.53}
$$

A 12+3 layered 16×4 scheme with rate $Q/T = 15/16$ can be constructed by applying the same orthogonal design (3.39) on both upper and lower layers. The lower layer consists of four elementary 4×4 blocks of the form (3.39). All in all, 12 symbols are encoded into these four blocks, and the processing delay is 16 symbol intervals. To construct the layered scheme, the zeros in the four lower-layer matrices are filled with rows from a upper-layer matrix, which is a orthogonal design of the form (3.39), encoding symbols x_{13}, x_{14}, x_{15},

$$
\mathbf{X}^{(\mathrm{u})} =
\left[
\begin{array}{cccc}
x_{13} & x_{14} & x_{15} & 0 \\
-x_{14}^* & x_{13}^* & 0 & -x_{15} \\
-x_{15}^* & 0 & x_{13}^* & x_{14} \\
0 & x_{15}^* & -x_{14}^* & x_{13}
\end{array}
\right] .
\tag{7.54}
$$

When pasting a row of the upper-layer matrix to the anti-diagonal of a lower-layer matrix, the signs of the two first elements of the row are changed, reflecting the anti-diagonal signs in (5.30). The resulting two-layer modulation matrix is of the form

$$
\mathbf{X}_{12+3} =
\left[\begin{array}{cccc}
x_1 & x_2 & x_3 & 0 \\
-x_2^* & x_1^* & 0 & -x_3 \\
-x_3^* & 0 & x_1^* & x_2 \\
0 & x_3^* & -x_2^* & x_1 \\
\hline
x_4 & x_5 & x_6 & 0 \\
-x_5^* & x_4^* & 0 & -x_6 \\
-x_6^* & 0 & x_4^* & x_5 \\
0 & x_6^* & -x_5^* & x_4 \\
\hline
x_7 & x_8 & x_9 & 0 \\
-x_8^* & x_7^* & 0 & -x_8 \\
-x_9^* & 0 & x_7^* & x_8 \\
0 & x_9^* & -x_8^* & x_7 \\
\hline
x_{10} & x_{11} & x_{12} & 0 \\
-x_{11}^* & x_{10}^* & 0 & -x_{12} \\
-x_{12}^* & 0 & x_{10}^* & x_{11} \\
0 & x_{12}^* & -x_{11}^* & x_{10}
\end{array}\right]
+
\left[\begin{array}{cccc}
0 & 0 & 0 & 0 \\
0 & 0 & x_{15} & 0 \\
0 & -x_{14} & 0 & 0 \\
-x_{13} & 0 & 0 & 0 \\
\hline
0 & 0 & 0 & -x_{15} \\
0 & 0 & 0 & 0 \\
0 & -x_{13}^* & 0 & 0 \\
x_{14}^* & 0 & 0 & 0 \\
\hline
0 & 0 & 0 & x_{14} \\
0 & 0 & x_{13}^* & 0 \\
0 & 0 & 0 & 0 \\
x_{15}^* & 0 & 0 & 0 \\
\hline
0 & 0 & 0 & x_{13} \\
0 & 0 & -x_{14}^* & 0 \\
0 & -x_{15}^* & 0 & 0 \\
0 & 0 & 0 & 0
\end{array}\right]
\tag{7.55}
$$

The four lower-layer blocks are indicated by vertical lines. The residual idle directions are the idle directions of the upper-layer modulation matrix. These can be used to transmit e.g. pilot symbols. Also, it is possible to construct a 12+4 layered 16×4 scheme with rate $Q/T = 1$ by transmitting a symbol x_{16} on the idle direction of the upper-layer rate 3/4 orthogonal design (7.54):

$$
\mathbf{X}_{12+4} = \mathbf{X}_{12+3} +
\left[\begin{array}{cccc}
0 & 0 & 0 & x_{16} \\
0 & 0 & 0 & 0 \\
0 & 0 & 0 & 0 \\
0 & 0 & 0 & 0 \\
\hline
0 & 0 & 0 & 0 \\
0 & 0 & x_{16} & 0 \\
0 & 0 & 0 & 0 \\
0 & 0 & 0 & 0 \\
\hline
0 & 0 & 0 & 0 \\
0 & 0 & 0 & 0 \\
0 & x_{16} & 0 & 0 \\
0 & 0 & 0 & 0 \\
\hline
0 & 0 & 0 & 0 \\
0 & 0 & 0 & 0 \\
0 & 0 & 0 & 0 \\
x_{16} & 0 & 0 & 0
\end{array}\right]
\tag{7.56}
$$

Note that, in this scheme, there is no interference between x_{16} and x_{13}, x_{14}, x_{15}, the scheme is still two-layered. The difference between x_{16} and x_{13}, x_{14}, x_{15} becomes actual if (5.29) is used for lower-layer interference cancellation.

Special IC Pre-stage for Non-homogeneous Schemes: Non-homogeneous layered schemes based on the rate 3/4 orthogonal design (5.17) have a very special interference cancelling pre-stage, where all lower-layer interference can be cancelled from the upper layers, without noise enhancement [156]. In the 3+1 layered scheme (5.27) there are two layers, one with three complex symbols x_1, x_2, x_3, and one with the symbol x_4. From (5.29) it follows that the twisted matched filter projection (5.31) is

$$\tilde{z}_4 = -\mathbf{h}^T \mathbf{B}^{(7)} \mathbf{y} = -\sum_{i=1}^{4} h_i^2 x_4 + \text{noise}. \tag{7.57}$$

Accordingly, $\tilde{z}_4$ gives an estimate of x_4 with *squared, non-coherently combined* channels, and *no interference from the lower layer*. This may be used as a pre-stage for iterative OSIC detection.

This pre-stage may be generalized to any layered scheme with extended block where the lowest layer is based on (5.17), most notably the 12+3 layered rate 15/16 scheme. With an analogue of pre-stage (7.57), the interference caused by the lower layer on the upper may be cancelled. Instead of (7.57), one gets the received symbols of the upper-layer matrix in *squared channels*. Thus one gets the full diversity advantage of the upper-layer modulation matrix after linear interference cancellation of the lower layer.

Generically, iterative IC detectors do not converge to a maximal likelihood result, and often fail even to reach the asymptotic diversity degree of a layered scheme. Judging from simulation results, there are two exceptions to this. For ABBA, with asymptotic diversity 2, PIC + iterations performs as well as maximal likelihood detection. The same holds for the 12+3 layered scheme (7.55) if iterative IC based on the pre-stage (7.57) is used.

Using this linear detector for the 12+3 scheme, a low complexity detector for the the 12+4 layered rate 1 scheme (7.56) may be constructed. This detector would split the 16 symbols into three sets, two corresponding to the layers of the 12+3 scheme, and the third consisting of the last symbol x_{16}. A simple iterative non-linear IC detector would involve a search over the points in the constellation of x_{16}, followed by iterative IC of the remaining code of the form (7.55) using pre-stage (7.57).

7.5 COMPARISON OF LAYERED SCHEMES FOR FOUR TX ANTENNAS

Here, the layered matrix modulators for $N_t = 4$ Tx antennas with rate $Q/T = 1$ discussed above are assessed in terms of the information and performance optimization criteria of Chapter 4, as well as in terms of the MAX-MIN-DET criterion. The schemes compared are

- The 3+1 layered 4×4 rate $R_s = 1$ scheme (5.27).

- The 2+2 layered 4×4 rate $R_s = 1$ ABBA scheme (7.1).

- The four-parameter constellation rotation optimized ABBA with a unitary matrix rotation parameterized by (7.51).

- The 6+2 layered 8×4 rate $R_s = 1$ scheme (7.53).

- The 4+4 layered 8×4 rate $R_s = 1$ $AB^2C^2D^2A$ scheme (7.52).

Table 7.6: Characteristics of layered schemes for four Tx antennas

Dimension	Scheme	Rank	MIN-DET	$\tilde{\mathcal{N}}$	$\frac{\mathcal{I}_2}{\mathcal{C}_2}$(dB)	XTRM	SH
4×4	3+1	2	0	3/8	-1.14	no	no
	ABBA	2	0	1/4	-0.79	yes	yes
	opt ABBA	4	1	1/4	-0.79	yes	yes
8×4	6+2	2	0	3/8	-1.14	no	no
	4+4	2	0	1/2	-1.46	yes	yes
16×4	8+8	2	0	1/2	-1.46	yes	yes
	12+3	4	1	$\frac{3}{10}$	-1.21	no	no
	12+4	4	1	3/8	-1.14	no	no

- The 8+8 layered 16×4 rate $R_s = 1$ $AB^2C^2D^2E^2F^2G^2H^2A$ scheme discussed in section 7.4.1.

- The 12+3 layered 16×4 rate $R_s = 15/16$ scheme (7.55).

- The 12+4 layered 16×4 rate $R_s = 1$ scheme (7.56).

7.5.1   Design Metrics

The characteristics of these schemes are collected in Table 7.6. All schemes are characterized by their layering, or nickname. The columns show minimum rank and determinant of codeword difference matrices, self-interference per symbol $\tilde{\mathcal{N}}$ (Equation (4.47)), the loss of second-order information, whether the scheme fulfils second-order information extremality (4.89), and whether it is symbol homogeneous (Equation (4.49)).

Normalization: All schemes have been considered with Gray-encoded QPSK symbols, and for numerical considerations they are normalized so that the total transmit power is 1/bit. If each symbol is transmitted with the same power, the basis matrices of a 4×4 MSD scheme with QPSK symbols are $\frac{1}{2}\mathbf{U}$ with $\mathbf{U}$ a unitary matrix. This gives $\mathrm{Tr}\mathcal{S}^{(kk)} = 2\,\mathrm{Tr}\,\mathbf{B}^{(k)\,\dagger}\mathbf{B}^{(k)} = \mathrm{Tr}\,\mathbf{U}^\dagger\mathbf{U}/2 = 2$, consistent with (4.73). The I- and Q-branch real (BPSK) symbols are ± 1 and the real symbol differences are $\Delta = \pm 2$. Thus the distance matrix related to an one-bit error event is $|\Delta|^2/4\,\mathbf{I}_4 = \mathbf{I}_4$, and its determinant is 1.

Regarding the two categories of extended block schemes, a slightly surprising result may be seen in Table 7.6. Extending 2+2 ABBA (7.1) to 4+4 and 8+8, self-interference is increased, and information is lost, whereas when extending 3+1 layered (5.27) to 6+2 and 12+4, self-interference and information remains constant. Also, schemes constructed by extending ABBA never have full rank, whereas extending a 3+1 layered scheme to a 16×4 matrix gets rid of singular error events. [1]

For all full-rank MSD schemes, MIN-DET is the one-bit error determinant 1. To get more insight into the spectrum of determinants of different error events, Table 7.7 enlists the MIN-DETs in subsets of all error events with a specified number of bit errors. For comparison,

---

[1]The self-interference measures for the extended layered schemes differ from the ones reported in [132] due to a different normalization.

Table 7.7: MIN-DETs for error events with one to eight bit errors

Dimension	Scheme	MIN-DET for bit-errors:							
		1	2	3	4	5	6	7	8
	MF bound	1	16	81	256	625	1296	N/A	N/A
4×4	3+1	1	0	1	0	41	144	457	1024
	ABBA	1	0	25	0	81	0	169	0
	opt ABBA	1	7.1	13.4	16	13.4	7.1	245	256
8×4	6+2	1	0	25	0	41	0	172	512
	4+4	1	9	25	0	25	121	81	0
16×4	8+8	1	9	25	64	144	361	729	0
	12+3	1	12	45	128	357	676	1408	2068
	12+4	1	12	45	128	357	676	1031	1936

these determinants are also tabulated for the rate 3/4 orthogonal design (3.39), which is the matched filter bound as discussed in section 5.1.1. For the best schemes (12+3 and 12+4), the minimal determinant is at least half of the theoretical maximum for each subset of error events.

From the determinants it should be noted that optimizing ABBA with constellation rotations changes MIN-DET protection from one category of error events to another, thus the protection against three-bit error events is worse for "opt ABBA" than for "ABBA", whereas the protection against two-bit errors is better. This is in complete agreement with section 7.3.2. Conversely, when extending the block length preserves information, as in the 3+1 ↦ 6+2 ↦ 12+4 sequence, the diversity protection does not become worse for any class of error events.

7.5.2 Performance

Due to the bit-interleaved coded space–time modulation paradigm applied in this book, the relevant performance measure of matrix modulation is the bit-error rate (BER).

For some schemes, performance results can be found in Table 7.8 and in Figures 7.9 and 7.10. The channels are block-fading Rayleigh with the channel assumed constant over the transmission of the modulation matrix. Perfect channel state information is assumed at the receiver, and ML (or equivalently, QOML, see section 6.3) detection is assumed.

It is illuminating to contrast performance with Tables 7.6 and 7.7. As discussed in Chapter 4, performance at low E_b/N_0 is almost completely explained by the self-interference per bit $\widetilde{N}$. At high E_b/N_0, the slope of the performance curves are explained by the rank [129], and the horizontal position of the slope by the determinant [19]. To see these properties more clearly, performance should be studied for larger E_b/N_0; here a cross-over from the $\widetilde{N}$ dominated region to a rank-and-det dominated region is discernible in the range 5–14 dB. In this intermediate region, both self-interference and the spectrum of determinants affect performance. Note that the longer the scheme, the higher E_b/N_0 the crossover is at. This can be understood in terms of the spectrum of determinants. Even though MIN-DET characterizes performance at high E_b/N_0, for an extended scheme the worst-case singular error events are so rare that their effect is not visible in the plotted range. Thus the 6+2

Table 7.8: Operation point in E_b/N_0 at various target BER, measured as distance from the matched filter bound provided by an orthogonal 4 Tx scheme.

scheme	E_b/N_0 at target BER			
	10^{-1}	10^{-2}	10^{-3}	5×10^{-4}
MF-bound	-0.11	6.23	10.06	11.04
	loss in dB due to non-orthogonality:			
12+4	0.92	0.66	0.39	0.27
opt ABBA	0.73	0.49	0.49	0.49
6+2	1.06	0.71	0.99	1.12
ABBA	0.71	1.03	2.04	2.43
3+1	1.25	1.54	2.50	2.86

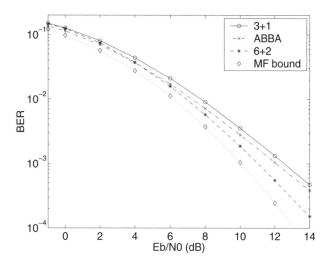

Fig. 7.9: BER of some rate 1 layered matrix modulators for 4 Tx antennas.

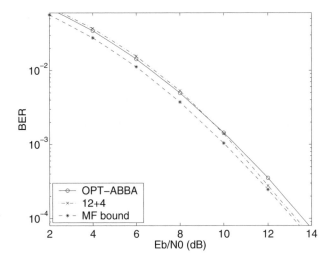

Fig. 7.10: BER of the best rate 1 layered matrix modulators found for 4 Tx antennas, compared to the matched filter bound.

layered 8×4 scheme (7.53) does not reach the asymptotic regime with a performance slope dominated by the true rank 2 in the plotted range.

Overall, the best performing schemes are optimized ABBA and the 12+4 layered 16×4 scheme (7.56), plotted in Figure 7.10. The former is better in terms of self-interference and worse in terms of determinant. Accordingly, their performance curves cross at $\approx$ 9 dB. Performance differences are slight, however. At high E_b/N_0, both perform within 0.5 dB of the theoretical maximum (see section 5.1.1) for a linear scheme with QPSK modulation, represented by the symbol rate $R_s = 3/4$ orthogonal design (3.39). The price of increasing the rate is a slight loss in performance and an increase in detection complexity.

The criterion of symbol homogeneity is more delicate. Of the two best schemes, one is SH, whereas the other is not. Here it should be noticed that the ABBA family (of which the optimized 2+2 scheme is a member) was found by minimizing $\tilde{\mathcal{N}}$ in the space of 4×4 rate 1 MSD schemes [132] (or by maximizing the mutual information [133]), and this family is SH. The non-SH 12+4 scheme was found by heuristic considerations. Combining this with the results of Chapter 4 leads to the heuristic rule that traceless self-interference and SH are prerequisites for minimal self-interference. In concrete terms, non-SH layered schemes do not fulfil the information extremality equation discussed in section 4.5. This indicates that the schemes discussed here are not likely to be information- nor performance-optimal extended block schemes.

7.5.3 Detection Complexity

Detecting a non-orthogonal matrix modulation is always more complex than detecting an orthogonal design. The main tradeoff in matrix modulation is between rate and complexity, with performance overtones: how much is one willing to compromise complexity in order to increase the rate, without losing too much in performance?

Table 7.9: Complexity per bit of some receiver algorithms for $N_t = 4$ space–time block codes with QPSK, compared to an orthogonal rate 3/4 transmission

Scheme	Detection	Real multiplications	Real additions n iterations	three iterations
orthogonal (5.17)	MF	8		8
ABBA (5.34)	PIC	11	$9 + 2n$	15
opt ABBA (7.32)	QOML	9		44
12+3 (7.55)	OSIC	13	$12 + 10n$	42
12+4 (7.56)	x_{16} search + OSIC	13	$13 + 42n$	139

For any linear scheme, detection starts with a matched filter, which provides the bulk of multiplicative complexity, and a significant fraction of additional complexity. IC pre-stages (decorrelating or LMMSE) introduce additional multiplications, whereas iterations and exhaustive searches introduce additional additions. In ML or QOML (see section 6.3), after matched filtering, only additions are needed. This is important when considering hardware implementation, where a multiplication is ~15 times as costly as an addition. Hard decision detection complexity of some schemes considered above can be found in Table 7.9. The number of operations is per bit, and rounded up to the nearest integer.

For ABBA (5.34), PIC is sufficient to reach ML performance. Optimized versions of ABBA (7.32) require ML or QOML for optimal performance. For the 12+3 layered scheme (7.55), OSIC with pre-stage (5.29) reaches ML performance within simulation accuracy. Similarly, for the 12+4 layered (7.56), a full search over x_{16} with iterative OSIC for the rest is also indistinguishable from ML. For iterative schemes, three iterations is enough to reach near saturation.

Constellation Rotation vs. Extended Block: Both applying constellation rotations and extending the block yields well performing schemes. The difference between these was seen above: with QPSK symbols, some of the latter yield full diversity gain with linear detectors, whereas the former always require non-linear (quasi-orthogonality assisted) ML-type detectors. Conjecturing that these properties hold for higher modulation as well, the extended block layered schemes are preferable to constellation-rotated due to lower detection complexity, at least for higher modulation and channel coherence times greater than the extended modulation block length.

7.6 WEIGHTED AND MULTIMODULATION NON-ORTHOGONAL MATRIX MODULATION

Looking at the expression (7.17) for the determinant of ABBA with error in symbols x_1, x_3, it is immediately clear that constellation rotations are not the only way to get rid of singular error events. An alternative method is to dilate the constellation of x_3 by applying a weight factor, or simply to choose the symbols from completely different modulation alphabets.

7.6.1   Weighted ABBA

In weighted ABBA, overlapping constellations are avoided by weighting symbols z_1, z_2 as compared to z_3, z_4. This can be done in addition to a possible constellation rotation. Thus, before transmitting, we map

$$x_1 \mapsto \sqrt{w}\, x_1, \quad x_2 \mapsto \sqrt{w}\, x_2, \quad x_3 \mapsto \sqrt{2-w}\, x_3, \quad x_4 \mapsto \sqrt{2-w}\, x_4 \qquad (7.58)$$

with w some real weight factor, $0 < w < 2$. The power-balanced case, with the powers of all symbols the same, is $w = 1$.

The reason why such dilatations were not considered in the previous sections is that they are not symmetries of information. Indeed, a dilatation breaks against "Equal average power for symbols" required from information optima in section 4.5.2, and against unitary equivalence of $\mathcal{S}^{(kk)}$ discussed in section 4.6.1. Nevertheless, weighting improves protection against multiple-symbol error events. The information (and low SNR performance) degradation comes about from the fact that protection against one-symbol errors in the symbols with smaller power is compromised.

For a given constellation rotation angle, one may search for the optimal weighting factor. As discussed above, linear detection methods are not capable of exploiting the performance benefits provided by constellation rotations. However, for a scheme with no constellation rotation and a suitable weighting, linear IC still works. With no constellation rotation, the union bound optimum weighting at high SNR is given by $w = 1.4$, giving the ratio $7/3$ to the symbol powers of different symbols. Second-order mutual information with this weighting can be found in Table 7.10. The loss in information is ≈ 0.4 dB. Performance with this weighting is plotted in Figure 7.11. Compared to ABBA, the performance is indeed better at high SNR; the true diversity degree is 4. However, the penalty of ≈ 0.4 dB due to lost information makes weighted ABBA worse than ABBA at low SNR, with a cross-over at $E_b/N_0 = 10$ dB. The best scheme at all values of SNR is the optimal constellation rotated ABBA with no weighting.

With milder weighting (w closer to 1), performance at low SNR would be better, at the price of compromised high SNR performance. The cross-over point moves to lower E_b/N_0. Thus with a given operation point, a weighting may be chosen that gives the maximum performance improvement at that point. With low E_b/N_0, however, this improvement is negligible.

7.6.2   Multimodulation ABBA

If one aims at higher data rates, looking at (7.17) immediately leads to the concept of multimodulation ABBA. The most straightforward way to have no overlapping points in the constellations of symbols z_1 and z_3 is to take them from different modulation alphabets altogether, generalizing the multimodulation concept discussed in section 5.3 to the non-orthogonal realm.

Consider bandwidth efficiency 3 bps/Hz, which may be realized by populating an ABBA block with four 8-PSK symbols. Alternatively, it may be realized by a multimodulation ABBA with z_1, z_2 in QPSK and z_3, z_4 in 16-QAM (with double average power). In the multimodulation scheme, no constellation rotation or weighting (apart from the doubling of the average power of 16-QAM) is needed; the constellations automatically avoid overlappings that cause singularities in the determinant and weaken performance. The corresponding values for Δ_1 and Δ_3 can be found in Figure 7.12. The multimodulation scheme has

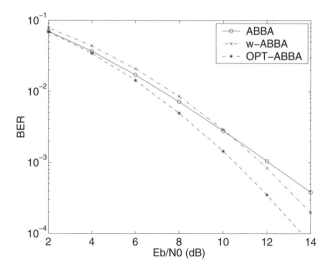

Fig. 7.11: Performance of ABBA, weighted ABBA ($w = 1.4$, no constellation rotation) and optimal (scalar) constellation-rotated ABBA.

Table 7.10: Second-order mutual information for ABBA, weighted ABBA with $w = 1.4$ and multimodulation ABBA

Scheme	$\mathcal{I}_2/\mathcal{C}_2$ (dB)	
	$N_r = 1$	$N_r = 2$
ABBA	-0.79	-2.22
Multimodulation ABBA, $|x_{3,4}|^2 = 2|x_{1,2}|^2$	-1.1	-2.59
Weighted ABBA, $|x_{3,4}|^2 = \sqrt{\frac{7}{3}}|x_{1,2}|^2$	-1.23	-2.74

better diversity protection. For multimodulation QPSK/16-QAM ABBA, MAX-MIN-DET is $\frac{256}{625} \approx 0.41$, compared to $\frac{81}{4}\left(10 - 7\sqrt{2}\right)^2 \approx 0.20$ for $\phi = \pi/8$ rotated 8-PSK ABBA.

At low SNR, one should again consider the second-order information. With multimodulation, the basis matrices transmitting 16-QAM symbols should be scaled by $\sqrt{2}$, resulting in an information penalty. The second-order information can be found in Table 7.10. With one Rx antenna, the loss is 0.3 dB, which leads to a small performance penalty at high SNR.

Regarding detecting these 3 bps/Hz examples, the same holds as for weighted schemes: the multimodulation scheme may be detected with linear IC detectors with practically optimal performance, whereas rotated 8-PSK ABBA requires non-linear detection.

7.7 SUMMARY

In this chapter, optimizing linear matrix modulation performance at high SNR was considered. The salient observations may be summarized as follows.

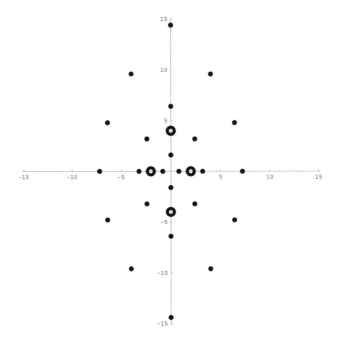

Fig. 7.12: Multimodulation QPSK/16QAM: values for Δ^2 for QPSK (circles) and 16QAM (dots).

- Non-orthogonal linear matrix modulation is often plagued by pathological error events that make the true diversity degree smaller than the symbolwise diversity degree. For such errors, the distance matrix is singular.

- It is possible to turn symbolwise diversity degree to true diversity degree using (possibly matrix-valued) symbol rotations. Sometimes this is possible by extending the block as well.

- For a given symbolwise diversity degree, orthogonal symbol rotations should only mix symbols within a quasi-orthogonal layer.

- The actual details of a symbol rotation are not that important, as long as the ensuing distance matrices are well away from singular points. The larger the distance from singular points, the bigger the MIN-DET.

- For higher modulation rates, the density of singular error events grows. The distance from singular errors becomes progressively smaller, and correspondingly the MAX-MIN-DETs smaller.

- Conversely, for higher modulation, the proportion of possibly singular error events from all error events becomes progressively lower, and the adverse effect of pathological error events on performance is pushed to higher SNR.

- Finally, it was noted that using weighting and/or multimodulation cures singular error events, with a information penalty at low SNR. These schemes may be favourably used with a linear detector.

8

Robust and Practical Open-loop Designs

Space-time block coding theory sets the requirements on modulation matrices for full orthogonality and diversity. Based on these requirements, several well-designed high rate and power efficient space–time modulation concepts have been developed in previous chapters. The practicality of these concepts, e.g. the applicability to 3G systems or their evolutions, depends on a number of issues. Among other things, a practical and robust design should also take into account the following issues:

- Space-time codes and matrix modulations are typically used as a part of concatenated encoding concept. Hence, ease of providing soft outputs (*a posteriori* probabilities) to the following receiver stage is crucial for overall performance.

- Preserving orthogonality of space–time block codes requires perfect channel knowledge. In practice, there are always channel estimation errors, and due to channel mismatch, the effective code after matched filtering is non-orthogonal.

- Receiver complexity typically increases when the number of transmit antennas is increased. However, there may be a hard limit on how much a receiver complexity can be increased without the need to completely redesign the receiver implementation.

- Additional diversity and rate improvements provided by new coding and modulation schemes are often incremental improvements to an existing system. Such new features should have a minimum effect on current system and terminal implementation, and to be backward-compatible to prior designs, if possible.

When addressing backward compatibility, one must consider the fact that the wireless system in question is likely to serve terminals that have different versions of the same standard. For example, WCDMA Release '99 terminals support transmit diversity using two transmit antennas, while future standard evolutions may introduce terminals that can exploit explicitly, e.g. four transmit antennas. Hence, terminals that support different standard versions may need channel estimates from different antennas. Additional pilot signals needed, for example with four transmit antennas, increase downlink interference. This reduces the performance (e.g. range or bit rates) of terminals that do not support four antenna transmission.

Another important aspect is to reuse the existing implementation solutions as much as possible, in concordance with the modular system design principle. Novel multi-antenna transmit diversity solutions should therefore reuse the key ingredients already present in a prevailing two-antenna transmit diversity solution.

These issues discussed above are important, although difficult to capture with elegant theory. Often, these targets constrain the set of all possible solutions. However, we can provide some examples of transmitter concepts in which the additional benefit of increasing the number of transmit antennas is still dramatic, but which avoid many of the design problems described above.

In this chapter we provide a number of multi-antenna modulation concepts that adhere to a large extent with the aforementioned targets and principles. We assume a quasi-static channel to explore symbol rate one multi-antenna open-loop methods. All of the proposed techniques are potentially applicable in future evolutions of 3G systems, and they all leverage STTD in one way or another, thus allowing at least partial reuse of existing hardware. We consider performance-enhancing concepts for quasi-static channels. In particular, we study the effect of channel ergodization via randomized space–time transmission methods.

8.1 RANDOMIZED MATRIX MODULATIONS

Wireless data services are often used in quasi-static channels, such as indoor channels, or some outdoor or indoor hot spots. In these environments the majority of users can be assumed to be either stationary or moving slowly. One simple channel model, suitable for such an environment, assumes that the link gains from each transmit antenna to each receive antenna are essentially constant during the encoded frame. The importance of designing space–time transmission schemes for quasi-static or block fading channel realizations becomes particularly relevant when the HSDPA system becomes operational. In the HSDPA system the Transport Time Interval (TTI) is reduced from 10 ms (Release '99) to only 2 ms. This time interval is in most cases within the channel coherence time, leading to a quasi-static channel model. Hence, quasi-static fading is an appropriate initial assumption when designing multi-antenna modulation and coding techniques. Due to lack of alternative diversity sources (e.g. time diversity or frequency diversity) multi-antenna modems can provide significant gains in these environments.

Randomized space–time codes and matrix modulations are attempts to ergodize the inherently static channel. In general, randomization can be implemented by applying time-varying left and right unitary transformations on the space–time modulation matrix. That is, regardless of whether the matrix modulation is orthogonal or non-orthogonal, all methods involve transformations applied on a modulation matrix $\mathbf{X}$,

$$\tilde{\mathbf{X}} = \mathbf{V}\mathbf{X}\mathbf{W}, \tag{8.1}$$

as discussed in (7.7). Clearly, appropriate randomization depends on the structure of the modulation matrix. As observed in Table 7.1, the right unitary transformation $\mathbf{W}$, acting directly on the matrix output, does not always preserve the information provided by a selected transmission scheme. In the signal model (1.1), a right unitary transformation can be interpreted as a beam-forming matrix. Thus, the randomized transmission methods discussed in this chapter can be considered randomized beam-forming methods.

In the following sections we study extensions of STTD (the Alamouti code), and develop several attractive symbol rate one transmit diversity methods. The proposed methods are described in detail, and their applicability to 3G systems is discussed. Multi-antenna transmission methods employing non-orthogonal and orthogonal space–time codes are proposed and refined for quasi-static channels via channel ergodization. In particular, we summarize and develop the following concepts:

- Combined space–time block code and orthogonal transmit diversity (STTD-OTD) [49]

- Randomized STTD-OTD

- Combined space–time block code and phase-hopping (STTD-PHOP), also known as Trombi) [24, 217]

- Randomized non-orthogonal matrix modulation [26].

Some of these techniques have been already proposed in 3G WCDMA or cdma2000 standardization meetings in 3GPP and 3GPP2. Evaluation work is still in progress in these forums, and the alternative concepts (STTD-OTD and STTD-PHOP) are seen as the most relevant candidates when extending STTD to four transmit antennas. Both of these concepts are symbol rate one designs.

8.1.1 Randomization for Non-orthogonal Schemes

Recall the ABBA scheme (5.34), defined for three or four transmit antennas as follows

$$\mathbf{X}_{ABBA}(x_1, x_2, x_2, x_4) = \begin{bmatrix} \mathbf{X}_A & \mathbf{X}_B \\ \mathbf{X}_B & \mathbf{X}_A \end{bmatrix} \tag{8.2}$$

where $\mathbf{X}_A = \mathbf{X}(x_1, x_2)$, $\mathbf{X}_B = \mathbf{X}(x_1, x_2)$.

8.1.1.1 Different Forms of ABBA
A host of different forms of ABBA may be constructed using the right (and left) unitary rotations. Part of the generality of these rotations may be fixed by combining the right rotation with a left one, and demanding that the real part of x_1 be encoded by $\mathbf{I}_4$. A three-dimensional family of schemes with the same performance and the same average information in Rayleigh fading channels, but different *instantaneous*, i.e. non-ergodic information, may be constructed as

$$\mathbf{X}_{UABBA} = \mathbf{I}_2 \otimes \mathbf{X}_A + \mathrm{j}\,\mathbf{U} \otimes \mathbf{X}_B\,. \tag{8.3}$$

Here $\mathbf{U}$ is a unitary 2×2 matrix with determinant 1. The symbol rotation $\mathbf{U}$ may be transformed to a pair of left and right unitary transformations using a 2×2 matrix $\mathbf{W}$ which transforms $\mathrm{j}\,\mathbf{U}$ into $\begin{bmatrix} 0 & 1 \\ 1 & 0 \end{bmatrix}$.[1] With such a matrix, we have

$$\mathbf{X}_{ABBA} = \left(\mathbf{W}^\dagger \otimes \mathbf{I}_2\right)\,\mathbf{X}_{UABBA}\,\left(\mathbf{W} \otimes \mathbf{I}_2\right)\,. \tag{8.4}$$

---

[1]In addition to being unitary, all 2×2 unitary matrices with unit determinant $\mathbf{U}$ are anti-Hermitian and may thus be diagonalized with a 2×2 unitary transformation $\mathbf{W}$. Transforming to the off-diagonal matrix is a rotated form of diagonalization. Note that j in (8.3) makes j $\mathbf{U}$ Hermitian.

This means that with different $\mathbf{U}$ in (8.3), we have different schemes that have the same average performance and the same average information, but different instantaneous performance and information.

Diagonal ABBA: As another example, consider $\mathbf{U} = \begin{bmatrix} \mathrm{j} & 0 \\ 0 & -\mathrm{j} \end{bmatrix}$. This gives the block-diagonal scheme

$$\mathbf{X}_{\text{diag ABBA}} = \begin{bmatrix} \mathbf{X}_A + \mathbf{X}_B & \mathbf{0}_2 \\ \mathbf{0}_2 & \mathbf{X}_A - \mathbf{X}_B \end{bmatrix}, \tag{8.5}$$

which has diagonal self-interference. This may be transformed to the ABBA form (8.2) with

$$\mathbf{W} = \frac{1}{\sqrt{2}} \begin{bmatrix} 1 & 1 \\ -1 & 1 \end{bmatrix} \tag{8.6}$$

using (8.4).

Phase-rotated Matrix Modulation: Another family of schemes with equivalent average performance but different instantaneous information is given by the right action of the matrix

$$\mathbf{W} = \begin{bmatrix} \mathbf{I}_2 & \mathbf{0}_2 \\ \mathbf{0}_2 & \Theta \end{bmatrix}, \tag{8.7}$$

where the phasor matrix

$$\Theta = \begin{bmatrix} e^{\mathrm{j}\theta_{t,3}} & 0 \\ 0 & e^{\mathrm{j}\theta_{t,4}} \end{bmatrix} \tag{8.8}$$

rotates symbols x_3, x_4. Applying this right action on the ABBA modulation matrix,

$$\mathbf{X}_{ABBA}\mathbf{W} \tag{8.9}$$

with the phasors changing from one matrix modulated block to the next, leads to simple ergodization and adaptation schemes in slowly fading channels, as discussed in [26] and below.

8.1.1.2 *Randomization and the Equivalent Channel Matrix*

The benefits achievable with channel ergodization are more transparent when viewed through the equivalent channel correlation matrix (4.57) or the equivalent correlation matrix (4.58). The equivalent channel matrix for ABBA in (8.2) is

$$\mathcal{H}_{ABBA}(h_1, h_2, h_2, h_4) = \begin{bmatrix} \mathcal{H}(h_1, h_2) & \mathcal{H}(h_3, h_4) \\ \mathcal{H}(h_3, h_4) & \mathcal{H}(h_1, h_2) \end{bmatrix}, \tag{8.10}$$

where $h_1, ..., h_4$ denotes the complex channel amplitudes between the four transmit antennas and the receive antenna. With N_r receive antennas, the effective channel matrix is obtained by stacking the effective channel matrices of each receive antenna on top of each other,

$$\mathcal{H} = [\mathcal{H}_{ABBA}^{(1)}, ..., \mathcal{H}_{ABBA}^{(N_r)}]^T, \tag{8.11}$$

where the superscript refers to the receive antenna index. The equivalent channel correlation matrix $\mathcal{H}^\dagger \mathcal{H}$ can be written as a sum of the (equivalent channel) diversity matrix

$$\mathcal{D}_{\mathbf{h}} = \sum_{j=1}^{N_r} \sum_{i=1}^{4} |h_{i,j}|^2 \mathbf{I}_4 \tag{8.12}$$

and the (equivalent channel) self-interference matrix

$$\mathcal{S}_{\mathbf{h}} = \sum_{j=1}^{N_r} 2\operatorname{Re}[h_{1,j}h_{3,j}^* + h_{2,j}h_{4,j}^*]\begin{bmatrix} 0 & 0 & 1 & 0 \\ 0 & 0 & 0 & 1 \\ 1 & 0 & 0 & 0 \\ 0 & 1 & 0 & 0 \end{bmatrix}, \tag{8.13}$$

where

$$a = \sum_{j=1}^{N_r}\sum_{i=1}^{4}|h_{i,j}|^2 \tag{8.14}$$

$$b = \sum_{j=1}^{N_r} 2\operatorname{Re}[h_{1,j}h_{3,j}^* + h_{2,j}h_{4,j}^*] \tag{8.15}$$

and where $h_{i,j}$ is the complex channel coefficient between transmit antenna i and receive antenna j. It is apparent from (8.15) that the factor

$$b = \sum_{j=1}^{N_r} 2\operatorname{Re}[h_{1,j}h_{3,j}^* + h_{2,j}h_{4,j}^*], \tag{8.16}$$

that determines the self-interference depends on both channel phases and channel powers. Coefficient b vanishes for certain phase profiles. On the other hand, correlation can be very high for other phase and power profiles. In a block fading or a quasi-static channel the correlation coefficient b is constant over the whole frame. This leads to a "block interference channel". Performance degradation due to block interference cannot be properly mitigated by bit-interleaving in a concatenated encoding chain where the non-orthogonal matrix modulation is used as the inner code. This leads to an error floor, essentially due to channel realizations that result in significant channel self-interference power.

However, the interference coefficient can be randomized to provide interference diversity [26], essentially in analogy with phase sweep or phase hop transmit diversity. One approach to achieve randomization is to weight at least one antenna output by a constant amplitude (complex) signal as suggested in (8.7). The phase of the weighting coefficient should change pseudo-randomly after each space–time block, and several times within a frame. This ergodization improves performance when combined with bit-interleaved coded modulation. The pseudo-random sequence of phases applied for antennas 3 and 4, $\{\theta_{t,3}\}$ and $\{\theta_{t,3}\}$, is assumed to be known *a priori* to the receiver in order not to exacerbate the channel estimation problem. Then, using right rotations of the form (8.7), the effect is to apply phasors $e^{j\theta_{t,3}}$ and $e^{j\theta_{t,4}}$ to columns 3 and 4 of ABBA. The effect of the rotations can be subsumed to the equivalent channel correlation matrix. More precisely, the correlation coefficient assumes the value

$$b[t] = 2\operatorname{Re}[e^{j\theta_{t,3}}h_1 h_3^* + e^{j\theta_{t,4}}h_2 h_4^*], \tag{8.17}$$

while the signal power a remains unaffected. With the phase evenly distributed in $(0, 2\pi]$, this clearly randomizes the interference without changing the effective correlation averaged over time and without sacrificing the diversity gain. In addition, transmit power in each antenna may be changed and the antenna indices may be permuted, to further enhance randomization. In fact, if g_n is the gain for antenna n, the correlation coefficient is given by

$$b[t] = 2\operatorname{Re}[e^{j\theta_t}g_1 h_1 g_3 h_3^* + e^{j\theta_t}g_2 h_2 g_4 h_4^*]. \tag{8.18}$$

However, unequal transmit power compromises the diversity gain. As an example, if $g_1 = g_2 = 0$ we get the two antenna STTD. If only $g_3 = 0$ we get three antenna transmission and different statistics for the interference coefficient. When randomization is used in conjunction with ABBA, we call the resulting scheme Randomized ABBA (RABBA).

8.1.2 Randomization for Orthogonal Space–Time Block Codes

With space–time block codes, the aforementioned randomizations do not change the performance at symbol level. Randomization is beneficial only if the orthogonal code is used as an inner code in a concatenated encoding chain.

8.1.2.1 Randomized STTD-OTD
The STTD-OTD concept above was the first four-antenna space–time transmission scheme proposed in 3G standardization. We demonstrate below that STTD-OTD can be further improved, when applied as an inner modulation matrix in a concatenated transmission chain. The improvement involves post-processing of the STTD-OTD signals using suitable unitary transformation matrices that essentially increase the number of effective channel states beyond two. Recall the transmission matrix for the combined STTD and Orthogonal Transmit Diversity (OTD) concept (4.22)

$$\mathbf{X} = \frac{1}{\sqrt{2}} \begin{bmatrix} \mathbf{X}_A & \mathbf{X}_B \\ \mathbf{X}_A & -\mathbf{X}_B \end{bmatrix}, \tag{8.19}$$

where time runs vertically and space (transmit antenna index) horizontally. This transmission method supports four transmit antennas. It is apparent from (8.19) that the two STTD blocks, each with different symbols, are transmitted in parallel using code-division multiplexing. An analogous transmission applying time-division multiplexing can be written as

$$\mathbf{X}(x_1, x_2, x_2, x_4) = \begin{bmatrix} \mathbf{X}_A & \mathbf{0}_2 \\ \mathbf{0}_2 & \mathbf{X}_B \end{bmatrix}, \tag{8.20}$$

as described in (4.21). Both designs above attain only diversity order two. This is identical to that provided by STTD alone. In fact, it easy to see that for a given antenna index configuration

$$\mathcal{D}_\mathbf{h} = \begin{bmatrix} \left(|h_1|^2 + |h_2|^2\right)\mathbf{I}_2 & \mathbf{0}_2 \\ \mathbf{0}_2 & \left(|h_3|^2 + |h_4|^2\right)\mathbf{I}_2 \end{bmatrix} \tag{8.21}$$

$$\mathcal{S}_\mathbf{h} = 0 \tag{8.22}$$

If $\mathbf{W}$ is a permutation matrix that shuffles indices $(1, 2, 3, 4) \to (i_1, i_2, i_3, i_4)$, we obtain

$$\mathcal{D}_{\mathbf{Wh}} = \begin{bmatrix} \left(|h_{i_1}|^2 + |h_{i_2}|^2\right)\mathbf{I}_2 & \mathbf{0}_2 \\ \mathbf{0}_2 & \left(|h_{i_3}|^2 + |h_{i_4}|^2\right)\mathbf{I}_2 \end{bmatrix} \tag{8.23}$$

$$\mathcal{S}_{\mathbf{Wh}} = 0 \tag{8.24}$$

There are six non-redundant permutations that lead to different $\mathcal{D}_\mathbf{h}$ matrices. In a randomized STTD-OTD concept, these can be sequentially applied, to partially ergodize the quasi-static channel and to improve performance in a concatenated coding scheme.

8.1.2.2   STTD-PHOP   The columns of STTD-OTD symbol matrix are orthogonal. When equipped with perfect channel knowledge, the receiver can demultiplex the transmitted symbols optimally using linear processing over four symbols.

An attractive alternative to STTD-OTD is to concatenate STTD and phase hopping. This approach, first proposed in [24], also retains full orthogonality and even simpler decoding, as the optimal detector operates only over two symbols. In the construction proposed in 3GPP [217] two identical copies of STTD are transmitted in parallel using four antennas, with pseudo-random phasing for different STTD blocks. Consider a modulation matrix

$$\frac{1}{\sqrt{2}} \begin{bmatrix} \mathbf{X}_A & \mathbf{X}_A \end{bmatrix}, \tag{8.25}$$

and the right action matrix

$$\mathbf{W} = \begin{bmatrix} \mathbf{I}_2 & \mathbf{0}_2 \\ \mathbf{0}_2 & \Theta \end{bmatrix}, \tag{8.26}$$

where the phasor matrix is

$$\Theta = \begin{bmatrix} e^{j\,\theta_{t,3}} & 0 \\ 0 & e^{j\,\theta_{t,4}} \end{bmatrix}. \tag{8.27}$$

Applying this right action on the transmission matrix (8.25) rotates the symbols in the third and fourth columns of the modulation matrix. The received signal is

$$\mathbf{y} = \mathbf{X}\mathbf{W}\mathbf{h} + \mathbf{n}. \tag{8.28}$$

After maximum ratio combining over antenna pairs 1 and 3, and 2 and 4, respectively, the equivalent channel correlation matrix becomes

$$\mathcal{D}_{\mathbf{h}} = \begin{bmatrix} |h_1 + h_3 e^{j\,\theta_{t,3}}|^2 & \mathbf{0}_2 \\ \mathbf{0}_2 & |h_2 + h_4 e^{j\,\theta_{t,4}}|^2 \end{bmatrix} \tag{8.29}$$

$$\mathcal{S}_{\mathbf{h}} = 0. \tag{8.30}$$

Clearly, the scheme is orthogonal, $\mathcal{S}_{\mathbf{h}} = 0$. In STTD-PHOP, the phasor matrix is fixed during two symbol periods, but changes pseudo-randomly after that. Phase hopping sequences are defined such that $\{\theta_{t,3}\}$ and $\{\theta_{t,4}\}$ sample the interval $(0, 2\pi]$ evenly over a time period that corresponds to one channel-coded block.

3GPP Proposal: One possible way to define the transmitted block is the following, which resembles the solution proposed in 3GPP [217]. Assume that the coded block spans 16 symbols (eight STTD blocks), the transmitted signal in one implementation is

$$\begin{bmatrix} \mathbf{X}_A & \mathbf{X}_A \mathrm{diag}(e^{j\,\theta_{1,3}}, e^{j\,\theta_{1,4}}) \\ \vdots & \vdots \\ \mathbf{X}_H & \mathbf{X}_H \mathrm{diag}(e^{j\,\theta_{8,3}}, e^{j\,\theta_{8,4}}) \end{bmatrix}. \tag{8.31}$$

Phase hopping sequence $\theta_{t,3}, t = 1, 2, ..., 8$ for antenna 3 over eight successive space–time blocks can follow the pattern given in Figure 8.3. For antenna four, $\theta_{t,4}, t = 1, 2, ..., 8$ is defined such that $e^{j\,\theta_{t,3}} = -e^{j\,\theta_{t,4}}, t = 1, .., 8$.

In the STTD-PHOP construction the terminal sees a linear combination of the channels. We assume that there are two different phase-hopping patterns, each randomizing one branch of STTD. Then, the effective received channels for two successive symbols (at the input of the STTD detector after signal combining) are given by

$$\tilde{h}_1 = h_1 + e^{j\theta_{1,3}} h_3, \tag{8.32}$$

$$\tilde{h}_2 = h_2 + e^{j\theta_{1,4}} h_4, \tag{8.33}$$

and these are used to decode one STTD block in the dedicated channel.

Implementation Options: There are several alternative STTD-PHOP implementations. The description above focused on an implementation that applies phase-hopping in conjunction with an orthogonal space–time block code. Alternatively, a continuous phase sweep can be utilized, in analogy with PSTD or frequency offset transmit solutions described in Chapter 3. Phase sweep with sufficiently small step size, or frequency offset transmit diversity, is (under certain constraints) transparent to the terminal. In principle, as long as the frequency offset remains within an implementation margin allowed by the specification, the BS can use four antenna elements and a frequency-offset based STTD-PHOP transmission. This improves the performance of all terminals within the cell. In this case the terminal measures only the effective channels $\tilde{h}_1$ and $\tilde{h}_2$, without requiring explicit knowledge of the frequency offset that is applied at the transmitter.

Furthermore, the different downlink pilot channels can be defined for each transmit antenna to enable efficient channel estimation. These are used to identify channels $h_1, h_2,$ h_3, h_4. Then, if the the complex phasors $\{e^{j\theta_{t,3}}\}$ $\{e^{j\theta_{t,3}}\}$ are known, the receiver can construct $\tilde{h}_1$ and $\tilde{h}_2$ in (8.32) and (8.33), respectively. The downlink pilot channels can be defined as common channels, available to all terminals in the cell, or antenna-specific training sequences can be multiplexed to the dedicated channels. The advantage of a solution, where channel estimation is decoupled from phase modulation, is that the phase modulation can be arbitrary without sacrificing channel estimation performance. Numerical simulations suggest that an eight-state hopping pattern is sufficient, see Figure 8.3. The transmitter structure is given in Figure (8.1). For comparison, the transmitter structure for STTD-OTD is depicted in Figure 8.2.

Finally, the transmitter architecture in STTD-PHOP directly supports closed-loop transmit diversity. The optimal phasor, one that maximizes received signal power, can be obtained by using a feedback link. We return to this option when discussing closed-loop transmit diversity solutions later in this book.

8.2 SPACE–TIME BLOCK CODE WITH ROTATED CONSTELLATIONS

The previous sections developed different multi-antenna transmission methods, where robustness to disadvantageous channel conditions is improved via randomization. These were seen to be effective in quasi-static channels when applied within a concatenated coding chain.

In this section we consider the problem to improve performance of the non-orthogonal matrix modulation schemes discussed in Chapters 5 and 7 in correlated channels, e.g. those subject to Ricean fading distribution. For this purpose we develop schemes in which the non-orthogonality coefficients are independent on channel phases, and the self-interference diminishes as the relative powers of the channels become similar. These characteristics are examined using the equivalent channel correlation matrix.

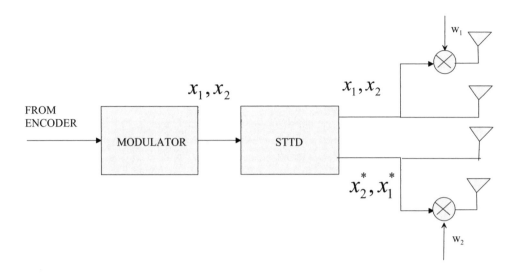

Fig. 8.1: Transmitter architecture for four Tx-antenna STTD-PHOP.

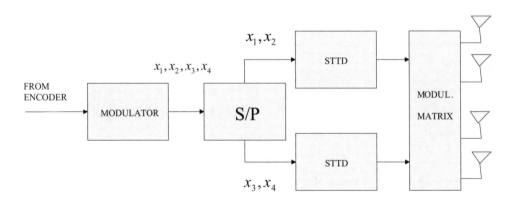

Fig. 8.2: Transmitter architecture for STTD-OTD.

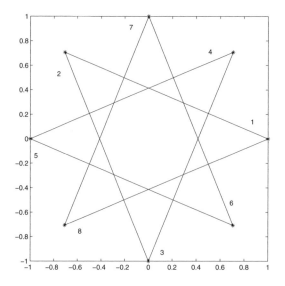

Fig. 8.3: State transitions for phase hopping in antenna three for STTD-PHOP.

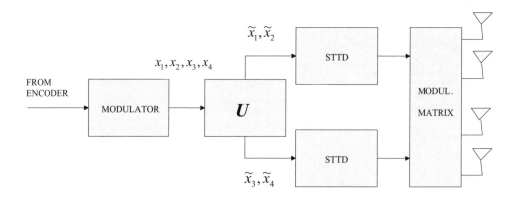

Fig. 8.4: Transmitter utilizing a space–time block code with symbol precoding.

The considered construction adopts orthogonal space–time block codes as constituents, subsumed within a matrix supporting multiple transmit antennas. Symbol precoding (rotated symbol constellations), as discussed in section 3.4 and Chapter 7, is used to improve the diversity order of the larger dimensional matrix, and accordingly performance at high SNR. The particular example discussed below may be called "transformed STBC-OTD", or "TrSTBC-OTD" for short. Here "STBC" refers to the constituent orthogonal STBC, and OTD is the chosen matrix extending the transmission to multiple antennas. The transmitter structure is depicted in Figure 8.4.

8.2.1 Transformed STBC-OTD Construction

Consider a transmitter with N_t transmit antennas and a square modulation matrix. Let $\mathbf{X}_A \in \mathbb{C}^{N_t/2 \times N_t/2}$ and $\mathbf{X}_B \in \mathbb{C}^{N_t/2 \times N_t/2}$ be arbitrary rate $R_s = Q/N_t$ orthogonal space–time block codes, and $\mathbf{U}$ be a unitary matrix of the form

$$\mathbf{U}(\mu, \nu) = \begin{bmatrix} \mu & \nu \\ -\nu^* & \mu^* \end{bmatrix} \otimes \mathbf{I}_{N_t/2}. \tag{8.34}$$

For simplicity, we apply parameters $\mu = \sqrt{\alpha}$ and $\nu = \sqrt{1-\alpha}\, e^{j\phi}$. A simple expression for a non-orthogonal matrix modulation is obtained by constructing a space–time matrix

$$\mathbf{X}_{tr} = \begin{bmatrix} \tilde{\mathbf{X}}_A & \tilde{\mathbf{X}}_B \\ \tilde{\mathbf{X}}_A & -\tilde{\mathbf{X}}_B \end{bmatrix} \tag{8.35}$$

where

$$\begin{aligned} \tilde{\mathbf{X}}_A &= \mathbf{X}_A(\tilde{x}_1, ..., \tilde{x}_{Q/2}) \\ \tilde{\mathbf{X}}_B &= \mathbf{X}_B(\tilde{x}_{Q/2+1}, ..., \tilde{x}_Q) \end{aligned} \tag{8.36}$$

with

$$(\tilde{x}_1, ..., \tilde{x}_Q) = (x_1, ..., x_Q)\mathbf{U}^T(\mu, \nu). \tag{8.37}$$

It should be stressed that each constituent $N_t/2 \times N_t/2$ space–time block code encodes $Q/2$ symbols. When specializing to four Tx antennas, the constituent STBC is STTD, and $Q = 4$. Thus the resulting matrix modulation has symbol rate $R_s = 1$. For $N_t = 4$, matrix (8.35) is otherwise identical to that used in STTD-OTD (8.19) except that here the symbols are linear combinations of each other. Thus, STTD-OTD is a special case, if we assume that μ is real and $\nu = 0$. For $N_t > 4$, we apply a higher-dimensional space-time block code, and the underlying scheme may be called STBC-OTD. With transformed symbols, the result is a TrSTBC-OTD.

It is instructive to recall that STTD-OTD and ABBA can be mapped to each other with symbol rotations that preserve the average information. The choice of $\alpha = 0.5$ in the precoding matrix, i.e.

$$\mathbf{U} = \frac{1}{\sqrt{2}} \begin{bmatrix} 1 & 1 \\ -1 & 1 \end{bmatrix} \otimes \mathbf{I}_2 \tag{8.38}$$

combined with

$$\tilde{\mathbf{X}}_{tr} = \sqrt{2} \begin{bmatrix} \tilde{\mathbf{X}}_A & \mathbf{0}_2 \\ \mathbf{0}_2 & \tilde{\mathbf{X}}_B \end{bmatrix}, \tag{8.39}$$

leads to diagonal ABBA (8.5). Thus precoding with $\mathbf{U}$ realizes a diversity-changing symbol rotation of the kind discussed in section 7.2.2, changing between symbolwise diversity two and four. This explains the observation discussed in section 4.7 that these two schemes provide the same (second-order) information and have identical performance at low SNR.

In (8.34) with $\alpha = 0.5$, the phase rotation part of the precoding matrix gives rise to a constellation rotation of the type discussed in section 7.3.1. More generic matrix-valued symbol rotations of the kind (7.32) may be realized by using a precoding matrix of the form

$$\tilde{\mathbf{U}} = \frac{1}{\sqrt{2}} \begin{bmatrix} 1 & 1 \\ -1 & 1 \end{bmatrix} \otimes \mathbf{U}_2 \tag{8.40}$$

where $\mathbf{U}_2$ is a 2×2 unitary matrix.

To complete the sequence of mappings between various space–time modulation schemes, note that the two forms of $\mathbf{X}_{tr}$ may be mapped to each other with the unitary left transformation

$$\mathbf{X}_{tr} = \frac{1}{\sqrt{2}} \begin{bmatrix} \mathbf{I}_2 & \mathbf{I}_2 \\ \mathbf{I}_2 & -\mathbf{I}_2 \end{bmatrix} \tilde{\mathbf{X}}_{tr} . \tag{8.41}$$

These transformations generalize in a natural way to higher-dimensional STBC-OTD and forms of ABBA with higher dimensional STBCs as constituent codes.

8.2.2 Effect on Equivalent Channel Correlation Matrix

In the parameterization of the precoding matrix $\mathbf{U}$ the parameter α interpolates between non-orthogonal transmission, where all symbols are transmitted homogeneously from all antennas, and orthogonal transmission, where each symbol is transmitted from half of the antennas to obtain diversity order two. Consider $N_t = 4$ and $\mathbf{X}_A, \mathbf{X}_B$ of the STTD form. When $\alpha = 0$ or $\alpha = 1$ we obtain the STTD-OTD concept of [49]. When $\alpha = 0.5$, $\mathbf{X}_{tr}$ is related to ABBA, as observed above. Due to the similar structure, the properties pertinent for performance in i.i.d. Rayleigh fading are the same as for ABBA (8.2). However, the form of the effective channel correlations is more favourable in correlated channels, as will be seen below.

The received signal is

$$\mathbf{y} = \mathbf{X}_{tr}\mathbf{h} + \mathbf{n}. \tag{8.42}$$

As before, we can write (8.42) using the effective channel matrix as

$$\tilde{\mathbf{y}} = \mathcal{H}\mathbf{U}\mathbf{x} + \tilde{\mathbf{n}},$$

where $\tilde{\mathbf{y}}$ is obtained from $\mathbf{y}$ using complex conjugations and linear transformations. With N_r receive antennas, N_r models are stacked according to (8.11). Direct calculation reveals that the equivalent channel correlation matrix of transformed STBC-OTD is

$$\mathbf{U}^\dagger \mathcal{H}^\dagger \mathcal{H} \mathbf{U} = \mathcal{D}_\mathbf{h} + \mathcal{S}_\mathbf{h}, \tag{8.43}$$

where

$$\mathcal{D}_\mathbf{h} = \begin{bmatrix} a_1 & 0 \\ 0 & a_2 \end{bmatrix} \otimes \mathbf{I}_{Q/2}, \tag{8.44}$$

and

$$\mathcal{S}_\mathbf{h} = \begin{bmatrix} 0 & b \\ b* & 0 \end{bmatrix} \otimes \mathbf{I}_{Q/2}, \tag{8.45}$$

and where

$$a_1 = p_1|\mu|^2 + p_2|\nu|^2 \tag{8.46}$$

$$a_2 = p_2|\mu|^2 + p_1|\nu|^2 \tag{8.47}$$

$$b = (p_2 - p_1)\mu^*\nu \tag{8.48}$$

$$p_1 = \sum_{j=1}^{N_r}\sum_{i=1}^{N_t/2} |h_{i,j}|^2 \tag{8.49}$$

$$p_2 = \sum_{j=1}^{N_r}\sum_{i=N_t/2+1}^{N_t} |h_{i,j}|^2. \tag{8.50}$$

This is written in a form that highlights the analogy to diversity transforms, discussed in section 3.4. The difference to Equation (3.18) is in statistical properties of the equivalent channel matrix. Here the difference of coefficients p_1 and p_2 is damped due to higher transmit diversity order, as they are the sums of $N_t N_r$ random variables.

The following issues are highlighted:

- In analogy with diversity transforms, the scheme is orthogonal e.g. if $|\nu| = 0$, in which case it retards to STBC-OTD, as explained above.

- If $|\mu| = |\nu|$ the diagonal elements of the equivalent channel correlation matrix are all identical, and each symbol obtains full diversity.

- The rate of transformed STBC-OTD is identical to the rate of the constituent space–time block codes $\mathbf{X}_A$ and $\mathbf{X}_B$.

When STTD is used as a constituent code, we have four transmission elements and symbol rate 1. On the other hand, if a rate 3/4 four antenna STBC is used as a constituent code, we obtain an eight-antenna transmission scheme with symbol rate 3/4, and so on. Note that when all channel coefficients are i.i.d. we obtain the same equivalent correlation structure with a transmitter employing four transmit and two receive antennas (4 Tx-2 Rx), and with a transmitter using eight transmit antennas and one receive antenna (8 Tx-1 Rx). Clearly, of these two cases the 4 Tx-2 Rx system enjoys a 3 dB performance gain due to receive antenna diversity, in addition to the symbol rate benefit.

8.2.3 Improved Performance in Correlated Channels

Now we may assess the benefits of the proposed scheme. The correlation factor b depends on the relative powers p_1 and p_2. When they are identical, the received signals are orthogonal. Hence, TrSTBC-OTD performs well in a channel with amplitude correlation between the antennas, since then p_1 and p_2 have similar expected magnitudes. Any of the transmit antenna correlation matrices discussed in section 2.1.2 would lead to increased amplitude correlation, and thus to a performance benefit of TrSTBC-OTD when compared to (constellation rotated) ABBA. A particular example of amplitude correlations is a channel with high Ricean factor K. If $K \to \infty$ for each transmit antenna, the multi-antenna channel approaches a Gaussian channel, and the transmission becomes orthogonal. Thus the transmission is more robust to changes in channel correlations. The drawback of applying transformed STBC-OTD, instead of rotated ABBA, is an increased peak-to-average power ratio.

Transformations Between Rate One Information Optimal Schemes: To understand in more detail where the performance gain comes from, we concentrate on $N_t = 4$ Tx antennas. Recall the chain of mappings between schemes discussed in section 8.2.1. Specifically, consider a precoding matrix expressed in terms of (8.34) as

$$\widehat{\mathbf{U}} = \begin{bmatrix} \mathbf{I}_2 & \mathbf{0}_2 \\ \mathbf{0}_2 & -e^{j\phi}\mathbf{I}_2 \end{bmatrix} \mathbf{U}\left(\frac{1}{\sqrt{2}}, \frac{e^{j\phi}}{\sqrt{2}}\right) = \frac{1}{\sqrt{2}}\begin{bmatrix} \mathbf{I}_2 & e^{j\phi}\mathbf{I}_2 \\ \mathbf{I}_2 & -e^{j\phi}\mathbf{I}_2 \end{bmatrix}. \tag{8.51}$$

Note that the additional diagonal precoding matrix just rotates all symbols within the orthogonally separable block $\widetilde{\mathbf{X}}_B$, and thus does not change performance. It is included to simplify notations below. As above, $\mathbf{X}_A$ and $\mathbf{X}_B$ are STTD blocks. The former encodes the symbols

x_1, x_2, whereas the latter encodes the constellation rotated symbols $e^{j\phi}x_3, e^{j\phi}x_4$. Furthermore, $\tilde{\mathbf{X}}_A$ and $\tilde{\mathbf{X}}_B$ are STTD blocks encoding symbols $\tilde{x}_1, \tilde{x}_2$ and $\tilde{x}_3, \tilde{x}_4$, respectively. These are constellation rotated and mixed by $\hat{\mathbf{U}}$:

$$\sqrt{2}\,\tilde{x}_1 = x_1 + e^{j\phi}x_3 \qquad \sqrt{2}\,\tilde{x}_2 = x_2 + e^{j\phi}x_4$$
$$\sqrt{2}\,\tilde{x}_3 = x_1 - e^{j\phi}x_3 \qquad \sqrt{2}\,\tilde{x}_4 = x_2 - e^{j\phi}x_4 \,.$$

Now the following identities connect ϕ-ABBA (7.33), diagonal ϕ-ABBA (8.5), diagonal TrSTBC-OTD (8.39) and TrSTBC-OTD (8.35):

$$\frac{1}{2}\begin{bmatrix} \mathbf{I}_2 & \mathbf{I}_2 \\ -\mathbf{I}_2 & \mathbf{I}_2 \end{bmatrix} \begin{bmatrix} \mathbf{X}_A & \mathbf{X}_B \\ \mathbf{X}_B & \mathbf{X}_A \end{bmatrix} \begin{bmatrix} \mathbf{I}_2 & -\mathbf{I}_2 \\ \mathbf{I}_2 & \mathbf{I}_2 \end{bmatrix} = \begin{bmatrix} \mathbf{X}_A + \mathbf{X}_B & \mathbf{0}_2 \\ \mathbf{0}_2 & \mathbf{X}_A - \mathbf{X}_B \end{bmatrix}$$

$$= \sqrt{2}\begin{bmatrix} \tilde{\mathbf{X}}_A & \mathbf{0}_2 \\ \mathbf{0}_2 & \tilde{\mathbf{X}}_B \end{bmatrix} = \frac{1}{\sqrt{2}}\begin{bmatrix} \mathbf{I}_2 & \mathbf{I}_2 \\ \mathbf{I}_2 & -\mathbf{I}_2 \end{bmatrix} \begin{bmatrix} \tilde{\mathbf{X}}_A & \tilde{\mathbf{X}}_B \\ \tilde{\mathbf{X}}_A & -\tilde{\mathbf{X}}_B \end{bmatrix} \tag{8.52}$$

As observed above, the symbol rotation connecting STTD-OTD and TrSTBC-OTD is of the type that changes the diversity degree, and thus performance properties. In (8.52), the different full diversity schemes are mapped to each other using *left- and right unitary transformations*. In section 7.1.2 these were identified as *symmetries of performance in i.i.d. Rayleigh fading*. For the left unitary transformations this extends to any channels; they never affect average performance. This is argued in Appendix A. Here, we shall extend the argument of Appendix A and show that right unitary transformations are not symmetries of performance for generic channel statistics.

Right Transformations and Channel Statistics: First consider i.i.d. Rayleigh fading channels. The fact that right transformations are symmetries of performance is seen from the probability density function (pdf) of the channel. The performance of a right unitary transformed matrix modulation $\mathbf{XW}$ in the channel $\mathbf{H}$ is the same as the performance of $\mathbf{X}$ in the channel $\mathbf{WH}$. When calculating the average performance, one averages over the channel statistics. For an i.i.d. Rayleigh fading MIMO channel, the channel pdf may be written as

$$f(\mathbf{H}) = \left(\frac{1}{2\pi\sigma^2}\right)^{N_t N_r} e^{-\,\text{Tr}\,\mathbf{H}^\dagger\mathbf{H}/2\sigma^2}, \tag{8.53}$$

which directly shows that $f(\mathbf{WH}) = f(\mathbf{H})$ for unitary $\mathbf{W}$. With non-trivial transmit covariance $\mathbf{R}_{\text{Tx}}$ the pdf becomes

$$f(\mathbf{H}) = \left(\frac{1}{\det \mathbf{R}_{\text{Tx}}}\right)^{N_r} e^{-\,\text{Tr}\,\mathbf{H}^\dagger\mathbf{R}_{\text{Tx}}\mathbf{H}}, \tag{8.54}$$

and in a right unitary transformation of the matrix modulation we get

$$f(\mathbf{WH}) = \left(\frac{1}{\det \mathbf{R}_{\text{Tx}}}\right)^{N_r} e^{-\,\text{Tr}\,\mathbf{H}^\dagger\mathbf{W}^\dagger\mathbf{R}_{\text{Tx}}\mathbf{WH}}. \tag{8.55}$$

The pdf, and thus performance, is invariant under this transformation only if $\mathbf{R}_{\text{Tx}}\mathbf{W} = \mathbf{W}\mathbf{R}_{\text{Tx}}$, which sets very stringent conditions on $\mathbf{W}$ for generic transmit covariance. In particular, we see immediately that the only covariance that is invariant under any $\mathbf{W}$ is the i.i.d. covariance proportional to the identity matrix.

Next consider a Ricean MIMO channel. Each channel matrix element h_{mn} has a line-of-sight component $\nu_{mn}e^{j\psi_{mn}}$, and the pdf is a Gaussian distribution around this first moment. The channel pdf may be written as

$$f(\mathbf{H}) = \left(\frac{1}{2\pi\sigma^2}\right)^{N_t N_r} e^{-\operatorname{Tr}\,(\mathbf{H}-\mathcal{V})^\dagger(\mathbf{H}-\mathcal{V})/2\sigma^2}, \tag{8.56}$$

where $\mathcal{V}$ is the matrix of first moments. The pdf is invariant under $\mathbf{H} \mapsto \mathbf{WH}$ only if $2\operatorname{Re}\operatorname{Tr}\mathcal{V}^\dagger(\mathbf{I} - \mathbf{W})\mathbf{H} = 0$. This again sets very stringent conditions on $\mathbf{W}$; the pdf is invariant under all $\mathbf{W}$ only if $\mathcal{V} = 0$, i.e. for i.i.d. Rayleigh.

Thus we have observed that for non-i.i.d. and/or non-Rayleigh fading, the right unitary transformation $\mathbf{W}$ is generically *not* a symmetry of performance. Thus ABBA, the first scheme in the chain (8.52), may have different performance from the remaining three schemes (the middle two of which are identical, the last one of which is equivalent up to a left unitary transformation). To see that the latter indeed perform better in correlated fading one may resort to simulations. Alternatively one may use upper- and lower-bounding of pairwise error probabilities, based on the Hadamard (A.14) and Schur (A.15) inequalities.

Finally, it should be noted that right unitary transformations are beam-forming matrices, and the analysis of this section shows that TrSTBC-OTD is an example of beam-forming based on information on the channel statistics only.

8.2.4 Combined Diversity Transform and STBC

We assumed above a transmitter with $N_t = 4$ antennas, and a block fading channel. However, the current WCDMA specifications support only two transmit antennas. Therefore, it is instructive to examine how the performance of a orthogonal two-antenna (or more generically N_t-antenna) transmission could be improved via rotated constellations. This can be accomplished transmitting the two parts of TrSTBC-OTD using different time slots. Time slot t_1 is allocated for $\tilde{\mathbf{X}}_A$ and another time slot at t_2 for $\tilde{\mathbf{X}}_B$, each transmitting STTD with rotated symbol constellations using two transmit antennas. Assume that the complex channel gains for antenna n at time t is $h_n[t]$. Then, the received signal can be modelled as

$$\mathbf{y} = \mathbf{X}_{\text{tr}}[h_1[t_1]\ h_2[t_1]\ h_1[t_2]\ h_2[t_2]]^T + \mathbf{n}, \tag{8.57}$$

where

$$\mathbf{X}_{\text{tr}} = \begin{bmatrix} \tilde{\mathbf{X}}_A & \mathbf{0}_2 \\ \mathbf{0}_2 & \tilde{\mathbf{X}}_B \end{bmatrix}, \tag{8.58}$$

and $\tilde{\mathbf{X}}_A$ and $\tilde{\mathbf{X}}_B$ are formed using (8.36) and (8.37), exactly as when $N_t = 4$. Hence, the equivalent channel correlation matrix is also identical to that in (8.43). Therefore, it is clear that in a time-invariant channel, where channel gains are identical at t_1 and t_2, the transmission is orthogonal and the diversity order is only two. At the other extreme, if the complex channel amplitudes are independent, we obtain a non-orthogonal transmission with diversity order four.

We consider the more interesting intermediate case, where channel coefficients are correlated in time. In particular, assume that the channel has the classical Doppler spectrum, where the maximum Doppler spread depends on the velocity of the terminal and the carrier frequency. The delay between the two transmissions is fixed at $\Delta t = t_2 - t_1 = 10$ ms. The performance of the proposed transmission scheme is depicted in Figure 8.5 for different

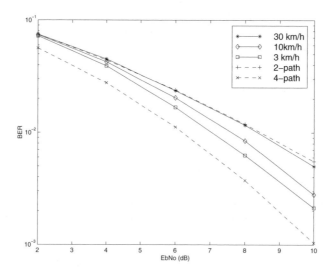

Fig. 8.5: Performance using combined STBC and diversity transform with 2 Tx antennas and with 10 ms delay. Performance achievable in the absence of the transform follows two-path curve.

terminal velocities. With higher velocities the temporal channel correlations are smaller and the scheme enjoys better diversity gain. This example shows that rotated constellations can be used to aggregate diversity from both space and time dimensions. Here, diversity order four performance can be obtained at the expense of increased latency.

8.3   PERFORMANCE EVALUATION

8.3.1   Performance without Channel Coding

In the absence of an outer code, it suffices to quantify the performance of ABBA without rotated constellations, STTD-OTD and transformed STBC-OTD. Phase-randomized counterparts of ABBA and STTD-OTD have all the same performance at symbol level as their deterministic siblings.

The simulated symbol rate 1 transformed STBC-OTD applies $\alpha = 1/2, \phi = \pi/4$ in (8.34)) to obtain full modulation diversity. Here each coordinate of $\mathbf{x}$ is QPSK modulated. The optimal constellation rotation for ABBA-type matrix modulations with QPSK deviates slightly from $\phi = \pi/4$ [211]. The aforementioned choice, related to set-partitioned 8-PSK, mitigates peak-to-average power ratio and is in practice indistinguishable from the optimal choice (see section 7.3). In each case we adopt Maximum Likelihood (ML) reception, although essentially the same performance can be achieved with interference cancellation receivers, as shown in [132] and in Chapter 6 in this book.

Figure 8.6 depicts the bit-error rates for three different transmission concepts in a Ricean fading channel with different K-factors. It is seen that the TrSTBC-OTD is superior to both STTD-OTD and to ABBA, in particular when the K-factor increases. ABBA performance

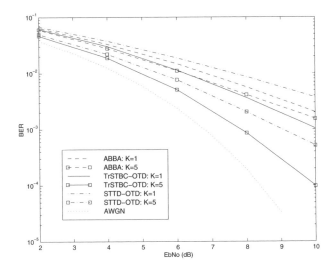

Fig. 8.6: Bit-error rates of ABBA, transformed (Tr) STBC-OTD and STTD-OTD with four
transmit antennas in a Ricean channel, with different K-factors.

is seen to improve only slightly with a larger K-factor, while STTD-OTD and TrSTBC-
OTD are more sensitive to the K-factor. The latter enjoys better performance due to higher
diversity order. This confirms the conjectures regarding robustness of TrSTBC-OTD in
correlated channels.

8.3.2   Performance with an Outer Code

We consider rate R_c=3/4 turbo-coded transmission with 378-bit frame size and random
interleaving. Turbo code polynomials are the same as in the cdma2000 standard, and symbols
are QPSK-modulated. The channel is block Rayleigh fading and the channel coefficients
are fixed for each frame, but change randomly for different frames. The turbo code is
concatenated with different four antenna transmission schemes ABBA, RABBA, STTD-
PHOP and STTD-OTD, for comparison with STTD, or with a single transmit antenna.

First, Figure 8.7 depicts the performance of turbo-coded transmission when concatenated
with ABBA and RABBA using $N_t = 4$ transmit antennas and a single receive antenna.
Randomized ABBA applies different phase rotation coefficients and a different antenna
shuffling pattern for each space–time coded block within a frame. That is, each unitary
transformation matrix $\mathbf{W}$ is a product of diagonal phasor matrix, and a 4×4 permutation
matrix. Hence, peak-to-average power ratio is one in each antenna element. In addition, with
this choice of matrix $\mathbf{W}$ the received signal power is identical for both ABBA and RABBA.
In all cases we applied the maximum likelihood decoder with ABBA and RABBA. However,
soft outputs transferred to the turbo decoder were calculated in a sub-optimum fashion using
one bit differences in the likelihood metric. This simplifies calculations, but induces a minor
performance loss, typically a fraction of a decibel. The reference curve (the matched filter
bound), with four-path diversity, is not realizable without bandwidth expansion. However, it
shows that we operate within one decibel from the absolute maximum performance. Figure

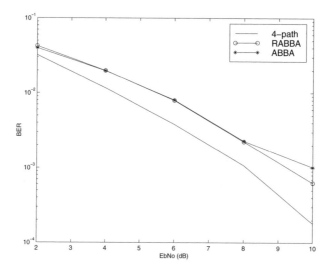

Fig. 8.7: Bit–error rates of turbo-coded ABBA and RABBA in a block fading channel. Turbo code rate is 3/4 with frame size 378. Matched filter bound for four-path diversity is shown as reference.

8.7 reveals that randomization improves performance only when the SNR is high. No attempt was made here to optimize the phase-hopping permutation sequence or the permutation matrices.

Figure 8.8 depicts the performance of different orthogonal symbol rate one transmission methods, using an identical simulation setup as before. We notice from Figure 8.8 that STTD-PHOP is always superior to STTD-OTD and STTD. At bit error rate 10^{-3} the difference is roughly 1 dB, when comparing performance with a single receive antenna. With two receive antennas, the performance difference between different schemes is smaller, as expected. Even two antenna transmission with STTD performs well. We conjecture that four element (open-loop) transmission is beneficial particularly when the number of receive antennas is small. The receiver complexity is comparable for all schemes in Figure 8.8. Thus, it is seen that simple transmission schemes can provide $2 - 4$ dB improvement over STTD in downlink performance.

Next, compare the coded and uncoded performance results. We notice that non-orthogonal concepts (ABBA and STBC with rotated constellations) outperform STTD-OTD and other diversity order two schemes in the absence of channel coding, as seen from Figures 7.8 and 8.6. However, when used with a turbo code of rate 3/4, RABBA performance is between STTD-PHOP and STTD-OTD. It can be conjectured that with very high channel coding rates (between 3/4 and 1) non-orthogonal codes also outperform STTD-PHOP, provided that optimal soft outputs are passed to the turbo decoder. On the other hand, Figure 8.7 shows that the performance for a virtual orthogonal transmission scheme with four-order diversity is only $1.0 - 1.5$ dB better than that of RABBA and STTD-PHOP.

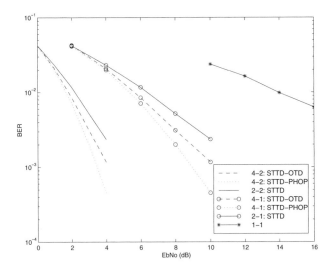

Fig. 8.8: Bit-error rate for selected symbol rate 1 orthogonal designs with different number of transmit and receive antennas. Channel coding rate is 3/4, using an eight-state turbo code with frame size 378 and random interleaving.

8.4 SUMMARY

In this chapter we have described various symbol rate one matrix modulation schemes, including concepts that have been proposed in 3GPP, such as STTD-OTD and STTD-PHOP. In HSDPA, these methods could be used to obtain peak data rates of 5.4 Mbps (with rate 3/4 turbo code) and 7.24 Mbps (without channel coding) with about 2 dB reduction in transmit power when compared to STTD using one receive antenna. It was shown that such extremely simple four antenna transmitter architectures can provide a significant performance gain when combined with channel coding. On the other hand, with two receive antennas the performance difference between STTD and the other schemes diminishes, when combined with an outer code.

With a very high channel coding rate, space–time block codes with rotated constellations perform well, although the peak-to-average ratio tends to increase with symbol rotations. It was shown that properly designed symbol rotations can be combined with an orthogonal space–time block code to make the code robust to channel fading statistics, in particular to the Ricean K-factor. If the peak-to-average ratio cannot be increased, symbol rotations are not applicable. Therefore, unless the channel coding rate is very high, one should apply randomizing unitary transformations that maintain the symbol constellation, e.g. permutations and phasing.

9

High-rate Designs for MIMO Systems

Most open-loop MIMO systems, starting from the first suggestions [15] and the methods used in the BLAST demonstrator [20, 21], to the bulk of more recent work, have been based on the simplest space–time modulation, namely unconstrained vector modulation. This means that a conventional modulation scheme is chosen, say QPSK, and an independent symbol in this scheme is transmitted separately from each antenna. Improving performance by spreading the Tx-diversity among the transmitted signals more evenly is performed indirectly by concatenated channel coding. Methods for this are D-BLAST [20], as well as horizontally or vertically encoded V-BLAST [21]. Taking into account the division between coding and modulation adopted in this book, multi-layered space–time architectures of the kind discussed in [218] belong to this category also. There are several reasons why more involved matrix modulation schemes should be considered in a MIMO setting.

Complexity: If the possible concatenated code has a high rate, sampling Tx diversity by the channel code is ineffective. From a complexity point of view it is beneficial to use as high-rate concatenated codes as possible, or equivalently, as small constellations as possible, to convey the information to be transmitted. Thus it is worthwhile to construct more involved and better performing matrix modulation schemes in a MIMO system, provided that a sufficiently simple detection algorithm exists.

Limitation on N_r in Downlink: In future cellular communication systems, regardless whether they are based on evolved 3G system or disruptive 4G technology, there is likely to be asymmetry between down- and uplinks. This will be seen both in traffic conditions and in hardware. The downlink direction is the projected bottleneck when it comes to data rates. In uplink, coverage is more important. Also, from implementation cost it follows that typical base stations will have more antennas than typical mobile stations.

In section 5.2.3 it was argued that the number of parallel streams(the symbol rate $R_s = Q/T$) should not exceed $\min(N_t, N_r)$. From this it follows that

- for typical downlinks, vector modulation is inefficient, even if mobile terminals have more than one Rx antenna,

- matrix modulation or beam-forming methods (including antenna selection) or both should be used for high data-rate downlink services.

Correlations and Rank Deficiency: Another reason for investigating constrained ($R_s < N_t$) schemes is that correlations between the Tx and/or Rx antennas often conspire to make the practical rank (Prank) of the channel less than $\min(N_t, N_r)$, see section 2.4. In such situations, the results of section 5.2.3 indicate that the number of parallel streams should not exceed Prank.

For these reasons, we shall investigate high-rate constrained schemes, with $1 < R_s \le N_t$. For such rates, dropping parts of a vector modulation leads to information deficient signalling and loss of performance at low SNR, as discussed in section 4.5. In this chapter, information-optimal constrained schemes will be discussed, along the lines of the principles discussed in Chapters 4 and 5, which perform well at low and medium SNR. High SNR performance may be optimized using the methods in Chapter 7.

9.1 SETS OF FROBENIUS ORTHOGONAL UNITARY MATRICES

Applying the principles of Chapters 4 and 5 should begin by constructing sets of basis matrices $\mathbf{B}^{(k)}$, having a chosen degree of transmit diversity. If a non-maximal symbolwise diversity degree is chosen, the basis should be a T-MSD (5.18) pseudo-unitary basis. Such bases may be constructed from unitary bases for fewer Tx antennas, as is done in (5.15). For this reason we shall discuss only square basis matrices in this section. These basis matrices should be unitary and Frobenius orthogonal.

Explicit matrix modulation schemes with lower diversity and rate $R_s > 1$ were constructed in [219]. As no transparent way of generalizing this work to various N_t, R_s is known, this approach is not discussed here.

Three alternatives for constructing Frobenius orthogonal unitary bases for $N_t \times N_t$ matrices have been discussed in the context of space–time code design.

9.1.1 Clifford Basis

This is discussed in section 5.4.1 and Appendix B, and is the fundament of constructing quasi-orthogonal layered matrix modulation schemes. A full-diversity Clifford basis may be constructed for any N_t that is a power of 2. For such schemes, $T = N_t$ and the schemes are MSD. When a full Clifford basis exists, it can be used to construct a modulation matrix employing N_t antennas with maximal transmit diversity N_t and symbol rate $R_s = N_t$ (the maximum number of parallel streams) [86]. With a lower transmit diversity, a Clifford basis may be constructed for any even N_t, with $T < N_t$ and T-MSD. A transmission with Tx diversity T and rate $R_s = N_t$ can be devised.

Interpretation in Terms of Quasi-orthogonal Layers with 2 Complex Symbols (STTD Blocks): As an example for $N_t = T = 4$, a rate $R_s = 4$ transmission in the Clifford basis may be described as a sum of eight parts (B.24)

$$
\mathbf{X}^{\text{Cliff}} = \begin{bmatrix} \mathbf{X}_A & \mathbf{X}_B \\ \mathbf{X}_B & \mathbf{X}_A \end{bmatrix} + \begin{bmatrix} \mathbf{X}_C & \mathbf{X}_D \\ -\mathbf{X}_D & -\mathbf{X}_C \end{bmatrix}
$$

$$
+ \, j \begin{bmatrix} \mathbf{X}_E & \mathbf{X}_F \\ \mathbf{X}_F & \mathbf{X}_E \end{bmatrix} + \, j \begin{bmatrix} \mathbf{X}_G & \mathbf{X}_H \\ -\mathbf{X}_H & -\mathbf{X}_G \end{bmatrix}, \tag{9.1}
$$

where $\mathbf{X}_A(x_1, x_2), \mathbf{X}_B(x_3, x_4), \ldots, \mathbf{X}_H(x_{15}, x_{16})$ are STTD blocks of the form (3.28), which describe the eight quasi-orthogonal layers of (9.1). Transmissions with lower rates may be constructed by removing some of the layers. The process of removing parts of a high-rate scheme to get a lower-rate one will be called "puncturing" in this chapter, for obvious reasons.

Interpretation in Terms of Maximal Orthogonal Design: An alternative way to view the Clifford basis is in terms of the maximal orthogonal design that is a part of the basis. For the 4×4 case this was done in (B.26). As a consequence, the basis matrix (9.1) may be written as

$$
\mathbf{X}^{\text{Cliff}} = \mathbf{X}_M(x_1, x_2, x_3) + \mathbf{X}_N(x_4, x_5, x_6) + \mathbf{X}_O(x_7, x_8, x_9)
$$
$$
+ \mathbf{X}_P(x_{10}, x_{11}, x_{12}) + \mathbf{X}_R(x_{13}, x_{14}, x_{15}, x_{16}), \tag{9.2}
$$

where $\mathbf{X}_M, \ldots, \mathbf{X}_P$ are rate 3/4 quasi-orthogonal layers (B.27), ...,(B.30) encoding three complex symbols each, and $\mathbf{X}_R$ of (B.31) encodes four complex symbols in four layers. It should be stressed that (9.1) and (9.2) are the very same matrix modulation; the symbols have only been divided into layers in a different way. The total number of layers in both divisions is 8; (9.1) is 2+2+2+2+2+2+2+2 layered, and (9.2) is 3+3+3+3+1+1+1+1 layered. At this stage, these two different divisions are superfluous. From the receiver point of view, the maximal quasi-orthogonality that may be used to simplify reception according to the lines of (6.3) is given by a layer with three complex symbols in (9.2).

The difference between the two layering counts shows up if layers are punctured to construct lower rate schemes, or if constellation rotations within layers are considered, in order to maximize diversity gain, as discussed in section 7.2.3.

9.1.2   Weyl Basis

In [133], a unitary Frobenius orthogonal basis was discussed for any N_t. This basis is constructed as follows. Take the two $N_t \times N_t$ matrices

$$
\mathbf{D} = \begin{bmatrix} 1 & 0 & \cdots & & 0 \\ 0 & e^{j\,2\pi/N_t} & 0 & & \\ \vdots & & \ddots & & \\ 0 & \cdots & & & e^{j\,2(N_t-1)\pi/N_t} \end{bmatrix}, \quad \mathbf{\Pi} = \begin{bmatrix} 0 & 1 & 0 & \cdots & 0 \\ 0 & 0 & 1 & 0 & \cdots \\ 0 & 0 & 0 & 1 & \cdots \\ \vdots & & \ddots & & \vdots \\ 0 & 0 & \cdots & 0 & 1 \\ 1 & 0 & \cdots & 0 & 0 \end{bmatrix} \tag{9.3}
$$

and construct the $N_t{}^2$ matrices

$$
\mathbf{B}_{\text{W}}^{(kl)} = \mathbf{D}^{k-1}\, \mathbf{\Pi}^{l-1}, k = 1, \ldots, N_t, \ l = 1, \ldots, N_t. \tag{9.4}
$$

Together with the matrices $\mathrm{j}\,\mathbf{B}_{\mathrm{W}}^{(kl)}$, these $2N_{\mathrm{t}}^2$ matrices form a basis for transmitting from N_{t} antennas with transmit diversity N_{t} and rate $Q/T = N_{\mathrm{t}}$ (the maximum number of parallel streams). The most fundamental property of the matrices $\mathbf{D}$ and $\mathbf{\Pi}$ is their commutation relation,

$$\mathbf{\Pi}\,\mathbf{D} = \mathrm{e}^{\mathrm{j}\,2\pi/N_{\mathrm{t}}}\,\mathbf{D}\,\mathbf{\Pi}\,. \tag{9.5}$$

Investigation of algebras with such commutation relations goes back to Hermann Weyl, so the basis (9.4) may be called the Weyl basis. In [133] it was shown that this basis reaches capacity for a transmission from N_{t} i.i.d. channels. As an example for $N_{\mathrm{t}} = 4$, $T = 4$, a Weyl modulation may be considered as a sum of four cyclic schemes of the kind of (4.29), as the matrix $\mathbf{\Pi}$ generates cyclic shifts:

$$
\begin{aligned}
\mathbf{X}^{\mathrm{Weyl}} = \quad &\mathbf{X}_{W\,1}(x_1, x_2, x_3, x_4) \; + \; \mathbf{X}_{W\,2}(x_5, x_6, x_7, x_8) \\
+ \; &\mathbf{X}_{W\,3}(x_9, x_{10}, x_{11}, x_{12}) \; + \; \mathbf{X}_{W\,4}(x_{13}, x_{14}, x_{15}, x_{16})\,.
\end{aligned} \tag{9.6}
$$

where

$$
\mathbf{X}_{W\,m}(x_1, x_2, x_3, x_4) =
\begin{bmatrix}
1 & 0 & 0 & 0 \\
0 & \mathrm{j} & 0 & 0 \\
0 & 0 & -1 & 0 \\
0 & 0 & 0 & -\mathrm{j}
\end{bmatrix}^m
\begin{bmatrix}
x_1 & x_2 & x_3 & x_4 \\
x_4 & x_1 & x_2 & x_3 \\
x_3 & x_4 & x_1 & x_2 \\
x_2 & x_3 & x_4 & x_1
\end{bmatrix}. \tag{9.7}
$$

The multiplying matrix is just the mth power of $\mathbf{D}$ in (9.3).

In (9.6), four cyclic layers been identified in $\mathbf{X}^{\mathrm{Weyl}}$. Lower-rate schemes may be constructed e.g. by dropping cyclic layers from (9.6). It should be stressed that these cyclic layers are not quasi-orthogonal layers. From the discussion in Chapter 4, it follows that each cyclic layer consists of four quasi-orthogonal layers. This is a generic feature of the Weyl basis. With the exception of the 2×2 case, which coincides with the 2×2 Clifford basis, the modulation matrices constructed according to (9.6) have quasi-orthogonality 1.

9.1.3 Hadamard Basis

Linear "threaded algebraic space–time codes" based on the threading prescription of [220] have been suggested in [144]. These have Frobenius orthogonal unitary bases. As we shall see, from (second-order) information point of view, the underlying Hadamard basis is equivalent to the Weyl basis.

The idea of threaded space–time coding is to take threads from a modulation matrix in the way that each thread is active during all symbol intervals used for transmission, and over time, uses each antenna as often. Essentially threading is a tail-biting version of D-BLAST. Consider a 4×4 example,

$$
\mathbf{X} =
\begin{bmatrix}
\tilde{x}_1 & \tilde{x}_5 & \tilde{x}_9 & \tilde{x}_{13} \\
\tilde{x}_{14} & \tilde{x}_2 & \tilde{x}_6 & \tilde{x}_{10} \\
\tilde{x}_{11} & \tilde{x}_{15} & \tilde{x}_3 & \tilde{x}_7 \\
\tilde{x}_8 & \tilde{x}_{12} & \tilde{x}_{16} & \tilde{x}_4
\end{bmatrix}, \tag{9.8}
$$

which is the usual vector modulation, considered over $T = 4$ symbol periods. In a threading interpretation, symbols $\tilde{x}_1, \ldots, \tilde{x}_4$ belong to the first thread, symbols $\tilde{x}_5, \ldots, \tilde{x}_8$ to the second, $\tilde{x}_9, \ldots, \tilde{x}_{12}$ to the third and $\tilde{x}_{13}, \ldots, \tilde{x}_{16}$ to the fourth. In [220], methods to choose convolutional codes acting on the threads so that the ensuing space time code has full rank were considered.

Diagonal Algebraic [52], or Khatri–Rao [53] space–time codes have symbol rate $R_s = 1$, and operate on one thread, say the diagonal. The transmission is made full rank by mixing constellation-rotated symbols with a Hadamard transform

$$
\mathbf{U}_{\text{Had}} = \frac{1}{2}
\begin{bmatrix}
1 & 1 & 1 & 1 \\
1 & -1 & 1 & -1 \\
1 & 1 & -1 & -1 \\
1 & -1 & -1 & 1
\end{bmatrix}.
\tag{9.9}
$$

In [144] this method is generalized to higher symbol rate by applying a Hadamard transform on rotated symbols on each thread. The corresponding set of basis matrices will be called a "Hadamard basis", following the basis naming convention in this Chapter. Additionally, to make the high-rate scheme full rank, algebraic ("Diophantine") rotations were suggested between the threads to make the matrix modulation full rank.

The unitary Hadamard rotation is a symmetry of the mutual information of (9.8) which changes the symbolwise diversity degree. It turns a vector transmission with no Tx diversity, of the type (9.8), to an MSD scheme. With this interpretation, a 4×4 threaded MSD modulation matrix following [144] would be

$$
\mathbf{X}^{\text{Thread}} = \mathbf{X}_{T1}(x_1, x_2, x_3, x_4) + \mathbf{X}_{T2}(x_5, x_6, x_7, x_8) \\
+ \mathbf{X}_{T3}(x_9, x_{10}, x_{11}, x_{12}) + \mathbf{X}_{T4}(x_{13}, x_{14}, x_{15}, x_{16}). \tag{9.10}
$$

where

$$
\mathbf{X}_{Tm}(\mathbf{x}) = \Pi^m \operatorname{diag}[\mathbf{U}_{\text{Had}}\, \mathbf{x}]. \tag{9.11}
$$

Here constellation rotations preserving the symbolwise diversity degree have been suppressed, as is usual when a short description of a matrix modulation is given. Lower-rate schemes may be constructed by dropping threads from (9.10).

From the appearance of the cyclic permutation operator Π in (9.11) it is evident that there is a straightforward mapping between the Weyl basis and the Hadamard basis. Indeed, the Weyl basis may be considered as a threaded basis, where instead of the Hadamard transform (9.9), a discrete Fourier transform

$$
\mathbf{U}_{\text{DFT}} = \frac{1}{2}
\begin{bmatrix}
1 & 1 & 1 & 1 \\
1 & j & -1 & -j \\
1 & -1 & 1 & -1 \\
1 & -j & -1 & j
\end{bmatrix}
\tag{9.12}
$$

is used to mix the symbols in a thread. This gives an alternative (threaded) way to divide (9.6) into four parts, and another natural way to decrease its rate. Similarly, (9.11) may be described in terms of cyclic matrices, where in (9.7), instead of powers of D, diagonal matrices with different rows of the Hadamard matrix (9.9) are used to multiply the cyclic matrices. As both threading and cyclic layers are to be found both in the Weyl (9.6) and threaded MSD (9.11) matrices, the latter will henceforth be called a "Hadamard scheme".

Both the Weyl basis and the Hadamard basis are thus based on threads, or on cyclic layers of the form (4.29), whereas the Clifford basis is based on quasi-orthogonal layers. In Chapter 4 it was shown that the cyclic scheme (4.29) fares badly at low SNR due to loss of information. With maximal rate $R_s = N_t$, all three bases discussed above provide the same information. Indeed they can all be constructed from a vector modulation extended in time (9.8) by different orthogonal symbol rotations. We shall see that when parts of the schemes are punctured, the schemes without quasi-orthogonality lose in information, and the loss grows progressively in puncturing. Due to the principles in section 4.6.2, a low SNR loss of information carries over to high SNR as loss in the maximum determinant distance.

9.2   OPTIMIZING RATE 2 MIMO-MODULATION FOR $N_T = T = 2$

As a warm-up, we optimize a scheme for $N_t = 2$ Tx antennas with symbol rate 2, extending over $T = 2$ symbol periods. This problem was briefly discussed in section 5.4.3. There, a Frobenius orthogonal MSD modulation matrix

$$
\begin{aligned}
\mathbf{X} &= \begin{bmatrix} x_1 & x_2 \\ -x_2^* & x_1^* \end{bmatrix} + \begin{bmatrix} 1 & 0 \\ 0 & -1 \end{bmatrix} \begin{bmatrix} x_3 & x_4 \\ -x_4^* & x_3^* \end{bmatrix} \\
&\equiv \mathbf{X}_A(x_1, x_2) + \begin{bmatrix} 1 & 0 \\ 0 & -1 \end{bmatrix} \mathbf{X}_B(x_3, x_4)
\end{aligned}
\tag{9.13}
$$

was constructed from the Clifford basis (5.20). Here $\mathbf{X}_A$ and $\mathbf{X}_B$ are 2×2 STTD blocks. Also, it was shown that the Weyl basis provides the same scheme (5.24).

It is easy to see that (9.13) suffers from pathological error events. Consider events where errors Δ_1 and Δ_3 are made in x_1 and x_3, and no errors are made in x_2, x_4. The codeword difference matrix is

$$
\mathbf{D}^{(\mathrm{ce})} = \begin{bmatrix} \Delta_1 + \Delta_3 & 0 \\ 0 & \Delta_1^* - \Delta_3^* \end{bmatrix},
\tag{9.14}
$$

and the determinant of the distance matrix is $|\Delta_1 + \Delta_3|^2 |\Delta_1 - \Delta_3|^2$. This clearly vanishes for some error events if the symbols x_1, x_3 have the same constellation. This kind of pathological errors may be cured by simple constellation rotations.

A more difficult set of pathological errors occurs when errors Δ_1 and Δ_4 are made in x_1 and x_4 and no errors are made in x_2, x_3. The codeword difference matrix is

$$
\mathbf{D}^{(\mathrm{ce})} = \begin{bmatrix} \Delta_1 & \Delta_4 \\ \Delta_4^* & \Delta_1^* \end{bmatrix},
\tag{9.15}
$$

and the determinant of the distance matrix is $\left(|\Delta_1|^2 - |\Delta_4|^2 \right)^2$. This clearly vanishes whenever errors with the same absolute difference may be made when detecting x_1, x_4. This kind of error cannot be cured by simple constellation rotations. One alternative is to dilate the constellations so that different symbols are transmitted with different powers. This, however, is information sub-optimal, as discussed in section 7.6.

Another way to cure these pathological errors is to apply matrix-valued constellation rotations. Thus we may apply the principles of Chapter 7 to optimize the performance of this scheme at high SNR. More specifically,

- Only symbol rotations within a quasi-orthogonal layer should be considered (7.29). For this the formulation (9.13) of the scheme is preferable over (5.24), as the two quasi-orthogonal layers are explicitly visible.

- As there are only two layers, only symbol rotations within one of the layers need to be investigated.

- STTD blocks have the special feature that an orthogonal symbol rotation acting on the symbols in a STTD block $\mathbf{X}_A$ may be expressed as a right and left unitary rotation (with determinant 1) acting on $\mathbf{X}_A$ itself, as in (7.32).

Thus for the rate 2 scheme (9.13), the MSD-preserving, performance-changing symbol rotations may be described as

$$\mathbf{X}(x_1, x_2, x_3, x_4) = \mathbf{X}_A(x_1, x_2) + \begin{bmatrix} 1 & 0 \\ 0 & -1 \end{bmatrix} \mathbf{V}\, \mathbf{X}_B(x_3, x_4)\, \mathbf{U}, \qquad (9.16)$$

where $\mathbf{V}$ and $\mathbf{U}$ are two 2×2 unitary matrices with unit determinant. This form was discussed in [51]. In section 7.1.1 it was argued that orthogonal symbol rotations should be restricted to unitary rotations that preserve complex symbols. As in (7.32), this means that one of the two matrices, say $\mathbf{V}$, should be of the form

$$\mathbf{V} = \begin{bmatrix} e^{j\rho} & 0 \\ 0 & e^{-j\rho} \end{bmatrix}. \qquad (9.17)$$

The other may be parameterized as (7.41) with three real parameters. Thus of the 28-dimensional set of orthogonal rotations mixing the four symbols in (9.16), only four need to be investigated to optimize performance.

In [51], the following unitary rotation was found to minimize the union bound of pairwise error probabilities at $E_b/N_0 = 10$ dB

$$\mathbf{V} = \mathbf{I}_2, \quad \mathbf{U} = \begin{bmatrix} e^{j\,7\pi/20}\cos\left(\frac{9\pi}{50}\right) & e^{j\,\pi/4}\sin\left(\frac{9\pi}{50}\right) \\ -e^{-j\,\pi/4}\sin\left(\frac{9\pi}{50}\right) & e^{-j\,7\pi/20}\cos\left(\frac{9\pi}{50}\right) \end{bmatrix} \qquad (9.18)$$

The MAX-MIN-DET optimum can be found in a similar way as it was found for ABBA in section 7.3.2. It provides optimal performance for infinite SNR, and is given by [86]

$$\mathbf{V} = \mathbf{I}_2, \quad \mathbf{U} = \frac{1}{\sqrt{7}} \begin{bmatrix} 1+j & 1+2j \\ -1+2j & 1-j \end{bmatrix}. \qquad (9.19)$$

Compared to the $R_s = 2\, N_t = 2$ schemes in [165, 213] with minimum product distances $3 - 2\sqrt{2} \approx 0.172$ and $1/\sqrt{10} \approx 0.316$, respectively, the scheme (9.16) with rotation (9.19) provides a minimum product distance $1/\sqrt{7} \approx 0.378$. This is calculated with the normalization that the total power transmitted during one symbol period is one. In addition, compared to the scheme in [213], where quasi-orthogonality is lost, (9.16) is 2+2 quasi-orthogonally layered, and has the ensuing simplification in detection discussed in section 6.3.

Performance of the optimized rate 2 matrix modulation for 2 Tx, 2 Rx antennas can be found in Figure 9.1. The symbols x_k are QPSK, the channel is block Rayleigh fading with block length 2, channel estimation is perfect and QOML detection (section 6.3) is used when applicable. For comparison, basic vector modulation and 16-QAM STTD are plotted. Comparing to Figure 5.4, it is clearly visible how the performance curve of the $R_s = 2$ matrix modulation has been bent to full fourth-order diversity by the diversity transform (9.19). Optimized $R_s = 2$ matrix modulation always outperforms 16-QAM STTD with ≈ 0.7 dB. This is of the same order as the second-order information loss of STTD when received with 2 Rx antennas (see Table 5.1), confirming the observations of section 4.7.

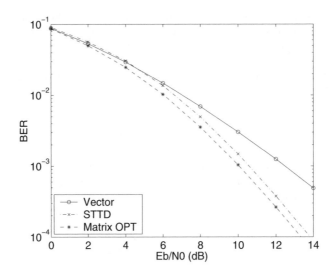

Fig. 9.1: Uncoded BER performance of more or less optimized schemes for $N_t = 2$, received with $N_r = 2$ Rx antennas, rate 4 bps/Hz.

9.3 FOUR TRANSMIT ANTENNAS, RATE 2

For transmission with $N_t = 4$ antennas, one may consider symbolwise Tx diversity one, two, three or four, with $T \geq 1, 2, 3, 4$ respectively. With judicious unitary precoding, the true diversity degree may be made equal to the symbolwise diversity degree, see Chapter 7.

Rate $R_s = 1$ schemes for $N_t = 4$ Tx antennas were thoroughly investigated in Chapters 4–8. A scheme with Tx-diversity 1 would be a scheme with antenna hopping. The optimal Tx-diversity 2 schemes were found to be STTD-OTD (4.21) and STTD-PHOP (8.25), with differences in slowly fading environments, see Chapter 8. Tx-diversity 3 was not discussed, but might be analysed along the lines of [219]. Among 4×4 schemes with symbolwise Tx-diversity 4, it was found that ABBA (4.35) was information-optimal, and could be constructed to perform well at high SNR as well. From the point of view of this chapter, it should be noted that ABBA is one example of puncturing the $R_s = 4$ scheme (9.1) to rate $R_s = 1$. It should be noted that not all puncturing schemes produce $R_s = 1$ schemes with the same mutual information.

Symbol rate 1 may not be sufficient for future evolutions of 3G systems or for systems beyond 3G. Below we discuss symbol rate two and higher enhancements of STTD-OTD and ABBA. Related high symbol rate designs, without the inherent quasi-orthogonal structure, have been shown to promise similar performance [52, 53] at high SNR.

For a transmission with symbol rate $R_s = 2$, the same alternatives exist for symbolwise transmit diversity. Tx diversity one could be provided by transmitting real modulation symbols in a vector modulation. This scheme fulfils the information extremality equations (4.89), but does not reach second-order capacity for any N_r. An alternative $R_s = 2$ transmission may be constructed by taking two threads from the vector modulation extended in time (9.8), e.g. $\tilde{x}_1, \ldots, \tilde{x}_8$.

9.3.1 Double STTD

An optimal rate 2, transmit diversity 2 scheme for $N_t = 4$ antennas can be constructed by transmitting two STTD blocks in parallel. The ensuing 2+2 layered scheme transmits $Q = 4$ symbols during $T = 2$ symbol periods with the modulation matrix

$$\mathbf{X} = \left[\; \mathbf{X}_A(x_1, x_2) \quad \mathbf{X}_B(x_3, x_4) \; \right] . \tag{9.20}$$

The underlying set of basis matrices is the rectangular Clifford basis (5.15) which fulfils the T-MSD principle discussed in section 5.4.1.

The philosophy behind this solution is analogous to that of multistream transmission, with added transmit diversity. Sufficiently complex detection is needed to attain spectral efficiency. Multiple receive antennas are required to reduce the signal correlations and enhance reception quality, as discussed in [16, 20]. Transmission of $N_t/2$ independent Alamouti blocks from $N_t/2$ pairs of different antennas over $T = 2$ symbol periods was discussed in [221]. The $N_t = 4$ antenna version with two independently coded parallel space–time codes was considered in [222] and in a 3GPP framework in [209], where the nickname DSTTD was suggested. In [223], applying different (multimodulation) symbol alphabets in the two parallel streams was discussed.

Reaching Capacity for $N_r = 1$:   The coefficient matrices of (9.20) fulfil the information extremality equations (4.89), and DSTTD *reaches second-order capacity for $N_r = 1$*. This is a remarkable fact. No constrained rate $R_s < N_t$ scheme can reach second-order capacity if $N_r > 1$, see (5.9). Also, constrained rate schemes that reach second-order capacity for $N_r = 1$ are few and far between. It is a delicate matter, what relevance such a fact has for a rate $R_s > 1$ scheme, which as argued in section 5.2.2 requires to be detected with $N_r > 1$ Rx antennas in order to keep complexity tractable. Comparing expressions (4.77) of second-order capacity and (4.82) of second-order information, quoted below,

$$\mathcal{C}_2 = -\frac{N_r(N_r + N_t)}{2N_t}$$

$$\mathcal{I}_2 = -\frac{N_r}{4T} \left(\sum_{k,l} \mathrm{Tr}\left(\mathcal{S}^{(kl)}\right)^2 + N_r \sum_{k,l} \left(\mathrm{Tr}\mathcal{S}^{(kl)}\right)^2 \right)$$

it should be noted that both have a linear and quadratic term in N_r. The linear term in N_r in the information is directly related to the total self-interference of the matrix modulation. The fact that these two quadratic functions are equal at $N_r = 1$ means that the term linear in N_r in second-information is optimal, i.e. that the scheme has *minimal self-interference*. This affects performance and information for multiple Rx antennas as well.

Optimizing Performance with Symbol Rotations: In the temporal 2×2 matrix $\mathbf{X}^\dagger \mathbf{X}$ there is no self-interference. As in the vector modulation example in section 7.1.3, there are no diversity transforms that preserve the diversity degree but change performance.

9.3.2 Double ABBA

With four-fold symbolwise transmit diversity, rate $R_s = 2$ schemes may be constructed by puncturing any of the four example schemes (9.1), (9.2), (9.6) and (9.10) above.

Considering the Hadamard and Weyl bases, two cyclic blocks, e.g. $\mathbf{X}_{W1}$ and $\mathbf{X}_{W2}$, may be punctured from the Weyl scheme (9.6), or similarly from the Hadamard scheme (9.10), or two threads may be punctured from either of the schemes. All of these methods result in the same second-order mutual information, which is worse than the one of DSTTD. With judicious unitary diversity transforms, these schemes may be constructed to have full rank, however, so at high SNR, performance would be better than for DSTTD.

A scheme with the same second-order information as DSTTD may be constructed by puncturing four quasi-orthogonal layers from (9.1). Here, different puncturing schemes result in different information. One example of a scheme with the same information as DSTTD is "double ABBA" (DABBA) which can be described as

$$
\mathbf{X}^{\text{Cliff}} = \begin{bmatrix} \mathbf{X}_A & \mathbf{X}_B \\ \mathbf{X}_B & \mathbf{X}_A \end{bmatrix} + \begin{bmatrix} \mathbf{X}_C & \mathbf{X}_D \\ -\mathbf{X}_D & -\mathbf{X}_C \end{bmatrix}, \tag{9.21}
$$

This scheme was discussed in [50]. It can be considered as double ABBA in two ways. One is the splitting above, another is in terms of a diagonal ABBA of the form (8.5), and another on the block anti-diagonal. Accordingly, the scheme inherits the correlation properties of ABBA for blocks $\mathbf{X}_A$ and $\mathbf{X}_B$ and blocks $\mathbf{X}_C$ and $\mathbf{X}_D$, whereas the correlation properties between $\mathbf{X}_A$ and $\mathbf{X}_C$ and blocks $\mathbf{X}_B$ and $\mathbf{X}_D$, are as for (8.5).

Note that BLAST [20] transmits multiple uncoded (unconstrained) symbol streams in parallel, DSTTD [209] transmits two orthogonal space–time codes in parallel, whereas Double ABBA transmits two non-orthogonal modulation matrices in parallel.

Interpreting the scheme in terms of a diagonal and off-diagonal ABBA is particularly illuminating. Using a complex diversity transform of the form (8.38), *which mixes quasi-orthogonal layers and thus changes the diversity degree*, one may interpolate between DSTTD (extended in time) and double ABBA. As this is a symmetry of information, it is no surprise that DSTTD and DABBA provide the same second-order information.

A sub-optimal diversity transformation (a local optimum) which renders the scheme full rank is given by a phase rotation $e^{j\pi/4}$ (45 degrees) acting on the symbols in $\mathbf{X}_C$, the matrix rotation $\mathbf{U}$ of (9.19) acting on $\mathbf{X}_D$ and $e^{j\pi/4}\mathbf{U}$ acting on $\mathbf{X}_B$ from the left. The minimum determinant distance is 0.021.

From the performance optimization point of view the version

$$
\mathbf{X} = \begin{bmatrix} \mathbf{X}_A + \mathbf{X}_C & \mathbf{X}_B + \mathbf{X}_D \\ j(\mathbf{X}_B - \mathbf{X}_D) & \mathbf{X}_A - \mathbf{X}_C \end{bmatrix} \tag{9.22}
$$

of (9.21), providing the same information, turns out to be better. Note the factor of j in the lower left corner. With a simple constellation rotation with the angle

$$
\phi = \arccos\left(2/\sqrt{5}\right) \tag{9.23}
$$

acting on the symbols in $\mathbf{X}_C$ and $\mathbf{X}_D$, the minimum determinant distance is 0.04.

Analogue of STTD for Fourth-order Tx Diversity: From an information point of view, these schemes may be considered as *the analogue of STTD for fourth-order transmit diversity*, in that they are the lowest rate schemes with fourth-order transmit diversity reaching capacity for one receive antenna.

Performance: Figure 9.2 compares the theoretical performance of Gray labelled QPSK and 16-QAM with two receive antennas and perfect Tx diversity (i.e. the symbol-rate 1

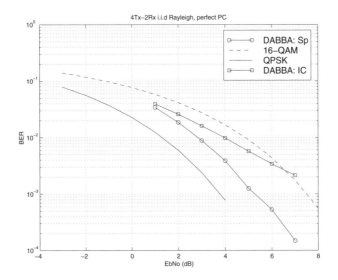

Fig. 9.2: Simulated bit-error rates of double ABBA with rotated symbols (sphere (Sp) and interference-canceling (IC) detectors) compared to single-stream matched-filter bound for 16-QAM and QPSK with four transmit antennas. Two receive antennas in a Rayleigh channel with perfect power control.

QPSK matched filter bound), with the simulated performance of double ABBA (9.21) with transformed symbols. For DABBA, a linear IC detector and the sphere detector are used. The channels are i.i.d. Rayleigh fading, and perfect power control is assumed. This means that the received power is one for all channel realizations. Then the 16-QAM bit error probability with Gray coding is given by [224]

$$P_b = \frac{3}{4}Q\left(\frac{2}{\sqrt{5}}\sqrt{E_b/N_0}\right) + \frac{1}{2}Q\left(\frac{6}{\sqrt{5}}\sqrt{E_b/N_0}\right) - \frac{1}{4}Q\left(\frac{10}{\sqrt{5}}\sqrt{E_b/N_0}\right). \tag{9.24}$$

The first term dominates when the signal to noise ratio is high. With multiple receive antennas and maximal ratio combining the signal power is N_r-fold. This is reflected in Figure 9.2, where $N_r = 2$, and the x-axis is E_b/N_0 per receive antenna. This idealistic simulation scenario highlights the inherent sub-optimality of single-stream 16-QAM modulation when compared to multistream transmission when a sufficient number of independent channels are available. Comparing to the $N_t = 2, N_r = 2$ case in Figure 9.1, the gain from applying a two-stream transmission is much increased.

Figure 9.3 compares double ABBA with double STTD at spectral efficiency 4 bps/Hz. With Maximum Likelihood decoding (ML) double ABBA follows diversity order 8 performance, with about 2 dB performance loss due to self-interference. DSTTD performance follows fourth-order diversity, with about 1 dB performance loss due to self-interference. As a reference, the performance of the LMMSE detector is depicted. The performance loss due to LMMSE detection, when compared to ML, is striking for both DABBA and DSTTD. Indeed, the slope of the BER performance curves suggests that only second-order diversity is obtained.

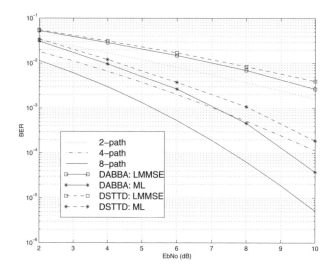

Fig. 9.3: Bit-error rates of transformed DABBA and DSTTD with four transmit antennas, two receive antennas in a Rayleigh fading channel, with 4 bps/Hz spectral efficiency. Theoretical 2, 4 and 8 order diversity curves are shown as reference. ML and LMMSE decoding.

We concentrated above on 4 bps/Hz spectral efficiency. A further data rate increase would be possible by replacing QPSK modulation e.g. with 16-QAM (8 bps/Hz). This would enable data rates beyond 20 Mbps in HSDPA. Further performance analysis can be found in [225, 226].

9.3.3 Rate 2 Scheme with Three-symbol Quasi-orthogonality

Double ABBA was based on puncturing the rate 4 Clifford modulation matrix in the form (9.1). When puncturing, the property of having a three-symbol quasi-orthogonality was lost, so the quasi-orthogonal layering is 2+2+2+2. Maximizing quasi-orthogonality is preferable for simple detection schemes. To keep this property, one should puncture the Clifford modulation matrix based on a division into three-symbol quasi-orthogonal layers, (9.2).

An example of a rate $R_s = 2$ scheme would be

$$\mathbf{X} = \mathbf{X}_M(x_1, x_2, x_3) + \mathbf{X}_N(x_4, x_5, x_6) + \mathbf{X}_R(x_7, x_8), \qquad (9.25)$$

where $\mathbf{X}_M$ and $\mathbf{X}_N$ are rate 3/4 quasi-orthogonal layers based on orthogonal designs (B.27) and (B.28) encoding three complex symbols each, and $\mathbf{X}_R$ is (B.31) with two complex symbols punctured. The layering of (9.25) is 3+3+1+1. This scheme provides less information than (9.21), and the coefficient matrices do not fulfil second-order information extremality (4.89). Detection, however, is simpler.

9.4   FOUR TRANSMIT ANTENNAS, RATE 3

With four-fold symbolwise transmit diversity, rate $R_s = 3$ schemes may be constructed by puncturing any of (9.1), (9.2), (9.6) or (9.10).

In particular, one cyclic layer or one thread may be punctured from the Hadamard (9.10) or Weyl schemes (9.6). These puncturings result in schemes with the same second-order mutual information.

Triple ABBA: A 2+2+2+2+2+2 layered scheme with two-symbol quasi-orthogonality may be constructed by puncturing two quasi-orthogonal layers from (9.1). Again, different puncturing schemes result in different information. It can be shown that with $T = N_t = 4, R_s = 3$ we are outside the region where simple intuition regarding Frobenius orthogonality, symbol homogeneity a.s.o. works. Thus the counterintuitive feature arises that it is information optimal to transmit different symbols with different power. This power offset depends on the number of receive antennas. The schemes with best mutual information found are

$$
\mathbf{X} = \begin{bmatrix} \mathbf{X}_A + \mathbf{X}_C & \mathbf{X}_B + \mathbf{X}_D \\ \mathrm{j}\,(\mathbf{X}_B - \mathbf{X}_D) & \mathbf{X}_A - \mathbf{X}_C \end{bmatrix}
$$

$$
+ \sqrt{\frac{N_r}{2N_r - 1}} \begin{bmatrix} 1 & \mathrm{j} \\ 1 & -1 \end{bmatrix} \otimes \mathbf{X}_E
$$

$$
+ \sqrt{\frac{N_r}{2N_r - 1}} \begin{bmatrix} -\mathrm{j} & -1 \\ \mathrm{j} & \mathrm{j} \end{bmatrix} \otimes \mathbf{X}_F , \tag{9.26}
$$

in terms of six STTD blocks (3.28).

Rate $R_s = 3$ Scheme with Three-symbol Quasi-orthogonality: To keep three-symbol quasi-orthogonality, one should puncture the Clifford modulation matrix (9.2) based on a division into three-symbol layers. Different puncturing schemes yield different information. A multi-stratum code constructed according to the principles of [143] would be 3+3+3+3 layered. It is information-equivalent to a scheme obtained by puncturing $\mathbf{X}_R$ from (9.2). The difference between a multi-stratum code and the punctured (9.2) would be that the latter has lower peak-to average power ratio. A 3+3+3+3 layered scheme, however, is not an information optimal three-symbol quasi-orthogonal scheme. Slightly more information is acquired with layering 3+3+3+1+1+1. An example of such a rate $R_s = 3$ scheme would be

$$
\mathbf{X} = \mathbf{X}_M(x_1, x_2, x_3) + \mathbf{X}_N(x_4, x_5, x_6) + \mathbf{X}_O(x_7, x_8, x_9) + \mathbf{X}_R(x_{10}, x_{11}, x_{12}) \tag{9.27}
$$

where $\mathbf{X}_M, \mathbf{X}_N, \mathbf{X}_O$ are rate 3/4 quasi-orthogonal layers (B.27)–(B.29) encoding three complex symbols each, and $\mathbf{X}_R$ is (B.31) with one complex symbol punctured.

9.5   FOUR TRANSMIT ANTENNAS, RATE 4

Transmit Diversity 2:   For $N_t = 4$, symbol rate 4 schemes that reach capacity for any N_r can be constructed with transmit diversity 1, 2 or 4, with $T = 1, 2$ and 4 respectively. It is an open problem whether capacity-reaching Tx diversity 3 schemes exist. With $T = 1$, one applies basic vector modulation. With $T = 2$, a capacity-reaching Tx diversity 2 scheme can be constructed by transmitting two codes of the form (9.16) in parallel,

$$\mathbf{X} = \begin{bmatrix} \mathbf{X}_A & \mathbf{X}_B \end{bmatrix} + \begin{bmatrix} 1 & 0 \\ 0 & -1 \end{bmatrix} \begin{bmatrix} \mathbf{X}_C & \mathbf{X}_D \end{bmatrix} \mathbf{U}. \tag{9.28}$$

This "2x Cliff 2" scheme is 2+2+2+2 layered quasi-orthogonal. In the 2×2 temporal self-interference matrix constructed from $\mathbf{X}^{\dagger}\mathbf{X}$, there is no interference between the layers transmitted from different antennas. Thus the two layers $\mathbf{X}_A, \mathbf{X}_C$ may be optimized separately from $\mathbf{X}_B, \mathbf{X}_D$, and the optimal diversity transform indeed operates as in (9.16) and is given by (9.19). A performance comparison of (9.28) and vector modulation may be found in Figure 9.4. Both schemes apply QPSK modulation, so the spectral efficiency is 8 bps/Hz. A performance gap opens up at around $E_b/N_0 = 5$ dB, after which the added Tx diversity becomes effective.

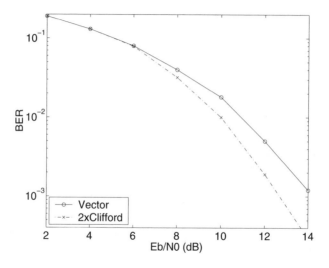

Fig. 9.4: BER performance of (9.28) and $N_r = 4$ vector modulation, received with $N_r = 4$ Rx antennas. Rate 8 bps/Hz.

Transmit Diversity 4: Rate $R_s = 4$ schemes with full Tx diversity may be constructed directly from the schemes (9.2), (9.6) and (9.10) that use the full Frobenius orthogonal unitary bases. All of these schemes reach capacity, and all may be rendered full rank with appropriate constellation rotations. Performance differences are slight. In this case, complexity arguments weight more heavily, and the choice between these schemes is almost wholly based on complexity. Thus the choice would fall on (9.2) which has three-symbol quasi-orthogonality, and accordingly the simplest detection, see Chapter 6. Even with a very high rate concatenated channel code, it is better to use the transmit diversity 2 scheme (9.28) due to the much lower complexity.

Table 9.1: The loss from second-order capacity in decibels

R_s	scheme	TxD	QO	XTRM	$N_r = 1$	$10\log_{10}(\mathcal{I}_2/\mathcal{C}_2)$ 2	3	4
1	STTD-OTD	2	2	yes	-0.79	-2.22	-3.01	-3.52
	ABBA	4	2	yes	-0.79	-2.22	-3.01	-3.52
	3-Clifford	4	3	no	-1.14	-2.43	-3.16	-3.64
	Weyl/Thread	4	1	yes	-2.04	-3.01	-3.59	-3.98
2	DSTTD	2	2	yes	0	-0.67	-1.09	-1.38
	Double ABBA	4	2	yes	0	-0.67	-1.09	-1.38
	3-Clifford	4	3	no	-0.41	-0.97	-1.33	-1.58
	Weyl/Thread	4	1	yes	-0.79	-1.25	-1.55	-1.76
3	Triple ABBA	4	2	yes	0	-0.28	-0.45	-0.57
	3-Clifford	4	3	no	-0.14	-0.35	-0.49	-0.59
	Multi-stratum	4	3	yes	-0.28	-0.46	-0.58	-0.67
	Weyl/Thread	4	1	yes	-0.28	-0.46	-0.58	-0.67
4	2 x Cliff 2	2	2	yes	0	0	0	0
	3-Clifford	4	3	yes	0	0	0	0
	Weyl/Thread	4	1	yes	0	0	0	0

9.6 THE INFORMATION PROVIDED BY THE SCHEMES

Here, the second-order information provided by the various puncturing schemes is considered. It should be noted that puncturing threads or cyclic layers from the Weyl (9.6) or Hadamard (9.10) bases yields schemes that are information extrema. These extrema, however, are likely to be minima. The best schemes constructed by puncturing (9.1) are also information extrema, but these extrema are likely to be maxima. Further, puncturing (9.2) typically does not yield information extrema, except in the symbol homogeneous 3+3+3+3 (multi-stratum) layered case. Investigating the mutual information of $R_s = 3$ schemes, the interesting observation may be made that the 3+3+3+3 layered scheme provides the same information as the punctured Weyl and Hadamard schemes, but it has quasi-orthogonality 3. In this case it is likely that there exist a symmetry of information that changes the quasi-orthogonality.

The mutual information data is collected in Table 9.1. "2x Cliff 2" indicates (9.28), and "3-Clifford" indicates the most informative schemes constructed by puncturing (9.2). For $R_s = 3$, the information of the 3+3+3+3 layered multi-stratum scheme is reported separately. The column "QO" indicates the maximal quasi-orthogonality, i.e. the number of symbols in the largest quasi-orthogonal layer. "XTRM" indicates whether the basis matrices fulfil second-order information extremality, and "TxD" is the symbolwise transmit diversity of the scheme.

From the numeric values it follows that at least for rate 3, it is likely that the minuscule differences in information between different puncturing schemes may be discarded, and the choice between schemes should be based on complexity arguments. In the full complexity analysis, the hard/software complexity of implementing the full set of schemes with adaptable rates should be considered.

9.7 SUMMARY

The design principles developed in Chapters 4, 5 and 7 can be used to construct matrix modulation schemes for MIMO systems with $N_r > 1$. In this chapter several examples of power-efficient transmission methods were constructed, where $R_s < N_t$. These solutions are able to increase both the symbol rate and the diversity order in a controlled manner. In particular, it was observed that rate-constrained schemes using quasi-orthogonal layers always provide more information than schemes based on other methods, e.g. those obtained by puncturing Weyl/Hadamard schemes. In addition, the detectors associated with quasi-orthogonal layers are less complex.

Part III

Closed-loop Methods

10

Closed-loop Methods:
Selected Multi-antenna
Extensions

Open-loop solutions, considered in previous chapters, are designed to operate without channel state information in the transmitter. In contrast, closed-loop concepts exploit channel state information that is provided to the transmitter using closed-loop signalling. The channel state information can be used to weight the signals transmitted from the BS antennas. The weighting should be such that the signals arrive co-phased in the receiver. Constructive signal combining increases the received signal power, and therefore link capacity increases.

The WCDMA system is currently the only wireless standard that contains explicit support for closed-loop transmit diversity (see Chapter 1). In the two closed-loop modes channel state information is calculated at the terminal, signalled to the transmitter, and finally used to control the beam-forming weights at different transmitting antenna elements. The feedback signal is determined using channel estimates obtained from two common pilot channels. The pilot channels are often mapped to two distinct antenna elements, but they also can be mapped to different beams. The WCDMA transmit diversity solutions explicitly support only two transmit antennas.

In this chapter we provide an overview of selected multi-antenna closed-loop concepts. We begin by summarizing the principles behind the two WCDMA feedback modes, with support for two transmit antennas. Then we discuss different ways to improve the two element concepts to provide increased robustness to feedback errors. Finally, we consider various multi-antenna extensions where more than two transmitting elements are used. We advocate a flexible solution that can cope with unreliable and low capacity feedback signalling, while still reaping a significant portion of the theoretical gains.

10.1 CLOSED–LOOP TRANSMIT DIVERSITY IN WCDMA

Transmit diversity techniques based on a feedback channel have been around for some time [89, 227, 228], but only recently have they found their way to commercial wireless systems. At first, it would seem that multi-antenna closed-loop concepts are plagued with certain inherent negative characteristics. For example, accurate control for a large number of antenna elements tends to require high feedback capacity. On the other hand, if the feedback capacity is fixed in the uplink frame, and the number of transmitting elements increases, feedback delay increases. Increased feedback delay deteriorates performance in rapid fading channels. However, these issues are not detrimental, although they in some cases limit the application to low mobility environments.

It was noticed in the 3G WCDMA standardization that even crude feedback signalling can be extremely useful in improving the downlink performance. The selective transmit diversity (STD) concept [14, 42] applied antenna or beam selection at slot frequency. Common pilot channel (CPICH) measurements were used in order to determine the transmit antenna weight that maximizes the signal power in the receiver. To support this operation, one additional feedback bit is signalled to the network at slot frequency, 15 times per 10 ms frame. The short control delay extends the applicability of closed-loop transmit diversity concepts beyond (pedestrian) low-mobility environments. The STD concept also incorporated so-called antenna verification to improve robustness to errors in feedback signalling. With antenna verification the terminal detects the actual transmit weight using received dedicated channel measurements. The purpose of antenna verification is to make sure that the terminal uses the correct channel estimate in detecting and despreading the dedicated physical channel.

Subsequent 3GPP proposals were based on co-phasing and weighting algorithms in the spirit of beam selection [1]. These are applicable when the delay profiles between each transmit antenna and the receive antenna(s) are identical within chip resolution. This requirement is typically satisfied even when the antenna elements are situated with about 10 wavelength separation from each other. In WCDMA this corresponds to about 1.4 metres. The signals transmitted via the two antennas can then be combined optimally at the transmitter to maximize received signal power. Field experiments related to different closed-loop modes have been conducted with different antenna and feedback mode configurations [229]. These results also partly validate the theoretical benefits from using simple closed-loop solutions. We summarize the relevant signal processing perspectives related to the feedback modes below.

10.1.1 Calculating the Feedback Weight

The signal model considered in this chapter is

$$\mathbf{y} = \mathbf{w}\mathbf{H}x + \text{noise}, \tag{10.1}$$

where the row vector $\mathbf{w}$ contains the beam-forming weights, $\mathbf{H}$ is the channel matrix, and x is the transmitted symbol. We assume initially that $N_t = 2$. The channel matrix is

$$\mathbf{H} = [\mathbf{h}_1, \mathbf{h}_2]^T, \tag{10.2}$$

where $\mathbf{h}_1$ is a vector with channel coefficients between transmit antenna 1 and all delay taps in the receive antenna 1. Similarly for $\mathbf{h}_2$. The transmit beam is determined at time t by coefficients $w_1[t]$ and $w_2[t]$, the complex weights for the two array elements at time t. The weight $\mathbf{w}[t] = (w_1[t], w_2[t])$ is normalized such that $||\mathbf{w}||^2 = 1$. The target is to determine the optimal linear combination of the two channel vectors, such that the (received) signal energy is maximized. Without loss of generality it can be assumed that w_1 is real. The solution, assuming two transmit antennas, is characterized by

$$w_2[t] \quad = \quad z[t]e^{j\,\phi[t]} \tag{10.3}$$

$$(z[t],\phi[t]) \quad = \quad \arg\max_{z\in\mathbb{A},\phi\in\mathbb{B}} ||(\sqrt{1-z^2}\mathbf{h}_1[t] + ze^{j\,\phi}\mathbf{h}_2[t])||^2.$$

where $\mathbb{A} = [0,1]$ and $\mathbb{B} = [0,2\pi)$. This yields

$$\phi[t] \quad = \quad \arg[\mathbf{h}_1[t]^\dagger\mathbf{h}_2[t]] \tag{10.4}$$

$$z[t] \quad = \quad \frac{||\mathbf{h}_2[t]||}{\sqrt{||\mathbf{h}_1[t]||^2 + ||\mathbf{h}_2[t]||^2}}. \tag{10.5}$$

Alternatively, the solution can be characterized using eigenvalue decomposition. Indeed, the solution is found as the dominant eigenvector of the channel correlation matrix $\mathbf{R}_{\mathrm{Tx}} = \mathbf{HH}^\dagger$. Note that with two transmit elements the solution is represented by a single complex coefficient. In the presence of multipath propagation this co-phasing/co-weighting coefficient maximizes the signal power at the output of a linear combiner. Analogously, in a MIMO channel, where a terminal has multiple receive antennas, it can operate essentially as above by treating the channel vectors in individual antenna elements as multipath components. In other words, with multiple receive antennas the feedback weight can be found as a dominant eigenvector corresponding to

$$\sum_{n=1}^{N_r} \mathbf{H}_{[n]}\mathbf{H}_{[n]}^\dagger, \tag{10.6}$$

where $\mathbf{H}_{[n]}$ is a matrix that contains the channel coefficients from N_t transmit antennas to receive antenna n. The transmitter structure for closed-loop transmit diversity is depicted in Figure 10.1. This structure supports N_t antennas. The WCDMA closed-loop modes are currently constrained for two element transmission.

10.1.2   Quantization and Feedback Signalling

As stated above, the feedback modes attempt to co-phase the transmitted signals so that they are combined coherently in the terminal. However, if the feedback word length is limited, co-phasing can be done only approximately, with finite resolution. Thus, once the desired feedback weight is calculated, it has to be quantized and converted to a feedback word. The bits within each word are multiplexed to the available signalling space in the uplink frame.

Quantization loss, signalling through a noisy feedback channel, and control delay are the primary error sources in feedback modes. These aspects need to be accounted for when defining the overall concept. The tradeoffs between different feedback signalling and quantization concepts are striking as the number transmitting elements (or the number of transmit beams) increases beyond two. These issues are addressed thoroughly in Chapter 11. As a prelude to Chapter 11, we describe below how the WCDMA concept resolves the quantization and signalling tasks in the presence of two transmit antennas.

In Mode 1 and Mode 2 closed-loop solutions, the optimal weight w_2 is quantized to different signal constellations with different quantization resolution. These constellations correspond to circular 16–QAM and QPSK, as shown in Figures 10.2 and 10.3, respectively.

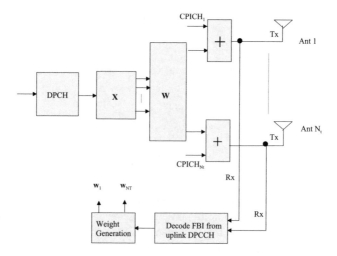

Fig. 10.1: Transmitter structure to support closed–loop transmit diversity with N_t transmit antennas.

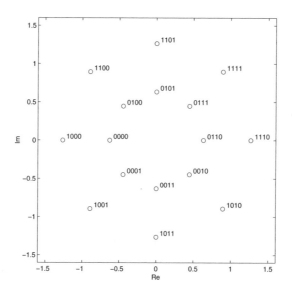

Fig. 10.2: Weight states (w_2) in Mode 2 with associated labels.

The feedback word corresponds to a Gray labelled signal state closest to w_2. The labels are transmitted to the network (base station) using the $FSMph$ field of the uplink signal. The uplink slot structure is shown in Figure 10.4. The Feedback Signalling Message (FSM) is a part of the FBI field of uplink Dedicated Physical Channel (DPCCH). Each message is of length $N_{po} + N_{ph}$ bits, and one bit is transmitted in each slot, resulting in a 1500 bps signalling overhead.

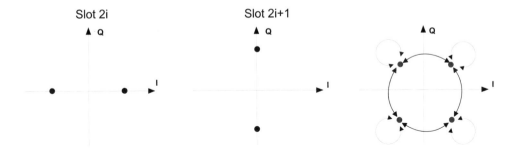

Fig. 10.3: Weight calculation at BS for Mode 1, with filtering over two slots.

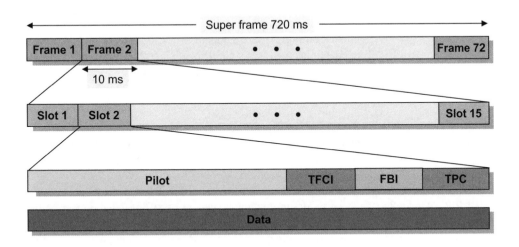

Fig. 10.4: Frame and slot structure for feedback mode transmit diversity.

10.1.2.1   *Mode 1*   In the WCDMA transmit diversity Feedback Mode 1, a terminal quantizes the real and the imaginary components of the current transmit weight w_2 in consecutive slots. Thus, each feedback bit designates either the real or the imaginary part of the feedback weight. The bits are sent in even and odd numbered slots, respectively, as described in Figure 10.3.

The BS combines two received feedback bits in order to construct a transmit weight for the diversity antenna [12, 230]. The base station interprets the bits that refer to the real and imaginary dimensions accordingly and combines them by filtering over two slots using a sliding window. More precisely, the weight for the diversity antenna is determined from

$$w_2[t] = 1/\sqrt{2}e^{j\,\phi[t]},\tag{10.7}$$

where

$$\phi[t] = \arg(j^{t \bmod 2}\mathrm{sgn}(y[t]) + j^{(t-1) \bmod 2}\mathrm{sgn}(y[t-1])),\tag{10.8}$$

where $y[t]$ denotes the feedback command received at the base station for slot t, and $w_2[t]$ is the complex weight applied in the diversity antenna for slot $t + 1$. Above, the sign function sgn is used to quantize each received feedback bit, and therefore the resulting weight constellation has four states. When a new feedback command is received, the transmit weight either remains unchanged, or it jumps to a neighbouring state, as shown in Figure 10.3.

Table 10.1: Feedback mode parameters

Mode	1	2
Phase bits per word (N_{ph})	1	3
Gain bits per word (N_{po})	0	1
Feedback bit rate	1500 bps	1500 bps
Update rate	1500 Hz	1500 Hz
Filtering at BS	yes (2 slots)	no

The gain information z determined by Equation (10.5) is not signalled to the transmitter in Mode 1. Instead, the weight amplitude in both transmit antennas is identical. This simplifies amplifier design at the base station, since the peak-to-average ratio per antenna element is minimized.

10.1.2.2   *Mode 2*   Mode 2 provides more accurate weight signalling to the BS transmitter with 16 possible weights, as shown in Figure 10.2. The transmit weight has eight phase states, which improves beam resolution at the expense of increased feedback delay when compared to Mode 1. Three feedback bits are used for phase adjustment and one for controlling the relative power between antennas 1 and 2. The relative transmit powers of antennas 1 and 2 are either $\{0.8, 0.2\}$ or $\{0.2, 0.8\}$ depending on the value of the $FSMpo$ field.

In analogy with Mode 1, feedback Mode 2 applies also time-varying quantization constellation for w_2. However, in contrast to Mode 1, the time-varying constellation follows from set-partitioning the 16-state weight constellation (see Tables 10.1.2.2 and 10.1.2.2), and the feedback word length is four bits. Each feedback bit is determined by conditioning on the other four bits. This enables slot-by-slot updating both at the terminal and at the base station [1, 13].

Table 10.2: Feedback power bits and corresponding relative transmit powers for Mode 2

FSM_{po}	Power A1	Power A2
0	0.2	0.8
1	0.8	0.2

Table 10.3: Feedback phase bits and corresponding phase differences for Mode 2

Phase	180	−135	−90	−45	0	45	90	135
FSM_{ph}	000	001	011	010	110	111	101	100

When the mobile speed is low and accurate channel estimates are available, Mode 2 is expected to be superior to Mode 1 due to its better transmit weight resolution. On the other hand, when the MS velocity is about 20 km/h or higher, Mode 1 typically outperforms Mode 2 [231, 232]. This is in part affected by performance degradation due to a longer signalling delay in Mode 2. Also, other system aspects affect the relative performance the transmit diversity concepts. Thus, in order to be able to use the best mode or concept, the network occasionally needs to switch dynamically between the two modes. Switching can be based on, for example, Doppler spread estimates at the base station.

10.1.2.3 *Weight Verification*   In the absence of feedback errors, MS could calculate the dedicated channel estimates for maximal ratio combining based on its own feedback commands transmitted to the uplink and CPICH estimates. Assume for simplicity that $N_t = 2$ and $N_r = 1$. The desired combined dedicated channel is

$$\mathbf{h}_d = \zeta \mathbf{w} \mathbf{H},$$

where $\mathbf{w}$ is a $1 \times N_t$ vector that contains the beam-forming coefficients determined by feedback and ζ is the amplitude ratio between the common channels and the power-controlled dedicated channel. In a flat fading channel the dedicated channel carrying the DPCH signal of a given user is given by

$$h_d = \zeta(z h_1 + \sqrt{1 - z^2} e^{j\phi} h_2). \tag{10.9}$$

If the MS knows z and ϕ, it can construct the dedicated channel estimate from (10.9), where h_1 and h_2 are estimated from (high-power) antenna-specific CPICHs. Note that the amplitude ratio ζ only scales the channel estimate.

However, realistic systems have feedback errors. Depending on the operation point of the system, the feedback error rate for each bit in the feedback word is in the range 4-20%.

The high error rate arises since the feedback bits are signalled uncoded to the base station, in order to reduce feedback delay and uplink signalling capacity. When a feedback error occurs, the diversity coefficient $w_2 = ze^{j\phi}$ and the associated vector $\mathbf{w}$ is different from that signalled by the terminal. This imposes performance degradation in two ways. First, the co-phasing or signal combining gain is reduced, reducing the received signal energy. Second, the common channel estimates h_1 and h_2 cannot be reliably combined to obtain a reliable estimate for the dedicated channel in (10.9). The latter problem, if not solved properly, would induce an error floor for link performance.

Mode 1 uses antenna-specific pilot patterns in DPCH. Using the dedicated pilot channel, the terminal can estimate the channel coefficient for the diversity antenna. This coefficient is the product of the antenna weight and the physical channel coefficient, $h_{2,d} = \zeta w_2 h_2$. Therefore, w_2 can be recovered by comparing the dedicated and common channel estimates. Indeed, instead of the four-hypothesis test per slot, a simplified binary weight verification procedure [2], operating separately for each feedback bit may determine the feedback phase as $\phi = \pi$ if

$$\frac{4}{\sigma^2} \operatorname{Re}\left(\zeta h_{2,d} h_2^*\right) > \ln\left(\frac{P(\phi = \pi)}{P(\phi = 0)}\right) \tag{10.10}$$

and $\phi = -\pi/2$ if

$$-\frac{4}{\sigma^2} \operatorname{Im}\left(\zeta h_{2,d} h_2^*\right) > \ln\left(\frac{P(\phi = -\pi/2)}{P(\phi = \pi/2)}\right), \tag{10.11}$$

during even and odd slots, respectively. Here $P(\phi)$ reflects the *a priori* probability for the corresponding feedback bit being received correctly at the base station and σ^2 is noise power in the feedback signalling channel in uplink. The *a priori* probability can be set such that it corresponds to the hypothesized feedback error rate. If the feedback signalling is considered reliable, the probability is set to 0 or 1. In this case the right hand side in (10.10) and (10.11) completely determines the verification result. On the other hand, if the *a priori* probability is $1/2$ for both states, the result is completely determined by the measurements in the terminal.

There are of course alternative formulations for the weight verification problem. In addition, verification procedures for Mode 2 have been proposed [233] even when Mode 2 does not utilize antenna-specific pilots in the dedicated channel. These verification techniques are deemed to be more complex than those in Mode 1. A default verification procedure for Mode 2 would require a multiple hypothesis test with 16 states. This is more error-prone than the binary test used in Mode 1. On the other hand, the longer feedback word of Mode 2 enables efficient multislot channel estimation, based solely on the dedicated channel pilot signal. At low MS speeds the detrimental effects of signalling and control delays are also less pronounced.

10.1.3   Enhancements

There are several ways to further improve robustness against feedback errors. In addition, more refined filtering techniques can improve the beam resolution, thus increasing the received signal power at terminal. Below, we summarize some recent developments.

10.1.3.1   *Improved Filtering for Mode 1*   First, we consider simple signal processing solutions that can be used in the base station without altering terminal operations. Such signal processing techniques need to be transparent to the terminal, for otherwise a standardization

effort is needed before they can be used. We focus on Mode 1, and assume that the feedback weight and signalling bits are determined as described above.

The WCDMA feedback Mode 1 relies on filtering at the BS. If two consecutive feedback bits were not filtered, each feedback bit would enable only crude beam-forming with 180 degree effective weight resolution using two possible weight states. With filtering over two slots the number of states is increased to four and only certain weight transitions are possible. In effect, Mode 1 filtering (averaging) imposes memory or trellis to the transmit weights, and the transmit weight is changed after each signalling bit. Other filtering techniques can further improve the weight resolution for Mode 1. In addition, proper filtering increases robustness to feedback errors.

One feasible solution is extend the filter length to cover more than only two feedback commands. The filter length should be dynamic in the sense that when Doppler spread is small a longer filter is used. Doppler spread can be assumed to be the same in uplink and downlink, and therefore uplink signal statistics can be used to adapt the filter length [234].

The base station can also make use of the reliabilities of the feedback signals $y[n]$. One solution is obtained by using the conditional mean estimate of the signalled weight. It is easy to see that the "Bayesian" weights (for a binary AWGN signalling channel) are given by [230]

$$w_{2,mmse}[t] = 1/N \sum_{k=t}^{t-N+1} j^{(k \bmod 2)/2} \tanh(|a[k]||z[k]/\sigma^2), \qquad (10.12)$$

where a is the amplitude of the feedback channel, and σ^2 is the noise variance at the base station. Thus, the transmit power of the diversity approaches zero when the signalling reliability decreases. In the absence of the feedback channel, the diversity antenna is automatically switched off. Various approximations to the tanh function can be used in practice.

Soft weights provided by (10.12) increase the peak-to-average ratio at the transmitter. If equal power transmission is desired, we need to quantize the weight $w_{2,mmse}$ to a desired MPSK constellation,

$$w_2[t] = \text{MPSK}(w_{2,mmse}[t]) \qquad (10.13)$$

When QPSK weight constellation is used, the verification algorithms proposed for Mode 1 are directly applicable.

Performance: Bit-error rate estimates of the closed-loop transmit diversity mode with new filtering concepts are presented in Figure 10.5. The example assumes rate $1/3$ convolutional coding defined for 8 kbps speech service in the WCDMA specification. Performance is depicted as a function of the feedback error rate in a single path Rayleigh fading channel without power control and assuming mobile velocity 10 km/h. Weight verification is not used. Instead, dedicated pilots according to Mode 2 are used for channel estimation purposes. Therefore, the comparison does not reflect the performance of WCDMA transmit diversity Mode 1 concept in every detail. On the other hand, the terminal operations including the weight quantization and feedback signalling aspects are identical to those used in Mode 1 in WCDMA regardless of the applied filter at the base station. Figure 10.5 shows that tanh filtering clearly provides the best robustness to feedback errors. This is seen especially when compared to Mode 1 at higher feedback error rates. In addition, overall performance is improved, which suggests that the longer filter length improves the weight resolution. Best performance is achieved when the Bayesian weight is projected to unit circle. However,

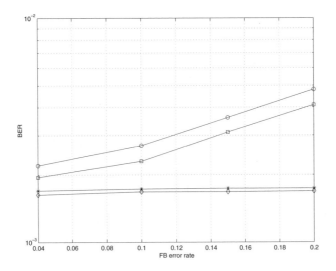

Fig. 10.5: Bit-error rates of FB mode 1 $(-\Box-)$, FB mode 2 $(- \circ -)$, tanh filter with QPSK weight constellation and $N = 4$ $(- * -)$; tanh filter output projected to complex unit circle with $N = 4$ $(- \diamond -)$. Single path channel, no power control, mobile speed is 10 km/h.

quantization to QPSK constellation according to (10.13) imposes negligible performance degradation.

10.1.3.2 *Soft-Weighted STBC*   We learned above that soft filtering at the base station provides robustness to feedback errors and improves performance. On the other hand, filters (10.12) deflate the amplitude of the diversity weight when the feedback bit is deemed unreliable. In extreme cases, when the feedback signal is completely unreliable, the diversity antenna is switched off. Clearly, this also reduces the transmit diversity order from two to one. This is undesirable from a performance point of view when operating at high signal-to-noise regions. Rather, it would be desirable that with unreliable signalling the transmit diversity concept should converge back to some full diversity open-loop concept.

The concept described below, Soft-Weighted STBC [25], satisfies these two goals. Consider the 2×2 Alamouti code, also called STTD. In SW-STTD different branches of a space–time block code are transmitted with different transmit powers. The branch of a space–time block code assigned to the desired antenna (or beam) has a larger weight than competing antenna(s) or beam(s). Soft-weighting refers to a form of soft antenna or beam selection. Assume that only one feedback bit is available per slot, and we would like to ensure that STTD is obtained as a special case of the SW-STTD concept. Assume that the terminal signals, in analogy with STD [14] the index of the desired antenna, and the base station uses this one bit to determine the relative weights for the two transmit antennas, each assigned to different columns of the STTD code. The received signal at the terminal can be written as

$$\begin{bmatrix} y_1 \\ y_2 \end{bmatrix} = \begin{bmatrix} a_1 x_1 & a_2 x_2 \\ -a_1 x_2^* & a_2 x_2^* \end{bmatrix} \begin{bmatrix} h_1 \\ h_2 \end{bmatrix} + \begin{bmatrix} n_1 \\ n_2 \end{bmatrix}, \quad (10.14)$$

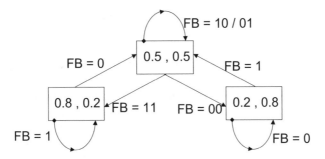

Fig. 10.6: Example of power transitions with SW-STTD.

where h_1 and h_2 refer to the complex channel coefficients between transmit antennas 1 and 2, and the receiver, respectively. Relative weighting factors a_1 and $a_2 = \sqrt{1 - a_1^2}$ are imposed on signals transmitted from antennas 1 and 2, but otherwise the signal respects the Alamouti code structure.

For optimal reception, to maintain orthogonality and to get full diversity, the receiver needs to know not only the channel coefficients but also the weights applied in the transmitter. This can be accomplished with a verification concept in the spirit of Mode 1. However, rather than devising a weight verification procedure for SW-STTD a simpler approach was proposed in [25]. In this solution the receiver assumes that $a_1 = a_2$. This is analogous to equal gain combining, and leads to a small performance degradation when compared to maximal ratio combining. However, if the relative transmit weights are changed based on two successive feedback commands according to Figure 10.6, rather than for each command independently, the robustness to feedback errors is increased. Two consecutive errors are needed to create a significant mismatch between the transmit weight and the downlink channel.

In Figure 10.6, we notice that when two consecutive feedback bits are detected as $[1\ 1]$ the transmit amplitudes are set to a_1 and $\sqrt{1 - a_1^2}$, where $a_1 > \sqrt{1 - a_1^2}$. When the successive feedback bits are different, equal power (STTD) transmission is applied. The actual values of the asymmetric weights can be fixed *a priori* to $\sqrt{0.8}$ and $\sqrt{0.2}$, which are the same amplitude weights that are applied in closed-loop Mode 2. Thus, we effectively have three different transmit weights for the STTD encoded signals. Such a concept may be suitable for future evolutions of the HSDPA or 1xEV-DV concepts since the terminal complexity is not increased. The relative performances of different concepts using 4% feedback error are shown in Figure 10.7. It is seen that SW-STTD achieves about $0.7 - 1.0$ dB performance improvement when compared to STTD, even when SW-STTD uses sub-optimal equal gain combining. Performance results given in [25] show that SW-STTD performance in not equally attractive in multipath channels, where the performance is similar to that of STTD. Mode 2 performance is very similar to Mode 1 in the current test case, and omitted from Figure 10.7.

10.2 MORE THAN TWO TRANSMIT ANTENNAS

Following the successful introduction of two-antenna closed-loop transmit diversity modes to 3GPP WCDMA, there was a considerable effort to develop closed-loop algorithms for $N_t > 2$ antennas, in particular for $N_t = 4$.

It is well-known that the optimum transmit weight vector is the eigenvector corresponding to the maximum eigenvalue of the channel autocorrelation matrix $\mathbf{R}_{Tx} = \mathbf{HH}^\dagger$, where $\mathbf{H} = [\mathbf{h}_1, \mathbf{h}_2, ..., \mathbf{h}_{N_t}]^T$, and $\mathbf{h}_m$ refers to the impulse response vector between the mth transmit antenna and the terminal, N_t is the number of transmit antennas and L refers to the number of paths in channel impulse response. In the WCDMA system, terminals can obtain $\mathbf{H}$ by measuring $\mathbf{h}_1$ and $\mathbf{h}_2$ from the common pilot channels. We assume that additional pilot channels are available when extending N_t beyond two. The correlation matrix $\mathbf{R}_{Tx}$ is based on instantaneous downlink channel measurements. Therefore it can be estimated by the terminal only (in FDD systems) and therefore some related side information must be signalled to the base station. Naturally, the signalling overhead should be minimized.

Some possible approaches to extending the transmit diversity modes are presented in Table 10.4. The first three cases in Table 10.4 are intuitive ones, while the case 4 is more general, consisting of the limited set of transmit weights and joint amplitude and phase quantization. While joint quantization schemes offer better performance than separate amplitude and phase quantization with the same amount of feedback bits, joint quantization is more difficult to analyse and more prone to errors in feedback signalling. Therefore, we will concentrate on separate quantization techniques when analysing different short-term feedback schemes in the next chapter. Case 5 uses long-term feedback based on structural properties of the channel, and effectively changes the coordinate system of the transmitted signal. Alternatively, this can be achieved without explicit feedback signalling by measuring the correlation matrix of the received signal in the base station and using frequency translation and transmit beam-forming. When there is no structure in the propagation environment, i.e. the channels are uncorrelated, long-term feedback alone does not bring in any performance improvement. To this end, long-term beam-forming can be combined with the short-term schemes 1-4 with the intention that the performance is robust with respect to different propagation environments. Case 7 provides another way to reduce feedback overhead signalling in correlated channels by utilizing the properties of the antenna array.

10.2.1 Extensions Using Fast Feedback Signalling

3GPP Contributions: Multi-antenna extensions have recently been developed and proposed in 3GPP. 3GPP contribution [235] proposed a direct extension of Mode 1 using phase-only feedback where phase differences of three antennas are independently quantized and signalled to the BS in a round-robin fashion. Filtering of consecutive feedback commands is analogous to that in Mode 1. Thus, the feedback scheme corresponds to case 2 in Table 10.4. A variation of the theme was proposed in [236] using two phase bits/antenna and progressive refinement of transmit weights, in analogy with Mode 2. Reference [237] proposed a feedback scheme similar to case 3 in Table 10.4 by selecting two or three out of four transmit antennas based on the feedback and quantizing the phase differences by two bits/antenna. References [235–237] only utilize short-term feedback and apply the same algorithm irrespective of the propagation environment. However, an alternative solution, that falls under case 6, was considered in [238]. This concept applied long-term eigenbeam-forming together

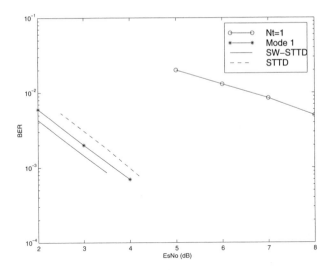

Fig. 10.7: Bit–error rates of single-antenna transmission $(-\circ-)$, Mode 1 $(-*-)$, SW-STTD
(-), and STTD (- -). Flat Rayleigh fading channel, mobile speeds 10 km/h, rate
1/3 convolutional coding.

Table 10.4: Different classes of feedback algorithms

1. Select the strongest transmit antenna:
$$\mathbf{w} \in \{\mathbf{e}_i \mid \mathbf{e}_i = (\overbrace{0, \cdots, 0}^{i-1}, 1, 0, \cdots, 0)^{\mathrm{T}}\}$$

2. Phase–only feedback:
$$\mathbf{w} \in \{\phi_i = (\{\frac{e^{j2\pi n_m/2^{N_{\mathrm{ph}}}}}{\sqrt{N_{\mathrm{t}}}}\}_{m=1}^{N_{\mathrm{t}}})^{\mathrm{T}} \mid n_m \in \{1, \cdots, 2^{N_{\mathrm{ph}}}\}\},$$
where phase differences are adjusted using N_{ph} bits.

3. A combination of 1 and 2 with independent phase adjustment and amplitude weighting.

4. Joint gain and phase adjustment:
$\mathbf{w} \in \{\mathbf{w}_1, \cdots, \mathbf{w}_K\}, \|\mathbf{w}_i\| = 1.$

5. Linear preprocessing in the transmitter, e.g. transmit beam-forming utilizing possible geometric properties of the channel.

6. Linear preprocessing in the transmitter together with feedback schemes 1–4.

7. Select the strongest beam using a parameterized array.

with short–term antenna/beam selection in case of micro and macro channel models [239]. A special case of the eigenbeam-forming concept was proposed in [240], where the transmitter consists of several spatially separated antenna arrays, where long-term feedback is applied to antenna and short-term feedback is applied to the beams of the arrays. We address some of these solutions below.

Open-loop and Closed-loop Systems— Numerical Results: Several efficient multi-antenna open-loop transmission methods were developed in Chapters 2–9. The benefit from closed-loop concepts with two transmit antennas is significant with sufficiently reliable feedback signalling. To appreciate the benefit from channel state information at the transmitter, it is necessary to compare certain efficient open-loop schemes and closed-loop schemes both using four transmit antennas. This continues the qualitative discussion between open-loop solutions and closed-loop solutions in Chapter 2.

In what follows, we provide some numerical results for a system that supports 4 bps/Hz in two different alternative transmitter–receiver architectures, each with a number of different MIMO modulation schemes. The MIMO modulation schemes considered here apply equal power transmission from all antenna elements using two parallel QPSK streams and the beam-forming matrix $\mathbf{W} = \mathbf{I}_4$. The closed-loop system uses Mode 1 or Mode 2 type feedback to control the weights of a single transmit beam $\mathbf{w}$. With closed-loop schemes a single 16-QAM stream is transmitted.

Figures 10.8 and 10.9 show the performance of several open-loop (OL) and closed-loop (CL) MIMO transmission concepts in spatially uncorrelated flat Rayleigh fading channel. All transmission methods obtain the same spectral efficiency of 4 bps/Hz. The closed-loop multi-antenna solution transmits one (uncoded) 16-QAM stream , whereas the symbol rate two space–time codes transmit two parallel (uncoded) QPSK streams.

The OL concepts shown in Figure 10.8 include single antenna transmission, STTD with one and two receive antennas, DSTTD, ABBA (7.1) with one and two receive antennas, transformed STBC-OTD (8.35) with one and two receive antennas, DABBA symbol rate two code, unconstrained vector modulation (BLAST) [21], as well as the optimized 2×2 matrix modulation (9.16). In the legends, the number of transmit and receive antennas are denoted by $(N_t \times N_r)$.

The CL concepts in Figure 10.9 loosely correspond to the closed-loop transmit diversity (Mode 1 and Mode 2) defined in 3GPP for WCDMA [235, 236], where the co-phasing information related to the dominant eigenvector of the channel correlation matrix is signalled to the transmitter. The simulations assume that two feedback bits per slot are used. This corresponds roughly to a feedback signalling overhead of 3000 bps.

As expected, the CL concepts with low mobile speed offer some gain compared to the corresponding OL concepts with the same number of transmit and receive antennas.

10.2.2 Linear Preprocessing Using Long-term (Structured) Feedback

If the channels are highly correlated, the required feedback overhead can be significantly reduced, since the same feedback information can be used to control more than one antenna element. Hence, the design of a feedback mode for correlated or structured channels is of great importance when defining future multi-antenna systems with a large number of antenna elements.

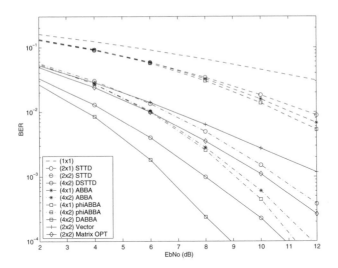

Fig. 10.8: Bit-error rates in uncorrelated Rayleigh channel for open-loop concepts using either one 16-QAM stream (dashed and dashed-dotted lines) or two QPSK streams (solid lines).

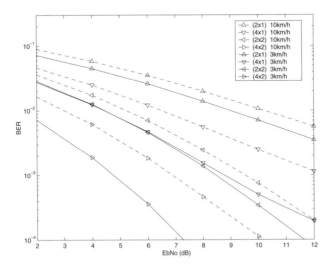

Fig. 10.9: Bit-error rates in uncorrelated Rayleigh fading channel for closed-loop concepts with one 16-QAM stream and with two different mobile speeds.

Linear preprocessing, corresponding to cases 5 and 6 in Table 10.4, can be used to reduce the required feedback capacity in spatially structured channels. The purpose of linear preprocessing is to change the coordinate system in the transmitter such that the number of dominant channel dimensions or eigenmodes is reduced. Dimension reduction via coordinate change is appropriate if the channel is spatially correlated due to the spatial geometric properties of the channel. Geometric properties arise when a small number of receive or transmit directions capture a significant portion of channel energy. In these case, linear preprocessing is similar to conventional beam-forming, with the possible exception that multiple beams may need to be used simultaneously.

When the geometric properties are invariant to frequency separation the beam-forming coefficients can be estimated either from uplink measurements, or or from downlink measurements using common pilot channels. The latter option is advocated in [89,91,94,228,241]. Below, we consider one particular example of obtaining the linear preprocessing using downlink measurements.

We assume that the terminal maintains an estimate of the averaged correlation matrix

$$\mathbf{R}_{\mathrm{Tx}}[t_0] = C \sum_{t=t_0-P}^{t_0} \mathbf{H}[t]\mathbf{H}^\dagger[t], \tag{10.15}$$

where $\mathbf{H}[t]$ constitutes the instantaneous channel matrix at slot t and C is a normalization coefficient. The integration window P should extend well over the channel coherence time in order to mitigate the effect of instantaneous fading realizations in the measurement result. In structured or correlated channels, this matrix has a small number of dominant eigenvalues. In this case, the dominant long-term beams are defined (at slot t_0) as the dominant eigenvectors of $\mathbf{R}_{\mathrm{Tx}}[t_0]$ using the eigenvalue decomposition

$$\mathbf{R}_{\mathrm{Tx}}[t_0]\mathbf{E} = \Lambda\mathbf{E}.$$

The dominant eigenvectors (or their parameters) are signalled to the base station and constitute the columns in the preprocessing matrix. The base station transmits then to these dominant long-term beams using any open-loop or closed-loop transmit diversity concept. The eigenvectors and the eigenvalues can be used to reconstruct the long-term channel correlation matrix at the base station.

When only a subset of eigenvectors/eigenvalues are signalled the BS can determine a low-rank approximation to the actual long-term channel correlation matrix. This approach bears some similarity to Karhunen–Loeve Transform (KLT)-based signal compression methods. Alternatively, one may feedback some other information information which enables the BS to reconstruct the entire correlation matrix. Alternative information may include quantized channel measurements e.g. on a slot-by-slot basis. These can be used to estimate the correlation matrix and long-term properties, even if a designated signalling channel for long-term information is missing. In the same vein, blind uplink measurements can be used, provided that the BS uses a calibrated array, and the uplink–downlink duplex separation is taken into account.

The beam-forming matrix at the base station can be set to $\mathbf{W} = \mathbf{E}$ and basically any open-loop method can be used to distribute the signals to the columns of $\mathbf{W}$, each column representing one eigenbeam. Further performance enhancement can be sought by combining short-term feedback appropriately. For example, the base station can request the index of the strongest beam, and transmit to the selected beam according to STD or case 5 in Table 10.4.

Another possibility is to utilize Mode 1 or Mode 2 co-phasing feedback in response to the effective channels

$$\mathbf{h}_{\mathbf{e}_j}[t] = \mathbf{H}^\dagger[t]\mathbf{e}_j, \ j = 1, ..., L. \tag{10.16}$$

where L long-term beams are used. When $L = 2$ these two effective channels should be used in place of $\mathbf{h}_1$ and $\mathbf{h}_2$ when calculating short-term feedback information using (10.16). Short-term feedback algorithms that are based on (10.16), adopted in Mode 1 or Mode 2, can be used to combine the "long-term" channels coherently. The transmit weight matrix reduces then to vector

$$\mathbf{w} = w_1\mathbf{e}_1 + w_2\mathbf{e}_2,$$

where we can set, without loss of generality, $w_1 = 1$ and optimize w_2 using this constraint, as in Mode 1 and Mode 2. More generally, we may write the instantaneous downlink beam as

$$\mathbf{w} = \sum_{l=1}^{L} w_l\mathbf{e}_l. \tag{10.17}$$

This is partially defined by the long-term properties of the channel $\{\mathbf{e}_l\}$, and partly by instantaneous feedback commands $\{w_l\}$.

Channel Estimation Issues: There are different ways to convey long-term channel information. The long-term correlation matrix can be estimated blindly from the signals received at the base station. Alternatively, both feedback information and blind processing can be applied. A responsible implementation should allow the transmitter to take all these into account when determining the long-term beams.

On the other hand, the UE needs to know the effective channel coefficients given by (10.16). If the basis vectors are arbitrary and unknown, the UE channel estimate can only be obtained using dedicated pilots. However, the problem can be avoided if it is possible to add overhead to the downlink. Then, the BS can signal to the UE the coefficients or parameters that characterize the long-term beams $\{\mathbf{e}_j\}$. Given that the basis vectors are known, fast feedback can be used to optimize w_2. Analogously, weight verification can follow in analogy with Mode 1.

10.2.3   Feedback Signalling and Array Parameterization

In correlated channels, the required feedback capacity can often be reduced, since the basis vectors of the channel can be defined appropriately. However, for example with eigenbeamforming the optimal basis (the set of eigenvectors) varies as the terminal position changes or the environment changes. Therefore, the eigenvectors need to be continuously updated.

Perhaps the most common array structure is the Uniform Linear Array (ULA). In a ULA the signals impinging on the array have a special structure. With $\lambda/2$ element spacing, where λ is the carrier wavelength, the received signal in different elements can be modelled with complex phasors, where the relative phase between neighbouring elements is the same, and varies depending on the Direction of Arrival (DOA) of the planewave. Therefore, in a ULA the channel coefficients modelling an impinging plane wave can be described using only one complex coefficient and the relative phase between the elements. The same parameterization can also be used when transmitting a signal towards a desired user. The problem is, however, that the desired transmit direction is difficult to estimate in multipath channels.

Below we propose a parameterized beam-forming concept that supports any number of transmit elements. The motivation stems from ULA parameterization, described above. Here the DOA estimation problem is avoided altogether. Rather, we let the terminal determine the Direction of Transmission (DOT), in response to the current channel realization in downlink. In particular, it is proposed that the terminal decides on the transmit direction or the relative phase between transmitting elements, as an extension to Mode 1 and Mode 2 closed-loop concepts.

Array Weights and the DFT Matrix: Assume that the base station has a discrete set of possible transmit weights. These may constitute the columns of an $M \times M$ "DFT" (Discrete Fourier Transform) matrix

$$
\mathbf{F} = \begin{bmatrix} 1 & 1 & 1 & \cdots & 1 \\ 1 & w & w^2 & \cdots & w^{(M-1)} \\ 1 & w^2 & w^4 & \cdots & w^{2(M-1)} \\ \vdots & \vdots & \vdots & \ddots & \vdots \\ 1 & w^{(M-1)} & w^{2(M-1)} & \cdots & w^{(M-1)(M-1)} \end{bmatrix}, \tag{10.18}
$$

where $w = e^{-\mathrm{j}\,2\pi/M}$. If $N_{\mathrm{t}} = M$, one of the M orthogonal columns is designated as the transmit weight vector, after normalization by dividing the vector with $\sqrt{M}$. If $N_{\mathrm{t}} < M$ the transmit vector is picked from the column vectors within an $N_{\mathrm{t}} \times M$ submatrix of $\mathbf{F}$, again normalized by $\sqrt{N_{\mathrm{t}}}$. More precisely, for given N_{t} and M the set of feasible beams is

$$
\left\{ \mathbf{w}_l = [1\,,w^l,\,...,\,w^{(N_{\mathrm{t}}-1)l}]^T / \sqrt{N_{\mathrm{t}}} \right\}_{l=0}^{M-1}. \tag{10.19}
$$

If the transmit elements form a Uniform Linear Array (ULA), each weight vector corresponds to a different transmit direction. This interpretation assumes that the array elements are separated by $d = \lambda/2$. If the element spacing in ULA is larger, beam pattern diverts from the classical directional beam-forming pattern, as sidelobes are created.

Feedback Signalling: In a closed-loop transmit diversity scheme the terminal selects the desired beam-forming vector and signals the corresponding index to the base station. In order to accomplish this task, we assume that the channel coefficients between each transmit antenna element and the receive antenna are estimated, using either common or dedicated pilot channels. The terminal then constructs a channel matrix $\mathbf{H} = [\mathbf{h}_1, \mathbf{h}_2, ..., \mathbf{h}_{N_{\mathrm{t}}}]^T$, where $\mathbf{h}_m$ is the impulse response between the mth array element and the terminal. The optimal DFT basis vector can be found as a solution to problem

$$
\hat{l} = \arg \max_{l \in \{0,...,(M-1)\}} \mathbf{w}_l^\dagger \mathbf{H}\mathbf{H}^\dagger \mathbf{w}_l. \tag{10.20}
$$

In the ULA parameterization the same relative phase is used between the neighbouring transmitting elements. Incidentally, this parameterization holds Mode 1 and Mode 2 phasors as a special case. The Mode 1 and Mode 2 diversity antenna weight is obtained when $M = 4$ and $M = 8$, respectively. The word length required for beam selection is $\log_2(M)$. However, in order to reduce the control delay, individual feedback bits can be transmitted sequentially to the base station, as in Mode 2. Alternatively, the relative phase can be determined directly from the phase of the feedback signal, possibly after filtering. Therefore, either Mode 1 or

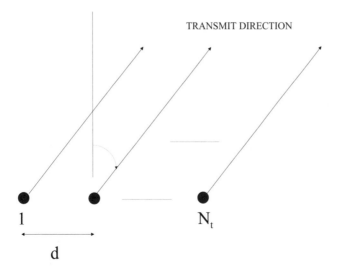

Fig. 10.10: Directional transmission using ULA with N_t antenna elements.

Mode 2 feedback can be used. Thus, in order to apply this concept in the WCDMA system only additional means (pilot signals) to estimate N_t channel impulse responses are needed. The most attractive characteristic of this scheme is the fact that the feedback signalling method need not change as the number of transmit antennas is increased.

10.3 PERFORMANCE

The dynamics of the received signal power for different closed-loop concepts are shown in Figure 10.11. It is apparent that optimal channel-matched beam-forming gives the largest received signal power. Other schemes, namely eigenbeam-forming and DOT-based solutions offer qualitatively similar receiver power.

The actual benefit from using multiple transmit antennas is ultimately evaluated from performance results when combined with a well motivated transmission chain. Below, we assume

- QPSK modulation with Gray labelling,

- 3GPP/3GPP2 turbo code, punctured to rate 3/4,

- Bit-Interleaved Coded Modulation (BICM) using a random interleaver,

- Frame size of 378 bits, and

- Two inter-element distances in a ULA with $N_t = 2, 4, 8$ or 16 elements.

The inter-element distances in ULA are $\lambda/2$ and 2λ. The MIMO channel model corresponding to case 2 in Chapter 2 is used to characterize a spatially correlated channel. Two inter-element spacings are used to construct scenarios where either the beam-forming gain or the diversity gain dominates performance. The number of antennas is varied as $N_t = 2, 4, 8$

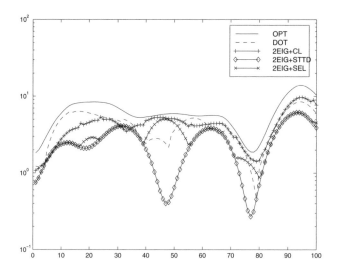

Fig. 10.11: Received powers for different closed-loop transmit diversity concepts with $N_t = 4$ and $N_r = 1$. Optimal channel-matched beam-forming (OPT) gives largest received power. Beam-space STTD (2EIG+STTD). Weighted eigenbeam combing (2EIG+CL) and Direction of Transmission feedback (DOT) offer similar performance.

or 16. However, the number of beams is in all cases fixed at $M = 16$. Therefore, the required feedback overhead is the same for all cases.

The target is to determine the potential performance improvement from using either (i) eigenbeam-forming or (ii) DOT/DFT feedback, when employing a different number of array elements in the base station. The number of receive antennas is always fixed at one. With eigenbeam-forming we choose to select two dominant eigenbeams corresponding to the long-term channel correlation matrix (10.15). The eigenbeams can either be determined via feedback from the terminal or via blind uplink measurements. Eigenbeams are represented with floating point accuracy, and the eigenbeams are combined with one of 16 DFT vectors, selected from a DFT matrix of size 2×16. With DOT feedback, the transmitter does not use the long-term channel correlation matrix or eigenbeams in any way.

Figures 10.12 and 10.13 depict the BER and FER performance of the eigenbeam-former when $N_t = 4, 8$ and 16. It is seen that the performance gain from increasing the number of transmitting elements is large when the inter-element spacing is $\lambda/2$. However, when the spacing is 2λ, performance degrades somewhat, especially when $N_t = 16$ elements are used. This phenomenon prevails largely due to the fact that only two eigenbeams are used. With $\lambda/2$ spacing, the channel energy can be captured with only a few eigenvectors. However, with 2λ spacing two eigenbeams are not sufficient, since the number of dominant eigenvalues in the long-term channel matrix increases with larger inter-element spacing. In fact, with 16-element transmission and 2λ inter-element spacing the average signal power loss (when using only two eigenbeams) is 5.3 dBs when compared to $\lambda/2$ spacing.

Figures 10.14 and 10.15 depict the BER and FER performance of the DOT array when $N_t = 2, 4, 8, 16$. It is seen that the performance gain from increasing the number transmitting

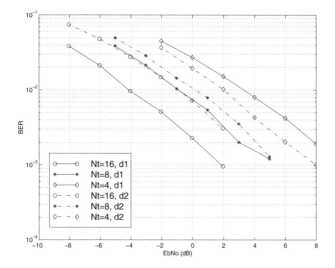

Fig. 10.12: Bit-error rate using eigenbeam-forming and closed-loop feedback. ULA with $N_t = 4, 8, 16$ antenna elements and $d1 = \lambda/2$ and $d2 = 2\lambda$ spacing. BICM using QPSK modulation and rate 3/4 turbo coding with 378 bits per frame.

elements is large regardless of inter-element spacing. As an example, $FER = 10^{-1}$ is achieved at $E_b/N_0 = -2$ dB with 16 elements, whereas with two elements the same quality of service requires $E_b/N_0 = 6.25$ dB or $E_b/N_0 = 9$ dB, with inter-element spacings of 2λ and $\lambda/2$, respectively.

In general with $\lambda/2$ inter-element spacing, the beam-forming gain due to increased received signal power is larger. On the other hand, the transmit diversity gain simultaneously reduces. This reflects the beam-forming–diversity tradeoff, which is apparent in all performance figures. When $N_t = 16$ the relative power gain due to narrow inter-element spacing is roughly 3.8 dBs.

10.4 SUMMARY

In this chapter we have given an overview of the two WCDMA transmit closed-loop modes. The support for these two modes in mandatory for all WCDMA terminals. In addition, the principles of a number of possible extensions, with support for $N_t > 2$ antenna elements, were discussed. The practical problem when designing multi-antenna extensions is to somehow limit the required feedback signalling capacity, which in a baseline solution increases linearly in N_t. Linear preprocessing or eigenbeam-forming, applied at the transmitter antenna array, can be used to reduce the number of dominant channel eigenmodes. Provided that these eigenmodes change infrequently, significant savings in feedback signalling can be achieved. Alternatively, one may use a parameterized array weight set that addresses multiple antenna elements with one or few feedback parameters. As an example, the receiver can determine the desired transmit direction using fast feedback. The numerical

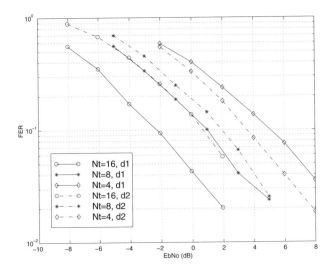

Fig. 10.13: Frame-error rate using eigenbeam-forming and closed-loop feedback. ULA with $N_t = 4, 8, 16$ antenna elements and $d1 = \lambda/2$ and $d2 = 2\lambda$ spacing. BICM using QPSK modulation and rate 3/4 turbo coding with 378 bits per frame.

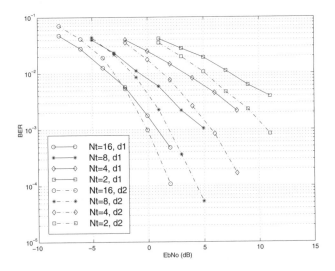

Fig. 10.14: Bit-error rate using DOT feedback. ULA with $N_t = 2, 4, 8, 16$ antenna elements and $d1 = \lambda/2$ and $d2 = 2\lambda$ inter-element spacing. BICM using QPSK modulation and rate 3/4 turbo coding with 378 bits per frame.

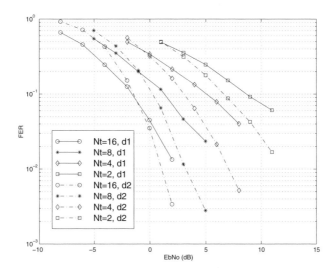

Fig. 10.15: Frame-error rate using DOT feedback. ULA with $N_t = 2, 4, 8, 16$ antenna elements and $d1 = \lambda/2$ and $d2 = 2\lambda$ spacing. BICM using QPSK modulation and rate 3/4 turbo coding with 378 bits per frame.

results provided in this chapter demonstrate that the gains achievable with feedback mode transmit diversity concepts are very significant, even with a single receiving antenna.

11

Analysis of Closed-loop Concepts

When uplink and downlink operate in different frequency bands the side information related to the downlink channel requires additional signalling, and the design of signalling formats optimizing some performance measure, e.g., received SNR in the mobile terminal while simultaneously minimizing the amount of uplink signalling makes the problem challenging. In practical cases, the feedback signal can be noisy, or due to the limited capacity of the feedback channel, only partial information of the channel state may be provided to the transmitter. Instead of vector coding in the transmitter as with the open-loop techniques, this chapter concentrates on a simpler scalar coding in MISO systems. Different design principles for the feedback algorithms presented in Table 10.4 are considered and analysed in detail utilizing explicit quantization schemes for the feedback message.

The performance of WCDMA closed-loop transmit diversity has been studied in [242, 243] without assuming quantization of the feedback. Rate distortion theory [244] has been applied to quantization in [245] assuming that channel and transmit weight vectors are jointly Gaussian-distributed. In an FDD system this would require a large number of feedback bits to quantize the channel state information. However, to minimize the load in uplink control channels and feedback latency it is necessary to use as short feedback words as possible to reach adequate performance. Therefore, we concentrate on quantization schemes where the transmit weight vector is not assumed to obey Gaussian distribution, and refer to the schemes as short-term feedback schemes. An information theoretic approach to the performance of MISO systems using two different partial channel state information strategies in the transmitter has been adopted in [92,93,246]. Mean feedback [92] assumes joint Gaussianity of channel and transmit weight vectors, which should be distinguished from the short-term feedback techniques discussed in this chapter. However, covariance feedback in [92,93,246] is similar to long-term feedback in 10.2.2.

11.1 GENERALIZED FEEDBACK SIGNALLING DESIGN

First we will present design principles for closed-loop transmit diversity using SNR gain as a performance measure. For simplicity, design principles are first presented in a single-path channel, and are further extended to multipath channels in section 11.8. Assuming the system model (2.38), the mathematical formulation for the problem of finding the best possible transmit weight $\mathbf{w}$ becomes

$$\text{Find } \mathbf{w}_* \in \mathbb{W} : \qquad |\mathbf{w}_* \, \mathbf{h}| = \max_{\mathbf{w} \in \mathbb{W}} |\mathbf{w} \, \mathbf{h}| \qquad (11.1)$$

where $\mathbb{W} = \{\mathbf{w} = [w_1, w_2, \cdots, w_{N_t}] : w_m \in \mathbb{C}\}$ and it has been assumed that $\|\mathbf{w}\| = 1$. Since the capacity of the feedback channel is limited, it is necessary to quantize $\mathbb{W}$ to reduce signalling overhead.

In the sequel we review low-complexity sub-optimal algorithms from Table 10.4 which can find a good solution for the problem (11.1). The performance of the algorithms is illustrated using the expected SNR gain

$$\gamma = \frac{\mathrm{E} \left\langle |\mathbf{w}\mathbf{h}|^2 \right\rangle}{\mathrm{E} \left\langle \|\mathbf{h}^2\| \right\rangle} N_t, \qquad (11.2)$$

where $\|\cdot\|$ is a norm in $\mathbb{C}^{N_t}$.

The first algorithm has been addressed in [90], where it was assumed that a feedback word of length $N_t - 1$ bits is provided to the base station having N_t antenna elements. Thus, it contains information about the phase differences between the reference antenna and other $N_t - 1$ antennas. A generalization utilizing N_{ph} feedback bits/antenna is given by the following algorithm.

Co-phase Algorithm: Assume that $(N_t - 1)N_{\mathrm{ph}}$ bits of side information is available. Then the transmit weights $\{w_{*,m}\}_{m=2}^{N_t}$ are chosen by using the condition

$$|h_1 + w_{*,m}h_m| = \max_{w_m \in \mathbb{W}} |h_1 + w_m h_m|, \qquad (11.3)$$

where $2 \leq m \leq N_t$ and $\mathbb{W} = \{e^{-j2\pi(n-1)/2^{N_{\mathrm{ph}}}}/\sqrt{N_t} : n = 1, 2, \ldots, 2^{N_{\mathrm{ph}}}\}$. That is, phases are independently adjusted against the phase of the first channel. The complexity of the algorithm increases linearly with additional antennas, i.e., complexity is proportional to $(N_t - 1)2^{N_{\mathrm{ph}}}$.

When $N_t = 2$ and $N_{\mathrm{ph}} = 2$ the co-phase algorithm resembles FDD WCDMA transmit diversity Mode 1. The only difference between Mode 1 and the given algorithm is that in Mode 1 the feedback word results from the interpolation between two consecutive one-bit feedback words. However, when feedback delay is not taken into account, the difference is irrelevant. The co-phase algorithm makes the elements of $\mathbf{w}$ dependent because the transmit weights are calculated with respect to a common reference, and consequently, the signals transmitted from different antennas become dependent even when the elements of $\mathbf{h}$ are independent. Applying joint phase quantization

$$\left| h_1 + \sum_{m=2}^{N_t} w_{*,m}h_m \right| = \max_{w_m \in \mathbb{W}} \left| h_1 + \sum_{m=2}^{N_t} w_m h_m \right|, \qquad (11.4)$$

would provide larger SNR gain than (11.3). However, joint phase quantization requires more complex signal processing in the receiver, and the resulting feedback word is more sensitive to feedback errors than that according to (11.3), where an error in w_m does not affect w_k, $m \neq k$. Parametrized feedback, discussed in Section 10.2.3, would calculate

$$\left| h_1 + \sum_{m=2}^{N_t} w_*^m h_m \right| = \max_{w \in \mathbb{W}} \left| h_1 + \sum_{m=2}^{N_t} w^m h_m \right| \tag{11.5}$$

assuming implicitely correlation between the transmit antennas.

A co-phase algorithm similar to the mean feedback [92] is given by

$$\left| w_{*,m} h_m \right| = \max_{w_m \in \mathbb{W}} \left| w_m h_m \right|, \ 1 \leq m \leq N_t, \tag{11.6}$$

which can be shown to give inferior performance to (11.3) in terms of SNR gain, assuming the same feedback word length. Assuming the signal model (2.38), the components of the transmit weight vector $w_m = \frac{1}{N_t} e^{j\psi_m}$ known by the transmitter are independent, where ϕ_m is uniformly distributed as $U(\arg(w_m) - \pi 2^{-N}, \arg(w_m) + \pi 2^{-N_{Ph}})$, but **h** and **w** are not jointly Gaussian.

A natural generalization of co-phasing is to weight transmission amplitudes in addition to adjusting phase differences. Intuitively, the stronger the channel, the more power should be transmitted through it. Bearing this in mind we give the following general feedback scheme.

Order and Co-phase Algorithm: Receiver ranks some or all $\{\alpha_m\}_{m=1}^{N_t} = \{|h_m|\}_{m=1}^{N_t}$ and adjusts the phase differences of the corresponding $\{h_m\}_{m=1}^{N_t}$ by applying the co-phase algorithm. Order and phase difference information is signalled to the transmitter which then chooses appropriate amplitude and phase weights from a finite quantization set.

This is a sub-optimum algorithm because amplitude and phase weights are determined separately. One of the crucial questions when applying the order and co-phase algorithm is the selection of a suitable quantizer. A uniform quantizer suffices for the quantization of phase differences, but finding a good quantizer for transmit powers is not a trivial task. If the number of transmit antennas is small then the quantization for amplitude weights can be found using simulations, but when the number of transmit antennas is large, it can become a cumbersome task. The problem is similar to the one in restricted polar quantization [247], where phase and magnitude are separately quantized. However, here we are not interested in minimizing mean square error but maximizing the received SNR.

Selection Algorithm: In this case the quantization consists of vectors $\mathbf{w}_i = [0, \cdots, \overbrace{0, 1}^{i-1}, 0, \cdots, 0], \ i = 1, \cdots N_t$, where the non-zero component in the ith position indicates the best channel in terms of received power. Hence

$$|\mathbf{w}_* \cdot \mathbf{h}| = \max\{|h_m| : 1 \leq m \leq N_t\}. \tag{11.7}$$

Since the quantization set has N_t points, $\lceil \log_2(N_t) \rceil$ feedback bits are needed.

Selection algorithm is very simple and requires a relatively low feedback capacity. However, the corresponding SNR gain is given by [60] $\gamma = \sum_{m=1}^{N_t} \frac{1}{m}$ when $E\langle |h_i|^2 \rangle = 1$ so that γ is proportional to $\log N_t$ for large N_t. We will see in section 11.5 that the selection algorithm is also sensitive to feedback errors.

Consider first an example where $N_t = 2$ and a single feedback bit is available. Let h_1 and h_2 be uncorrelated zero-mean complex Gaussian variables corresponding to the channel coefficients of the first and the second antennas. Interestingly, under these assumptions selection and the co-phasing algorithm are equivalent. Let us denote $k_{\pm} = h_1 \pm h_2$. Then it is easily seen that $\mathrm{E}\langle k_+ k_-^* \rangle = \mathrm{E}\langle |h_1|^2 \rangle - \mathrm{E}\langle |h_2|^2 \rangle = 0$. Thus, k_+ and k_- are uncorrelated zero-mean complex Gaussian random variables, and $\mathrm{E}\langle \max\{|h_1|, |h_2|\}\rangle = \mathrm{E}\left\langle \max \frac{1}{\sqrt{2}}\{|k_+|, |k_-|\}\right\rangle$.

The following sections analyse the performance of the example algorithms using SNR gain as a performance measure. The analysis can be also applied to predetection diversity in handsets [248], where the receiver has several antennas but only one receiver channel. The combining of the signals is performed in radio frequency, and therefore, phase shifters and attenuators are heavily quantized which makes the scheme analogous to quantized feedback signalling.

11.2 ANALYSIS OF SNR GAIN OF THE CO-PHASE ALGORITHM

In this section we review the performance analysis of the co-phase algorithm with different channel models presented in [38, 249] assuming zero feedback latency and perfect channel state information in the mobile station. Before adjustments, the phases of the coefficients $h_m = \alpha_m e^{j\phi_m}$, $m = 1, 2, \cdots, N_t$ are uniformly distributed on $(-\pi, \pi)$. Therefore the adjusted phases $\Psi_m = \phi_m + \arg(w_m)$ are uniformly distributed on $(-\frac{\pi}{2^{N_{\mathrm{ph}}}}, \frac{\pi}{2^{N_{\mathrm{ph}}}})$. Amplitudes α_m and phases ϕ_m are mutually independent, and assuming the signal model (2.38)

$$\mathrm{E}\langle (w_m h_m)^* w_k h_k + w_m h_m (w_k h_k)^* \rangle = 2/N_t \cdot \mathrm{E}\langle \alpha_m \alpha_k \rangle \cdot \mathrm{E}\langle \cos(\Psi_m - \Psi_k) \rangle, \quad (11.8)$$

where we can set $\Psi_1 = 0$ without losing generality. Now we can write

$$|\mathbf{w}\,\mathbf{h}|^2 = \sum_{m=1}^{N_t} |w_m|^2 \alpha_m^2 + 2 \sum_{m=2}^{N_t} \sum_{k=1}^{m-1} |w_m| \alpha_m |w_k| \alpha_k \cos(\Psi_m - \Psi_k). \quad (11.9)$$

where $\Psi_m = \phi_m + \arg(w_m)$ and we can set $\Psi_1 = 0$ without losing generality. If $m \neq k$ and no adjustments are done the expectation (11.8) is equal to zero. Furthermore,

$$\mathrm{E}\langle \cos(\Psi_m - \Psi_k) \rangle = \begin{cases} c_{N_{\mathrm{ph}}}, & k=1 \text{ or } m=1 \\ c_{N_{\mathrm{ph}}}^2, & \text{otherwise} \end{cases}, \quad m \neq k, \quad (11.10)$$

where $c_{N_{\mathrm{ph}}} = \mathrm{sinc}(2^{-N_{\mathrm{ph}}}) = \frac{2^{N_{\mathrm{ph}}}}{\pi} \sin \frac{\pi}{2^{N_{\mathrm{ph}}}}$. Since all channel parameters are equally distributed we can reject the dependence from indices and denote $a = \mathrm{E}\langle \alpha_m^2 \rangle$ and $b = \mathrm{E}\langle \alpha_m \alpha_k \rangle$, $m \neq k$. Then, combining (11.9) and (11.8) and setting $|w_m| = \frac{1}{\sqrt{N_t}}$ we obtain

$$\mathrm{E}\langle |\mathbf{w}\mathbf{h}|^2 \rangle = a + (N_t - 1)\left(1 + \frac{N_t - 2}{2} c_N\right)\frac{2c_N}{N_t} \cdot b. \quad (11.11)$$

The SNR gain is given by

$$\gamma = 1 + (N_t - 1)\left(1 + \frac{N_t - 2}{2} c_{N_{\mathrm{ph}}}\right)\frac{2c_{N_{\mathrm{ph}}}}{N_t} \cdot \frac{b}{a}. \quad (11.12)$$

Thus it remains to recall the value of b/a for Rayleigh, Nakagami and Rice distributions (2.1) and (2.2). In the first two cases

$$\frac{b}{a} = \left(\frac{\Gamma(\kappa + \frac{1}{2})}{\sqrt{\kappa}\,\Gamma(\kappa)}\right)^2. \quad (11.13)$$

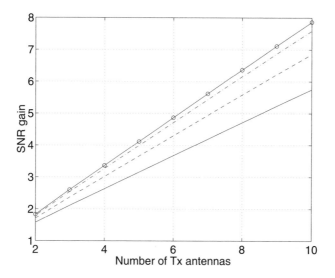

Fig. 11.1: SNR gains (linear scale) as a function of the number of transmit antennas using the co-phase algorithm when $N_{\mathrm{ph}} = 2$, Nakagami fading is assumed and $\kappa = \frac{1}{2}$ (—), $\kappa = 1$ (- -), $\kappa = 2$ (- ·), $\kappa = 3$ (-o-).

Figure 11.1 displays γ when $N_{\mathrm{ph}} = 2$ for different values of the fading figure κ in (2.2), where $\kappa = 1$ corresponds to Rayleigh fading. It is noticed that γ increases with κ, which results from the fact that for larger κ the probability of a deep fade is lower than in the case of small κ (as shown in Figure 2.1). Thus, the loss due to transmitting with equal power from all the transmit antennas becomes smaller with increasing κ.

When α follows Rice distribution

$$\mathrm{E}\langle \alpha \rangle = \sqrt{\frac{\pi}{2}} \sigma e^{-\frac{1}{2}K}\left((1+K)I_0(\tfrac{1}{2}K) + KI_1(\tfrac{1}{2}K)\right), \qquad (11.14)$$

where $I_0(\cdot)$ and $I_1(\cdot)$ are modified Bessel functions of order zero and one respectively. This equation can be deduced using integration by parts and properties of Bessel functions. Hence, in the case of Ricean fading

$$\frac{b}{a} = \frac{\pi}{2}\frac{e^{-K}\left((1+K)I_0(\tfrac{1}{2}K) + KI_1(\tfrac{1}{2}K)\right)^2}{2(1+K)}. \qquad (11.15)$$

The extreme case of a static channel is obtained when $K \to \infty$. Using the asymptotic formula

$$I_l(z) = \frac{e^z}{\sqrt{2\pi z}}\left(1 + O_l(\tfrac{1}{z})\right), \qquad |\arg z| < \frac{\pi}{2}, \; l = 0, 1 \qquad (11.16)$$

(see [250], (9.7.1)) we find that

$$\lim_{K \to \infty}\frac{b}{a} = \lim_{K \to \infty}\frac{(1+2K)^2}{4K(1+K)} = 1. \qquad (11.17)$$

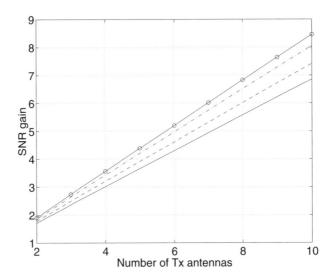

Fig. 11.2: SNR gains (linear scale) as a function of the number of transmit antennas using the co-phase algorithm when $N_{\mathrm{ph}} = 2$, Ricean fading is assumed and $K = -\infty$ dB (—), $K = 3$ dB (- -), $K = 9$ dB (- ·), $K = \infty$ dB (-o-).

When $N_{\mathrm{t}} = 2$ we have $\gamma = 1 + c_{N_{\mathrm{ph}}}$. Hence, when $N_{\mathrm{ph}} \to \infty$ (phase difference of the two channels is perfectly known in the transmitter) SNR gain attains the theoretical upper limit.

Figure 11.2 displays SNR gain γ for different values of K. It is noticed that γ increases with K, which is again due to the fact that the channel becomes more static (see Figure 2.2) and the probability of a deep fade is small when K is large. The power ν^2 in (2.3) of the fixed-path component is equal for all channels since it is assumed that all antennas are located in the same antenna mast.

If the signal has a strong static component co-phasing can be done in the base station based on uplink measurements. If mobile feedback is neglected, the achievable SNR gain is given by

$$\gamma = 1 + \frac{(N_{\mathrm{t}} - 1)K}{1 + K}, \tag{11.18}$$

where it has been assumed that the phase differences between the static signal components are known. Now the SNR gain is purely due to downlink beam-forming and calibration of the antenna array is required.

>From (11.9) we find that the absolute upper bound (denoted by γ_*) for the SNR gain γ is achieved when all phases Ψ_m are identical. Thus, in the case of Rayleigh fading channel

$$\gamma \le \sum_{m=1}^{N_{\mathrm{t}}} \mathrm{E}\left\langle |\alpha_m|^2 \right\rangle + 2 \sum_{m=2}^{N_{\mathrm{t}}} \sum_{k=1}^{m-1} \mathrm{E}\left\langle \alpha_m \right\rangle \mathrm{E}\left\langle \alpha_k \right\rangle$$

$$= 2N_{\mathrm{t}}\sigma^2 + \pi\sigma^2 \frac{N_{\mathrm{t}}(N_{\mathrm{t}} - 1)}{2} = 1 + \frac{\pi}{4}(N_{\mathrm{t}} - 1) =: \gamma_*, \tag{11.19}$$

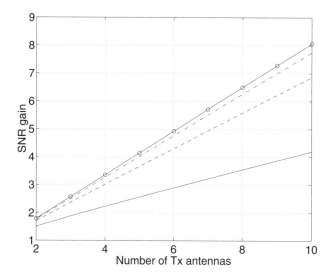

Fig. 11.3: SNR gains (linear scale) of the co-phase algorithm as a function of the number of transmit antennas, $N_{\rm ph} = 1$ (—), $N_{\rm ph} = 2$ (- -) $N_{\rm ph} = 3$ (- ·), upper bound γ_*, (-o-).

which is the SNR gain of equal gain combining. This bound cannot be reached in practice in FDD systems because it requires unquantized weight vector **w** to be communicated to the base station.

It can easily be seen that $\gamma \to \gamma_*$ when $N_{\rm ph} \to \infty$ in (11.12). Thus, the SNR gain γ corresponding to the co-phase algorithm is consistent with the absolute upper bound γ_*. This is also illustrated in Figure 11.3 which shows theoretical SNR gains in a flat Rayleigh fading environment as a function of the number of the transmit antennas for values $N_{\rm ph} = 1, 2, 3$ with upper bound γ_*. The performance is very near to the optimal, unquantized case when the number of feedback bits per antenna is three. Figure 11.4 shows SNR gains of the co-phase algorithm as a function of required feedback bits. One bit/phase/antenna provides the best performance, but on the other hand the number of transmit antennas is three times larger than that with $N_{\rm ph} = 3$. Comparing Figures 11.3 and 11.4 suggests that $N_{\rm ph} = 2$ provides the best overall performance in terms of number of antennas and feedback bits.

Similar analysis has been also conducted in [251], where the lower bound of SNR gain for the co-phase algorithm has been obtained in flat Rayleigh fading channels for different $N_{\rm ph}$. The lower bounds follow from (11.12) by setting $c_{N_{\rm ph}} \mapsto |\Psi_1 - \Psi_k| = \frac{\pi}{2^{N_{\rm ph}}}, N_{\rm ph} = 1, 2, \cdots$.

11.3 ANALYSIS OF SNR GAIN OF THE ORDER AND CO-PHASE ALGORITHM

Let us now consider the general case of $N_{\rm t}$ transmit antennas, single path channels and the order and co-phase algorithm. Since the number of different permutations of $\{\alpha_m\}_{m=1}^{N_{\rm t}}$ is

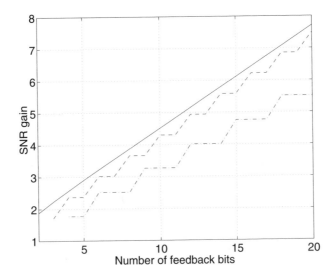

Fig. 11.4: SNR gains (linear scale) of the co-phase algorithm as a function of the number of the feedback bits, $N_{\mathrm{ph}} = 1$ (—), $N_{\mathrm{ph}} = 2$ (- -), $N_{\mathrm{ph}} = 3$ (- ·).

$N_{\mathrm{t}}!$, the number of feedback bits becomes $\lceil \log_2(N_{\mathrm{t}}! \cdot 2^{(N_{\mathrm{t}}-1)N_{\mathrm{ph}}}) \rceil$, where N_{ph} is the number of phase bits/antenna. In a similar manner as in the co-phase algorithm we obtain [38]

$$\gamma = \sum_{m=1}^{N_{\mathrm{t}}} u_m^2 \mathrm{E}\left\langle \alpha_{(m)}^2 \right\rangle + 2 \sum_{m=2}^{N_{\mathrm{t}}} \sum_{k=1}^{m-1} u_m u_k \mathrm{E}\left\langle \alpha_{(m)}\,\alpha_{(k)} \right\rangle \mathrm{E}\left\langle \cos(\Psi_m - \Psi_k) \right\rangle = \mathbf{u}^T \mathbf{C} \mathbf{u},$$

(11.20)

where $\mathbf{u} = (u_1, u_2, \cdots, u_{N_{\mathrm{t}}})^T$, $u_m = |w_m|$ and the instantaneous amplitudes of the channels have been ordered as $\alpha_{(1)} \geq \cdots \geq \alpha_{(N_{\mathrm{t}})}$. The matrix $\mathbf{C}$ is given by

$$\mathbf{C} = \begin{pmatrix} \sigma_1^2 & c_{N_{\mathrm{ph}}}\sigma_{12} & c_{N_{\mathrm{ph}}}\sigma_{13} & \cdots & c_{N_{\mathrm{ph}}}\sigma_{1,N_{\mathrm{t}}} \\ c_{N_{\mathrm{ph}}}\sigma_{12} & \sigma_2^2 & c_{N_{\mathrm{ph}}}^2\sigma_{23} & \cdots & c_{N_{\mathrm{ph}}}^2\sigma_{2,N_{\mathrm{t}}} \\ & & \ddots & & \\ c_{N_{\mathrm{ph}}}\sigma_{1,N_{\mathrm{t}}} & c_{N_{\mathrm{ph}}}^2\sigma_{2,N_{\mathrm{t}}} & \cdots & c_{N_{\mathrm{ph}}}^2\sigma_{N_{\mathrm{t}}-1,N_{\mathrm{t}}} & \sigma_{N_{\mathrm{t}}}^2 \end{pmatrix}$$

(11.21)

where the elements of $\mathbf{C}$ are defined as

$$\begin{aligned} c_{m,m} &= \sigma_m^2 = \mathrm{E}\left\langle \alpha_{(m)}^2 \right\rangle \\ c_{m,1} &= c_{N_{\mathrm{ph}}} \mathrm{E}\left\langle \alpha_{(m)}\alpha_{(1)} \right\rangle = c_{1,m} \\ c_{m,k} &= c_{N_{\mathrm{ph}}}^2 \mathrm{E}\left\langle \alpha_{(m)}\alpha_{(k)} \right\rangle = c_{k,m},\ m \neq k,\ k \neq 1. \end{aligned}$$

(11.22)

Instead of the spatial domain, the correlation matrix $\mathbf{C}$ is defined in the order domain, where the element $c_{1,1}$ refers to the second moment of the strongest antenna instead of the second moment of the antenna in the first position in the antenna array. The largest eigenvalue of the matrix $\mathbf{C}$ is equal to γ and the amplitude weight vector $\mathbf{u}_*$ maximizing the expected SNR gain is the corresponding eigenvector. This is a general design principle which is not restricted to uncorrelated single-path Rayleigh fading channels, as further discussed in section 11.8. Furthermore, it is immediately seen that if co-phasing is not applied so that $c_{N_{\mathrm{ph}}} = c_0 = 0$ the SNR gain is maximized by the selection algorithm, which only transmits from the antenna with the largest received SNR. The order and co-phase algorithm combines short-term and long-term feedback where the latter information is contained in amplitude weights $\mathbf{u}$ corresponding to the short-term order information received from the feedback channel. Furthermore, the feedback schemes can operate on top of any coordinate system (case 6 in Table 10.4 on page 221) so that they can be combined with correlation/covariance/eigenbeam-forming/long-term feedback which transmits the signal to the directions of the eigenvectors corresponding to the largest eigenvalues [252] as discussed in section 10.2.2.

Instead of using $\alpha_{(1)}$ as a reference for co-phasing we may use a fixed reference antenna for phase differences. In the case of i.i.d. channels all feedback messages are equally probable so that

$$
\begin{aligned}
c_{m,m} &= \mathrm{E}\left\langle \alpha_{(m)}^2 \right\rangle \\
c_{m,k} &= \frac{1}{N_{\mathrm{t}}}(c_{N_{\mathrm{ph}}} + (N_{\mathrm{t}} - 1)c_{N_{\mathrm{ph}}}^2)\mathrm{E}\left\langle \alpha_{(m)}\alpha_{(k)} \right\rangle = c_{k,m}, \ m \neq k.
\end{aligned}
\tag{11.23}
$$

This leads to a slightly smaller SNR gain but in practical feedback scenarios the fixed reference antenna has the advantage of making the signalling of order and phase difference independent, which gives improved performance when the feedback channel is subject to errors. If the phase reference varies according to the strongest channel and there is an error in signalling the order information, all phase adjustment commands within the current feedback word are erroneously decoded in the base station.

In case of spatial correlations between transmit antennas $N_{\mathrm{t}}!$ orderings are not all equally probable. Furthermore, $\mathrm{E}\left\langle h_{i,(j)}^* h_{k,(l)} \right\rangle$ depends on the corresponding indices in the order and spatial domain, where $h_{i,(j)}$ refers to the channel coefficient of the ith antenna in the spatial domain, whose amplitude has rank j in the order domain. Instead of using a single $\mathbf{C}$ we can define $N_{\mathrm{t}}!$ correlation matrices $\mathbf{C}(\mathbf{p}_i)$, $i = 1, \cdots, N_{\mathrm{t}}!$ where a permutation $\mathbf{p}_i$ defines a vector $(h_{p_{i,1},(1)}, \cdots, h_{p_{i,N_{\mathrm{t}}},(N_{\mathrm{t}})})^{\mathrm{T}}$. Thus, for each permutation there is a different $\mathbf{C}(\mathbf{p}_i)$ and consequently a different transmit weight vector maximizing the corresponding SNR gain which depends on the order and spatial domains. However, it turns out that the additional SNR gain is marginal and does not justify additional complexity when compared to using a single $\mathbf{C}$ in case of macro and micro spatial channel models [239].

The most difficult problem when calculating γ is to find moments of order statistics $\mathrm{E}\left\langle \alpha_{(m)}^2 \right\rangle$ and $\mathrm{E}\left\langle \alpha_{(m)}\alpha_{(k)} \right\rangle$. Fortunately, in the case of single-path Rayleigh fading channels this problem has been solved in [253] resulting in

$$
\begin{aligned}
\mathrm{E}\left\langle \alpha_{(k)}^2 \right\rangle &= 2\sigma^2 \sum_{i=0}^{k-1} K_k(i), \ K_k(i) = \frac{(-1)^i k \binom{N_{\mathrm{t}}}{k}\binom{k-1}{i}}{(N_{\mathrm{t}} + i - k + 1)^2} \\
\mathrm{E}\left\langle \alpha_{(k)}\alpha_{(l)} \right\rangle &= \sigma^2 \sum_{i=0}^{l-1}\sum_{j=0}^{k-l-1} K_{k,l}(i,j)\Psi_{k,l}(i,j), \ l < k,
\end{aligned}
\tag{11.24}
$$

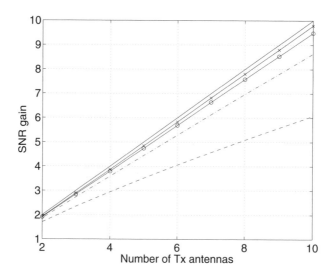

Fig. 11.5: SNR gains (linear scale) of the order and co-phase algorithm as a function of the number of transmit antennas when the number of feedback bits is $\lceil \log_2(N_t! \cdot 2^{(N_t-1)N_{\mathrm{ph}}}) \rceil$, $N_{\mathrm{ph}} = 1$ (- -), $N_{\mathrm{ph}} = 2$ (- ·) and $N_{\mathrm{ph}} = 3$ (–o–), $N_{\mathrm{ph}} = \infty$ (-x-), where N_{ph} refers to the number of phase bits/antenna. Solid line corresponds to the ideal case where the transmitter has complete knowledge of the channel.

where coefficients $K_{k,l}(i,j)$ are of the form

$$K_{k,l}(i,j) = \frac{(-1)^{i+j} N_t! \binom{l-1}{i} \binom{k-l-1}{j}}{(l-1)!(k-l-1)!(N_t-k)!} \tag{11.25}$$

and $\Psi_{k,l}(i,j) = \psi(k-l+i-j, N_t-k+j+1)$, where $\psi(\cdot,\cdot)$ is defined by

$$\psi(t,u) = (tu)^{-\frac{3}{2}} \left(\arcsin\left(\sqrt{\frac{t}{t+u}}\right) - \frac{1}{4}\sin\left(4 \cdot \arcsin\left(\sqrt{\frac{t}{t+u}}\right)\right)\right). \tag{11.26}$$

It is worth noticing that in [253] $\psi(\cdot,\cdot)$ is expressed in terms of Beta function since the main results are given for Weibull density function (2.4). In case of Rayleigh distribution it is easy to see that $\psi(\cdot,\cdot)$ has also the simpler expression (11.26) [38].

Figure 11.5 displays SNR gains when applying the order and co-phase algorithm and $\lceil \log_2(N_t! \cdot 2^{(N_t-1)N_{\mathrm{ph}}}) \rceil$ feedback bits. It is noted that the SNR performance of the algorithm is very near to optimal when there are three phase adjustment bits/antenna. On the other hand, the performance is seriously degraded when $N_{\mathrm{ph}} = 1$ showing the importance of co-phasing. The performance of the order and co-phase algorithm with complete phase information shows that quantization based on order information is sufficient, and there is no reason to increase the number of bits in gain quantization. Instead, we should think about, how to decrease the amount of quantization bits assigned to order information.

In cases of Nakagami fading with the probability density function (2.2) it is possible to find an upper bound for γ when $N_t > 2$ and κ is the integer by utilizing the fact that r^2

in (2.2) obeys Gamma distribution. When $E\langle\alpha^2\rangle = E\langle r^2\rangle = 1$, the diagonal entries in C become [254]

$$c_{1,1} = \frac{N_t}{\Gamma(\kappa)}^{(\kappa-1)(N_t-1)} \sum_{n=0} a_n(\kappa, N_t - 1)\frac{\Gamma(\kappa + n + 1)}{N_t^{\kappa+n+1}}$$

$$c_{k,k} = \frac{N_t!}{(k-1)!(N_t-k)!} \sum_{p=0}^{k-1}(-1)^p \binom{k-1}{p} \times$$

$$\frac{c_{1,1}|N_t \rightarrow N_t - k + p + 1}{N_t - k + p + 1},$$

where $a_n(\kappa, p)$ is the coefficient of t^n in the expansion of $(\sum_{i=0}^{\kappa-1} t^i/i!)^p$ satisfying

$$a_n(\kappa, 1) = \frac{1}{n!}$$

$$a_n(\kappa, p) = \sum_{i=0}^{\kappa-1}\frac{1}{i!}a_{n-i}(\kappa, p - 1)$$

Next we utilize the relation $\sqrt{E\langle\alpha_{(k)}^2\alpha_{(l)}^2\rangle} \geq E\langle\alpha_{(k)}\alpha_{(k)}\rangle$, where

$$E\langle\alpha_{(k)}\alpha_{(l)}\rangle = \frac{C}{\Gamma(\kappa)^2} \sum_{a,b,p,t,i}(-1)^{a+b}\binom{k-1}{a}\binom{l-k-1}{b} \times$$

$$a_p(\kappa, q - 1)a_t(\kappa, s - 1)\frac{\Gamma(t + \kappa + 1)\Gamma(p + \kappa + i + 1)}{s^{t+\kappa-i+1}i!(s + q)^{p+\kappa+i+1}}$$

where $C = N_t!/((k-1)!(l-k-1)!(N_t - l)!)$, $q = a + l - k - b$, $s = N_t - l + b + 1$, $l > k$, and the summation is carried over $0 \leq a \leq l-1, 0 \leq b \leq l-k-1, 0 \leq p \leq (\kappa-1)(q-1)$, $0 \leq t \leq (\kappa-1)(s-1), 0 \leq i \leq t+\kappa$. Now $c_{k,l} = c_{N_{ph}}^{\min\{2,k,l\}}\sqrt{E\langle\alpha_{(k)}\alpha_{(l)}\rangle}, k \neq l$ so that the maximum eigenvalue of C provides an upper bound of γ. This upper bound becomes tighter with increasing κ, because the range $\alpha_{(1)} - \alpha_{(N_t)}$ decreases.

Figures 11.6 and 11.7 show analytical upper bounds for SNR gains and simulated SNR gains in a flat Nakagami fading environment with $\kappa = 1, 2, 3$ when $N_{ph} = 1$ and $N_{ph} = 2$, respectively. (The case $\kappa = 1$ corresponds to Rayleigh fading, and therefore it is possible to calculate the SNR gain exactly.) It is noticed that the order and co-phase algorithm is robust with respect to the fading figure κ, and the upper bound for γ is reasonably accurate when compared with the simulated values. Interestingly, with $N_{ph} = 1$ the SNR gain decreases slightly when κ is increasing. Comparing Figures 11.1 (co-phase algorithm with $N_{ph} = 2$) and 11.7 (order and co-phase algorithm with $N_{ph} = 2$) shows that κ has a smaller effect on the latter algorithm.

11.3.1 Reducing the Amount of Feedback Signalling

As noticed before, the capacity of the feedback channel is limited in practical FDD systems, and therefore it is important to design schemes that provide good SNR gains with a small number of feedback bits. Short feedback words are also necessary from the feedback latency

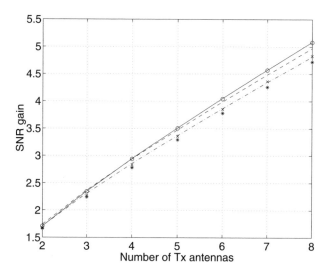

Fig. 11.6: SNR gains (linear scale) as a function of the number of transmit antennas using the order and co-phase algorithm with $N_{\mathrm{ph}} = 1$ in flat Nakagami fading channels: analytical upper bounds, $\kappa = 1$ (—), $\kappa = 2$ (- -), $\kappa = 3$ (-·), simulated $\kappa = 1$ (o), $\kappa = 2$ (x), $\kappa = 3$ (⋆).

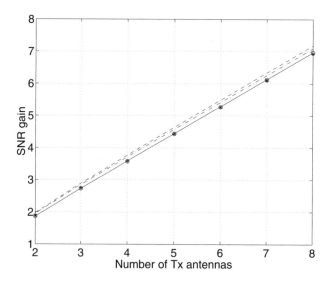

Fig. 11.7: SNR gains (linear scale) as a function of the number of transmit antennas using the order and co-phase algorithm with $N_{\mathrm{ph}} = 2$ in flat Nakagami fading channels: analytical upper bounds, $\kappa = 1$ (—), $\kappa = 2$ (- -), $\kappa = 3$ (-·), simulated $\kappa = 1$ (o), $\kappa = 2$ (x), $\kappa = 3$ (⋆).

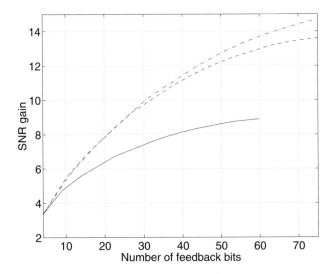

Fig. 11.8: SNR gains (linear scale) as a function of the number of feedback bits of the order and co-phase algorithm when K out of $N_t = 16$ antennas are chosen, and the number of feedback bits is $\lceil \log_2(N_t!/(N_t - K)! \cdot 2^{(K-1)N_{ph}}) \rceil$, where $N = 1$ (–), $N = 2$ (- -), $N = 3$, (- ·).

point of view, in the sense that any delay in signalling the feedback word should not exceed the channel coherence time.

Without assuming any spatial structure of the underlying channel, there are two possible ways to decrease the number of feedback bits in the short-term feedback schemes analysed in sections 11.2 and 11.3. First, one may dedicate fewer bits to phase adjustments or second, one may reduce the number of bits dedicated to order information. Figure 11.8 shows that γ deteriorates rapidly when the number of phase bits decreases. Therefore we concentrate on reducing the number of feedback bits corresponding to order information, which translates into reducing the dimension of $\mathbf{C}$ in (11.21). In the following we present two examples of such reduced feedback schemes.

In the first scheme we assume that K out of N_t antennas are selected so that $\mathbf{w} = [w_1, \cdots, w_K, 0, \cdots, 0]$. Now the corresponding amplitude weight vector $\mathbf{u}_*$ is the eigenvector corresponding to the largest eigenvalue of $\mathbf{C}_K$, where $\mathbf{C}_K$ is the $K \times K$ principal submatrix of $\mathbf{C}$. This is similar to the scheme in [237] for four transmit antennas, where two or three strongest antennas are selected and their relative phases are adjusted. However, [237] applies equal amplitude weight to the selected antennas. Figure 11.8 shows the SNR gains for different numbers of feedback bits in flat Rayleigh fading environment when applying the order and co-phase algorithm and selecting K out of $N_t = 16$ antennas so that the number of feedback bits becomes $\lceil \log_2(N_t!/(N_t - K)! \cdot 2^{(K-1)N_{ph}}) \rceil$. The value of K in Figure 11.8 then depends on the number of feedback bits available. It is noticed that when the number of phase adjustment bits is large, the SNR gain is near optimum. Comparing the curves $N_{ph} = 2$ and $N_{ph} = 3$ shows that when the number of feedback bits is low, it is more efficient to increase the number of order bits than the number of phase bits when N_t is fixed.

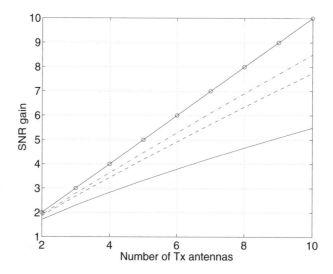

Fig. 11.9: SNR gains (linear scale) as a function of the number of transmit antennas when applying the order and co-phase algorithm and $\lceil \log_2(N_t \cdot 2^{(N_t-1)N_{ph}}) \rceil$ feedback bits, $N_{ph} = 1$ (–), $N_{ph} = 2$ (- -) and $N_{ph} = 3$ (- ·), ideal case with perfect channel state information in the transmitter (-o-).

In the second example, only the index of the strongest antenna is signalled to the transmitter so that $\mathbf{u} = [u_1, \overbrace{u_2, \cdots, u_2}^{N_t-1}]^T$ and phase differences with respect to the strongest antenna are adjusted with N_{ph} bits/antenna so that $\mathbf{C}$ becomes a 2×2 matrix denoted by $\tilde{\mathbf{C}}_2$. The entries of $\tilde{\mathbf{C}}_2$ are given by

$$\tilde{c}_{1,1} = c_{1,1}, \; \tilde{c}_{1,2} = \frac{1}{\sqrt{N_t - 1}} \sum_{m=2}^{N_t} c_{1,m} = \tilde{c}_{2,1},$$

$$\tilde{c}_{2,2} = \frac{1}{N_t - 1} \Big(\sum_{m=2}^{N_t} c_{m,m} + 2 \sum_{m=3}^{N_t} \sum_{k=2}^{m-1} c_{m,k} \Big), \tag{11.27}$$

so that $\lceil \log_2(N_t \cdot 2^{(N_t-1)N_{ph}}) \rceil$ feedback bits are required. When N_t is large the number of feedback bits is far lower than in the scheme ordering N_t antennas. Figure 11.9 depicts SNR gains when the last scheme is applied. Comparing Figures 11.5 and 11.9, we find that the degradation in SNR gain is small.

Finally, we emphasize that one may design several different schemes based on the order and co-phase algorithm. One more option is to vary the number of phase bits according to the rank of the corresponding antenna so that stronger antennas are assigned more phase quantization bits than weaker ones. The idea is similar to unrestricted polar quantization [255], where the phase step size depends on the quantized magnitude. In each case the amplitude weight vector $\mathbf{w}$ can be obtained as the eigenvector corresponding to the largest eigenvalue of suitably modified $\mathbf{C}$.

Another way to reduce the feedback rate is to exploit the temporal correlation of the feedback messages. For example, suppose that the base station maintains the list of the order information signalled by MS. Some options for updating the ordering and the corresponding amplitude weights are presented in Table 11.1

Table 11.1: Updating alternatives for order feedback

1.	MS signals the index of the strongest antenna to BS, and BS swaps the new and old indices.
2.	MS signals the index of the strongest antenna to BS, and BS updates antenna ordering by moving the previous index to the second highest position.
3.	MS alternatively signals the indices of the strongest and the weakest antenna and BS moves the previous index to the second strongest/weakest position.
4.	Generalization of the signalling schemes 1–3, where MS signals the index of $\|\mathbf{h}_{(i)}\|$ in round-robin fashion.

When $N_t = 4$ the last alternative in Table 11.1 effectively results in full ordering of the transmit antennas using two bits to signal the order information at the time while full order information requires five bits. The resulting SNR gain of different update strategies will be further examined in section 11.9 when studying the performance of feedback algorithms in spatially correlated transmit antennas.

There are many ways to reduce the amount of order feedback. However, order and co-phase signalling sets a strict upper limit of $\lceil \log_2 N_t \rceil$ bits of order information. If we want to increase the number of feedback bits allocated to transmit weights then, instead of signalling the order, it is necessary to quantize amplitudes or differences of amplitudes of the channels. However, order signalling together with co-phasing results in SNR gains which are very close to optimal ones, as seen in Figure 11.5.

11.3.2 Analysis of Two Transmit Antennas

It turns out that in the case of two transmit antennas it is possible to deduce analytical results for the SNR gain for single-path Nakagami and Ricean fading channels, while the SNR gain for Nakagami fading for $N_t > 2$ on section 11.3 provides an upper bound for γ. This stems from the fact that the calculation of cross correlations of order statistics can be avoided because $E \langle \alpha_{(1)} \alpha_{(2)} \rangle = E \langle \alpha_1 \alpha_2 \rangle$ when $N_t = 2$. Suppose that u_1 and u_2 are real and positive amplitude weights such that $u_1^2 + u_2^2 = 1$. Then we have

$$\gamma = u_1^2 E \left\langle \alpha_{(1)}^2 \right\rangle + u_2^2 E \left\langle \alpha_{(2)}^2 \right\rangle + 2c_{N_{\mathrm{ph}}} u_1 u_2 E \left\langle \alpha_{(1)} \alpha_{(2)} \right\rangle, \tag{11.28}$$

where $c_{N_{\mathrm{ph}}} = \mathrm{sinc}(2^{-N_{\mathrm{ph}}})$. Since the order of the antennas is signalled to the base station we may state that $\alpha_{(1)}^2 = \max\{\alpha_1^2, \alpha_2^2\}$ and $\alpha_{(2)}^2 = \min\{\alpha_1^2, \alpha_2^2\}$.

Let us deduce analytic expressions for elements of $\mathbf{C}$ when $N_t = 2$. Since

$$E \left\langle \|\mathbf{h}_{(\cdot)}\|^2 \right\rangle = E \left\langle \|\mathbf{h}\|^2 \right\rangle = N_t \cdot E \left\langle \alpha_1^2 \right\rangle \tag{11.29}$$

there holds

$$
\begin{aligned}
c_{m,m} &= \frac{\mathrm{E}\left\langle \alpha_{(m)}^2 \right\rangle}{\mathrm{E}\left\langle \alpha_1^2 \right\rangle} \\
c_{2,1} &= \frac{c_{N_{\mathrm{ph}}} \mathrm{E}\left\langle \alpha_1 \alpha_2 \right\rangle}{\mathrm{E}\left\langle \alpha_1^2 \right\rangle} = c_{N_{\mathrm{ph}}}\frac{b}{a},
\end{aligned}
\tag{11.30}
$$

where b/a is given by (11.13) and (11.15) with Nakagami and Ricean fading, respectively. Moreover, there holds $\mathrm{E}\left\langle \alpha_{(2)}^2 \right\rangle = \mathrm{E}\left\langle \|\mathbf{h}\|^2 \right\rangle - \mathrm{E}\left\langle \alpha_{(1)}^2 \right\rangle$, and when α_1 and α_2 follow Nakagami distribution [249]

$$
\mathrm{E}\left\langle \alpha_{(1)}^2 \right\rangle = \frac{\Gamma(2\kappa)}{2^{2(\kappa-1)}\Gamma(\kappa)}\left(\frac{\sqrt{\pi}}{\Gamma(\kappa+\frac{1}{2})} + \frac{1}{\Gamma(\kappa+1)}\right)\sigma^2.
\tag{11.31}
$$

The SNR gains for order and co-phase algorithm in the case of Nakagami fading with $\kappa = \frac{1}{2}, 1, 2, 3$ are listed in Table 11.2 when $N_{\mathrm{ph}} = 3$, which corresponds to closed-loop Mode 2. It is also noticed that the algorithm is robust against the change of the fading figure κ.

Table 11.2: SNR gains (linear scale) of the order and co-phase algorithm when $N_{\mathrm{ph}} = 3$ in a Nakagami fading environment

κ	1/2	1	2	3
γ	1.890	1.914	1.939	1.950

In the case of Ricean fading the expectation for the power of the weaker channel can be expressed using the first-order Marcum Q-function given by

$$
Q_1(a,b) = \int_b^\infty x e^{-\frac{x^2+a^2}{2}} I_0(ax)dx,
\tag{11.32}
$$

where $I_0(\cdot)$ refers to the modified Bessel function of order zero. After normalising $\mathrm{E}\left\langle \alpha_k^2 \right\rangle = \nu^2 + 2\sigma^2 = 1$, there holds

$$
c_{2,2} = \int_0^\infty Q_1\left(\sqrt{2K}, \sqrt{2(1+K)r}\right)^2 dr,
\tag{11.33}
$$

where the Ricean factor $K = \nu^2/2\sigma^2$. A closed-form solution for this integral is not available, and therefore we evaluate it numerically. The SNR gains corresponding to the order and co-phase algorithm assuming Ricean fading are given in Table 11.3 for $N_{\mathrm{ph}} = 3$.

Table 11.3: SNR gains (linear scale) of the order and co-phase algorithm when $N_{\mathrm{ph}} = 3$ in a Ricean fading environment

K (dB)	$-\infty$	3	9	∞
γ	1.914	1.929	1.957	1.974

In single-path Rayleigh fading channels [38]

$$
\mathbf{C} = \begin{pmatrix} \pi c_{N_{\mathrm{ph}}}^{\frac{3}{2}} & \frac{\pi c_{N_{\mathrm{ph}}}}{4} \\ \frac{\pi c_{N_{\mathrm{ph}}}}{4} & \frac{1}{2} \end{pmatrix}
\tag{11.34}
$$

so that

$$u_{1,2}^2 = \frac{1}{2}\left(1 \pm \frac{1}{\sqrt{1 + \left(\frac{\pi c_{N_{ph}}}{2}\right)^2}}\right). \tag{11.35}$$

Setting $N_{ph} = 3$ is analogous to the FDD WCDMA transmit diversity Mode 2 and $\mathbf{u}_{*,1} = [\sqrt{0.7735}, \sqrt{0.2265}]^T$, while the 3GPP WCDMA standard specifies $\mathbf{u}_{*,2} = [\sqrt{0.8}, \sqrt{0.2}]^T$. The corresponding SNR gains become $\gamma_1 = 1.9142$ and $\gamma_2 = 1.9123$, respectively. Thus, although the amplitude weights for Mode 2 were originally obtained by Monte Carlo simulations they are close enough to the optimal ones maximizing the received SNR in case of independent flat Rayleigh fading channels. However, it should be noted that the transmit weights maximizing the received SNR gain change as a function of the propagation environment and the number of feedback bits assigned to co-phasing. Furthermore, when fast power control is applied to the dedicated channel, the channel statistics and optimal transmit weights change as well.

11.4 SNR GAIN IN MULTIPATH RAYLEIGH FADING CHANNELS

So far the analysis of SNR gain has considered flat fading environments. Here we consider the performance of the co-phase algorithm and the order and co-phase algorithm in multipath Rayleigh fading channels when $N_t = 2$. With a proper scaling of SNR values the results also apply to flat Rayleigh fading channels and several receive antennas when $N_t = 2$.

In case of multipath channels, the cost function of the co-phase algorithm (11.3) becomes

$$\|\mathbf{h}_1 + w_{*,m}\mathbf{h}_m\| = \max_{w_m \in \mathbb{W}} \|\mathbf{h}_1 + w_m\mathbf{h}_m\|, \tag{11.36}$$

where $2 \leq m \leq N_t$ and $\mathbb{W} = \{e^{-j\pi(n-1)/2^{N_{ph}-1}}/\sqrt{N_t} : n = 1, 2, \cdots, 2^{N_{ph}}\}$. That is, each $\mathbf{h}_m$ is independently adjusted against $\mathbf{h}_1$.

When $N_t = 2$, $w_1 = 1/\sqrt{2}$ and $w_2 = e^{j\psi}/\sqrt{2}$, where $\psi \in \mathbb{W}$. We obtain

$$\|\mathbf{Hw}\|^2 = \frac{1}{2}\left(\|\mathbf{h}_1\|^2 + \|\mathbf{h}_2\|^2\right) + |\mathbf{h}_1^\dagger\mathbf{h}_2|\cos(\phi - \psi), \tag{11.37}$$

where $\phi = \arg(\mathbf{h}_1^\dagger\mathbf{h}_2)$ is uniformly distributed on $(-\pi, \pi)$. After the phase adjustments the angle $\phi - \psi$ is uniformly distributed on $\left(-\frac{\pi}{2^{N_{ph}}}, \frac{\pi}{2^{N_{ph}}}\right)$ and $E\langle\cos(\phi - \psi)\rangle = c_{N_{ph}} = \frac{2^{N_{ph}}}{\pi}\sin\frac{\pi}{2^{N_{ph}}}$. Employing the previous two equations results in

$$E\langle\|\mathbf{Hw}\|^2\rangle = 2\sum_{l=1}^{L}\sigma_l^2 + c_{N_{ph}}E\langle|\mathbf{h}_1^\dagger\mathbf{h}_2|\rangle. \tag{11.38}$$

The SNR gain γ is now given by

$$\gamma = 1 + \frac{c_{N_{ph}}E\langle|\mathbf{h}_1^\dagger\mathbf{h}_2|\rangle}{2\sum_{l=1}^{L}\sigma_l^2}, \tag{11.39}$$

where $E\langle\|\mathbf{h}_m\|^2\rangle = 2\sum_{l=1}^{L}\sigma_l^2 = 2\|\boldsymbol{\sigma}\|^2$, which results in [256]

$$\gamma = 1 + \frac{\pi c_{N_{ph}}}{4\sum_{l=1}^{L}\sigma_l^2}\sum_{l=1}^{L}\sigma_l^2\prod_{k=1, k\neq l}^{L}\frac{\sigma_l^4}{\sigma_l^4 - \sigma_k^4}, \tag{11.40}$$

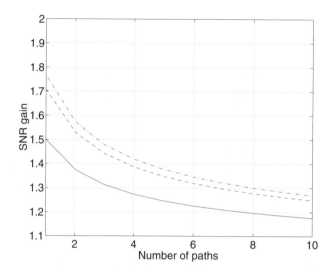

Fig. 11.10: SNR gains (linear scale) of the co-phase algorithm as a function of the number of paths when the path gains are equal, $N_t = 2$, and $N_{ph} = 1$ (—), $N_{ph} = 2$ (- -) and $N_{ph} = 3$ (- ·).

assuming that $\sigma_l \neq \sigma_k$ if $l \neq k$.

When the expected powers of all channel taps are equal

$$\gamma = 1 + \frac{c_{N_{ph}}}{2L!}\Gamma(\frac{1}{2})\Gamma(L + \frac{1}{2}), \qquad (11.41)$$

where $\Gamma(\cdot)$ is the gamma function.

Figure 11.10 shows the SNR gain γ when the co-phase algorithm is employed in a multipath channel with equal σ_l for all $l = 1, 2, \cdots, 10$ when $N_t = 2$. As expected, γ is inversely proportional to the number of paths, because given one spatial complex coefficient w_2 it is not possible to optimally adjust the phase differences of different propagation paths in the time domain.

In multipath channels, instead of $\{|h_m|\}_{m=1}^{N_t}$ a generic order and co-phase receiver ranks some or all $\{\|\mathbf{h}_m\|\}_{m=1}^{N_t}$ and adjusts the phase differences of the corresponding $\{\mathbf{h}_m\}_{m=1}^{N_t}$ according to (11.36). Consider next the quantization points for amplitude weights in the order and co-phase algorithm. Let $\mathbf{h}_{(1)}$ and $\mathbf{h}_{(2)}$ be such that

$$\begin{aligned} \|\mathbf{h}_{(1)}\| &= \max\{\|\mathbf{h}_1\|, \|\mathbf{h}_2\|\}, \\ \|\mathbf{h}_{(2)}\| &= \min\{\|\mathbf{h}_1\|, \|\mathbf{h}_2\|\}. \end{aligned} \qquad (11.42)$$

Moreover, let the transmit weights be u_1 and $u_2 e^{j\psi}$, where u_1, u_2 are real and non-negative, $\|\mathbf{u}\| = 1$ and $\psi \in \{\pi(n - 1)/2^{N_{ph}-1} : 1 \leq n \leq 2^{N_{ph}}\}$. After phase adjustment, the SNR gain for a weight vector $\mathbf{u}$ can be written in the form

$$\gamma(\mathbf{u}) = \frac{E\left\langle \|u_1 \mathbf{h}_{(1)} + u_2 e^{j\psi}\, \mathbf{h}_{(2)}\|^2 \right\rangle}{2\|\boldsymbol{\sigma}\|^2} = \mathbf{u}^T \mathbf{C} \mathbf{u}, \qquad (11.43)$$

where components of matrix $\mathbf{C} = (c_{i,j})^2_{i,j=1}$ are given by

$$c_{m,m} = \frac{\mathrm{E}\left\langle \|\mathbf{h}_{(m)}\|^2 \right\rangle}{2\|\sigma\|^2}$$

$$c_{1,2} = c_{2,1} = \frac{c_{N_{\mathrm{ph}}} \mathrm{E}\left\langle |\mathbf{h}^{\dagger}_{(1)} \mathbf{h}_{(2)}| \right\rangle}{2\|\sigma\|^2}. \tag{11.44}$$

Since the goal is to maximize $\gamma(\mathbf{u})$ we set $\mathbf{u} = \mathbf{u}_*$, where $\mathbf{u}_*$ is the eigenvector corresponding to the largest eigenvalue $\lambda_{\max}$ of matrix $\mathbf{C}$, and the SNR gain is then equal to $\lambda_{\max}$. It remains to find expressions for elements of $\mathbf{C}$. From (11.40) and (11.41) the off-diagonal elements of $\mathbf{C}$ become

$$c_{2,1} = c_{1,2} = \frac{\pi c_{N_{\mathrm{ph}}}}{4\|\sigma\|^2} \sum_{l=1}^{L} \sigma_l^2 \prod_{k=1, k\neq l}^{L} \frac{\sigma_l^4}{\sigma_l^4 - \sigma_k^4}, \tag{11.45}$$

when $\sigma_k \neq \sigma_l$ whenever $k \neq l$, and in case of $\sigma_l = 1$ for all $l = 1, 2, \cdots, L$

$$c_{2,1} = c_{1,2} = \frac{c_{N_{\mathrm{ph}}}}{2L!} \Gamma(\frac{1}{2}) \Gamma(L + \frac{1}{2}). \tag{11.46}$$

Variables $\|\mathbf{h}_m\|^2$, $m = 1, 2$ follow the central chi-square distribution (whether the expected powers of separate paths are equal or not). We simplify the notation by denoting

$$\pi_l = \prod_{k=1, k\neq l}^{L} \frac{\sigma_l^2}{\sigma_l^2 - \sigma_k^2}. \tag{11.47}$$

If $\sigma_k \neq \sigma_l$ whenever $k \neq l$ then we find that

$$c_{1,1} = \frac{2}{\|\sigma\|^2} \sum_{l=1}^{L} \sum_{k=1}^{L} \pi_l \pi_k \sigma_l^2 \left(1 - \frac{\sigma_k^2}{(\sigma_k^2 + \sigma_l^2)^2}\right),$$

$$c_{2,2} = \frac{2}{\|\sigma\|^2} \sum_{l=1}^{L} \pi_l \sigma_l^2 - c_{1,1}. \tag{11.48}$$

On the other hand, if $\sigma_l = 1$ for all $l = 1, 2, \cdots, L$ then there holds

$$c_{1,1} = 2 - c_{2,2}$$

$$c_{2,2} = \sum_{l=1}^{L-1} \frac{1}{2^{L+l}} \binom{L+l}{l} \tag{11.49}$$

Figure 11.11 depicts the SNR gain γ of the order and co-phase algorithm in a multipath channel when all σ_ls are equal. Gains are larger than in Figure 11.10 but the behaviour as a function of propagation paths is similar. Furthermore, Figure 11.12 depicts SNR gains when applying co-phase algorithms and order and co-phase algorithms to cases 1 (pedestrian) and 3 (vehicular) in Table 2.1 on page 25 with the exception that feedback latency due to mobile speed is ignored.

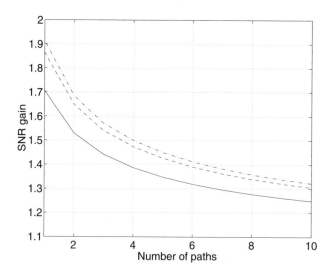

Fig. 11.11: SNR gains (linear scale) of the order and co-phase algorithm as a function of the number of paths when the path gains are equal, $N_t = 2$, and $N_{ph} = 1$ (—), $N_{ph} = 2$ (- -) and $N_{ph} = 3$ (- ·).

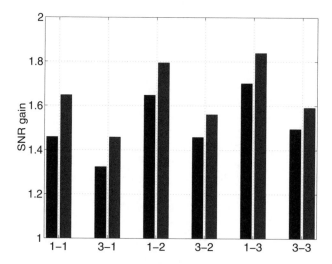

Fig. 11.12: SNR gains (linear scale) for 3GPP test channel profiles 1 and 3 from Table 2.1 (neglecting feedback latency), where $1 - N_{ph}$ and $3 - N_{ph}$ refer to 3GPP multipath test channels 1 and 3 with N_{ph}-bit phase adjustment and with (rightmost bars in the pairs) or without (leftmost bars) gain adjustment when $N_t = 2$.

11.5 ERRORS IN FEEDBACK SIGNALLING

Multipath fading decreases the SNR gains of closed-loop algorithms when compared to single-path channels. In addition, there are other sources of performance degradation like feedback bit errors and delay when signalling the feedback from MS to BS. In this section we study the effect of feedback errors to the received BEP using selection algorithm and co-phase algorithms as examples. For that purpose we first recall the well known BEP of the Selection algorithm for BPSK modulation (see, for example [257]),

$$P_{sa}(0) = \frac{1}{2}\left(1 - \sum_{m=1}^{N_t} \binom{N_t}{m}(-1)^{m-1}\sqrt{\frac{\gamma}{m+\gamma}}\right), \qquad (11.50)$$

where $\gamma = E_b/N_0$ is the SNR per bit, and argument 0 emphasizes that the given BEP is valid when no feedback bit errors occur in antenna selection. It is known that the selection algorithm provides full diversity, i.e., the slope of the asymptotic BEP curve in a logarithmic scale is N_t with N_t transmit antennas. Intuitively the co-phase algorithm should also give full N_t–fold diversity, which it indeed does, shown in [258] and further discussed in section 11.7. However, in the following we show that this is not true in the case of feedback bit errors.

We assume that the feedback bit error probability is constant and bit errors are uniformly distributed in time. The model can be considered to be approximately valid in FDD WCDMA since fast power control is applied to the uplink control channel carrying the feedback information. The assumption does not hold with high mobile speeds when the delay of the feedback loop exceeds the coherence time of the channel. However, the assumption is well justified within low mobility environments.

Consider next the effect of feedback bit errors to the BEP performance when $N_t = 2$. Let $P_{sa}(p)$ be the BEP of a two-antenna selection algorithm when the probability of a feedback bit error is p. Then $P_{sa}(p) = (1-p)P_{sa}(0) + p \cdot P_{sa}(1)$, where $P_{sa}(0)$ and $P_{sa}(1)$ refer to error-free and erroneously received feedback bits, respectively. Under the assumption that the feedback bits are equally probable there holds $P_s = \frac{1}{2}P_{sa}(0) + \frac{1}{2}P_{sa}(1)$, where P_s is the BEP of the single antenna transmission depending on the received SNR. The two-antenna system performance is reduced to that of a single transmit antenna if each feedback bit is randomly selected. Combining the two equations gives

$$P_{sa}(p) = (1 - 2p)P_{sa}(0) + 2p \cdot P_s. \qquad (11.51)$$

Thus, it is seen that the asymptotic diversity of the co-phase algorithm is only one when $p > 0$.

Figure 11.13 displays the BEP curve of two-antenna selection/co-phasing with $N_{ph} = 1$. It is seen that the term $2p$ in (11.51) defines the asymptotic difference between the single antenna and the two-antenna bit error probabilities. In practice, the loss of asymptotic diversity is not critical in the WCDMA feedback modes since the effect of feedback bit errors is not serious at low SNR values where WCDMA is usually operating. However, when studying extensions of the closed-loop modes to more than two antennas, this phenomenon should be taken into account.

Let us next consider a more general feedback scheme with quantization $\mathbb{W}_K$ such that all K feedback words are equally likely. Moreover, we assume that the optimal feedback word

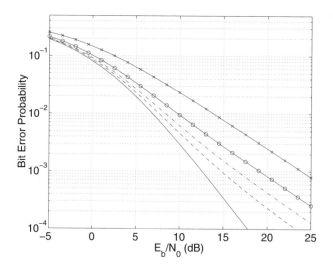

Fig. 11.13: Bit error probability curves as a function of SNR for two-antenna selection/co-phasing with $N_{\mathrm{ph}} = 1$ when $p = 0.00 (-), p = 0.04 (- -), p = 0.08 (- \cdot), p = 0.16$ (-o-), $p = 0.50$ (-x-).

$\mathbf{w}_*$ is arbitrary but fixed. The single antenna BEP is given by

$$P_s = \frac{1}{K} \sum_{\mathbf{w} \in \mathbb{W}_K} P_g(\mathbf{w}), \tag{11.52}$$

where $P_g(\mathbf{w})$ is the BEP of the general algorithm provided that the feedback word $\mathbf{w}$ is used. Hence, irrespective of the optimal feedback word, the performance of the multi-antenna system is equal to the performance of a single antenna when the feedback word is randomly selected. Let us denote by P_g the BEP of the general feedback algorithm in the presence of feedback errors and, finally, let $q_{\mathbf{w}}$ be the probability of a feedback word $\mathbf{w}$ on the condition that $\hat{\mathbf{w}}$ is the optimal (selected) feedback word. Then there holds [259]

$$\begin{aligned} P_g &= \sum_{\mathbf{w} \in \mathbb{W}_K} q_{\mathbf{w}} P_g(\mathbf{w}) \\ &= \sum_{\mathbf{w} \in \mathbb{W}_K} \left(q_{\mathbf{w}} - \min_{\mathbf{w} \in \mathbb{W}_K} \{q_{\mathbf{w}}\} \right) P_g(\mathbf{w}) + K \cdot \min_{\mathbf{w} \in \mathbb{W}_K} \{q_{\mathbf{w}}\} P_s. \end{aligned} \tag{11.53}$$

Since the difference in the sum term is always non-negative, we find that P_g admits the lower bound

$$P_g \geq K \cdot \min_{\mathbf{w} \in \mathbb{W}_K} \{q_{\mathbf{w}}\} P_s. \tag{11.54}$$

This bound shows that the asymptotic BEP of the general feedback method is always reduced to one in the presence of feedback errors. The assumption that all feedback words occur with equal frequency in the long run is not too restrictive and can easily be removed. The same conclusion has also been reached in [245] based on Chernoff bound of pairwise codeword error probabilities.

Consider the lower bounds of BEP of selection algorithm and co-phase algorithm. There holds [259]

$$P_{sa}(p) \geq \frac{N_t}{N_t - 1}\left(1 - (1 - p)^{\log_2(N_t)}\right)P_s = C_{sa}P_s$$

$$P_{cp}(p) \geq (2p)^{N_{ph}(N_t - 1)}P_s = C_{cp}P_s.$$

The bounds are equal when $N_t = 2$ and $N_{ph} = 1$ as expected, because the algorithms provide the same expected SNR gain as noticed on page 236. Comparing the bounds, e.g., when $p = 0.04$, $N_t = 8$ and $N_{ph} = 1$ shows $C_{sa} = 0.13$ while $C_{cp} = 2 \cdot 10^{-8}$. This difference is only a hint concerning the asymptotic performance of the two algorithms, because the coefficients refer to the lower bounds of BEP instead of exact expression, and they therefore do not necessarily give the difference in asymptotic curves as was the case in (11.51) when $N_t = 2$.

In selection algorithm $q_{\mathbf{w}_*} = (1 - p)^{\log_2(N_t)}$ while for other weights there holds $q_{\mathbf{w}} = 1/(N_t - 1)(1 - (1 - p)^{\log_2(N_t)})$. Hence

$$q_{\mathbf{w}} - \min_{\mathbf{w} \in \mathbb{W}_K} \{q_{\mathbf{w}}\} = \begin{cases} q_{\hat{\mathbf{w}}} - \min_{\mathbf{w} \in \mathbb{W}_K} \{q_{\mathbf{w}}\}, & \mathbf{w} = \mathbf{w}_* \\ 0, & \mathbf{w} \neq \mathbf{w}_*. \end{cases} \qquad (11.55)$$

Thus, the BEP of the selection algorithm becomes

$$P_{sa}(p) = \left(1 - \frac{q}{N_t - 1}\right)P_{sa}(0) + \frac{q}{N_t - 1}P_s, \qquad (11.56)$$

where $q = 1 - (1 - p)^{\log_2(N_t)}$ is the probability of a feedback word error.

Figure 11.14 displays the BEP of the selection algorithm with different feedback bit error probabilities when $N_t = 8$. It is seen that at the feedback bit error level $p = 0.04$, which is the nominal BEP used in simulations in 3GPP standardization contributions, the effect of the errors is significant, and the slope is similar to that of a single transmit antenna ($p = 0.50$).

Let us study the effect of feedback word labelling. When different feedback adjustment options are randomly labelled, a single feedback bit error makes the selection of the feedback word random, and there holds

$$P_g = \left(q_{\mathbf{w}_*} - \min_{\mathbf{w} \in \mathbb{W}_K} \{q_{\mathbf{w}}\}\right)P_g(\mathbf{w}_*) + K \cdot \min_{\mathbf{w} \in \mathbb{W}_K} \{q_{\mathbf{w}}\}P_s. \qquad (11.57)$$

This cannot be avoided in the selection algorithm, but in the co-phase algorithm Gray coding can be used when $N_{ph} > 1$. Even when $N_{ph} = 1$, $N_t > 2$ error probabilities of the feedback words of the co-phase algorithm are not uniformly distributed, because feedback words with a different number of feedback bit errors give rise to different bit error probabilities. Figure 11.15 displays the BEP of the co-phase algorithm when $N_{ph} = 1$, $N_t = 8$ and $p = 0.00, 0.04, 0.08, 0.16$, and 0.50. Comparing Figures 11.14 and 11.15 shows that when $p = 0$, the BEP performance of the co-phase algorithm is almost the same as the performance of the selection algorithm, and with erroneous feedback the co-phasing scheme outperforms the selection algorithm. However, the comparison is not necessarily a fair one, because when $N_t = 8$ the number of feedback bits is three for the selection algorithm and seven for the co-phase algorithm with $N_{ph} = 1$. However, the goal was not to compare the performance of different feedback algorithms but to illustrate the detrimental effects of feedback errors to closed-loop transmit diversity schemes.

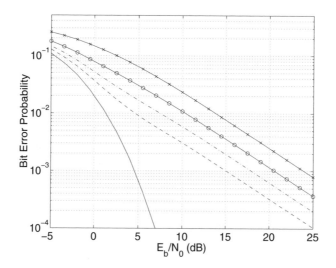

Fig. 11.14: Bit error probability curves as a function of SNR for the selection algorithm when $N_t = 8$ and $p = 0.00$ (–), $p = 0.04$ (- -), $p = 0.08$ (- ·), $p = 0.16$ (-o-), and $p = 0.50$ (-x-).

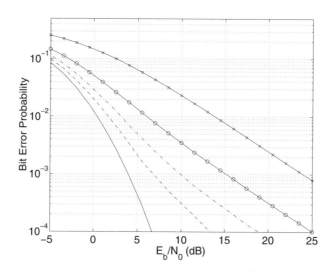

Fig. 11.15: Bit error probability curves as a function of SNR for the co-phase algorithm when $N_{ph} = 1$, $N_t = 8$ and $p = 0.00$ (–), $p = 0.04$ (- -), $p = 0.08$ (- ·), $p = 0.16$ (-o-), and $p = 0.50$ (-x-).

11.6 FEEDBACK LATENCY

It has been observed how SNR gain in closed-loop algorithms decreases in multipath propagation, and how feedback bit-errors reduce the asymptotic diversity order to one. In this
section we further examine the effect of feedback delay to SNR gain using the co-phase
algorithm as an example in a flat Rayleigh fading environment. In order to simplify the
analysis we first assume that all feedback bits are chosen based on the channel state at time
$t - \tau$ and they are applied at time t, where τ is the feedback delay. Let us first rewrite (11.9)

$$|\mathbf{wh}(t)|^2 = \sum_{m=1}^{N_t} \sum_{k=1}^{N_t} w_m h_m(t) w_k^* h_k^*(t) =: A_1 + A_2 + A_3, \tag{11.58}$$

where the channel has been expressed as a function of time, and where

$$A_1 = \sum_{m=1}^{N_t} |w_m|^2 \alpha_m(t)^2$$

$$A_2 = \sum_{k=2}^{N_t} 2\mathrm{Re}\left(w_1 w_k^* h_1(t) h_k^*(t)\right) \tag{11.59}$$

$$A_3 = \sum_{m=3}^{N_t} \sum_{k=2}^{m-1} 2\mathrm{Re}\left(w_m w_k^* h_m(t) h_k^*(t)\right).$$

Now $\mathrm{E}\langle A_1 \rangle = 1$, because each sample $\alpha_m(t)$ follows Rayleigh distribution and the weight
vector $\mathbf{w}$ is normalized such that $\|\mathbf{w}\| = 1$. Moreover, $w_1 = 1/\sqrt{N_t}$ and amplitudes $\alpha_m(t)$
have no effect on the phase adjustments given by (11.3). Therefore

$$\mathrm{E}\langle w_1 w_k^* h_1(t) h_k^*(t) \rangle = \frac{\pi}{4N_t} \mathrm{E}\left\langle e^{j(\phi_1(t) - \phi_k(t) - \psi_k)} \right\rangle. \tag{11.60}$$

Here $\phi_k(t)$ is the phase of $h_k(t)$ and $w_k = 1/\sqrt{N_t} e^{j\psi_k}$. Now we write

$$\phi_1(t) - \phi_k(t) - \psi_k = \Delta\phi_1(t) - \Delta\phi_k(t) + \varphi_k =: \psi_k(t), \tag{11.61}$$

where $\Delta\phi_k(t) = \phi_k(t) - \phi_k(t - \tau)$. Since ψ_k is chosen by using (11.3) at time instant $t - \tau$,
the variable $\varphi_k = \phi_1(t - \tau) - \phi_k(t - \tau) - \psi_k$ is uniformly distributed on $\left(-\frac{\pi}{2^{N_{ph}}}, \frac{\pi}{2^{N_{ph}}}\right)$
and there holds

$$\mathrm{E}\langle \sin \varphi_k \rangle = 0$$

$$\mathrm{E}\langle \cos \varphi_k \rangle = \frac{2^{N_{ph}}}{\pi} \sin \frac{\pi}{2^{N_{ph}}} = c_{N_{ph}}. \tag{11.62}$$

Thus we have

$$\mathrm{Re}\left(\mathrm{E}\left\langle e^{j(\phi_1(t) - \phi_k(t) - \psi_k)} \right\rangle\right) = \mathrm{E}\langle \cos(\Psi_k(t)) \rangle$$

$$= c_{N_{ph}} \mathrm{E}\langle \cos \Delta\phi_1(t) \cos \Delta\phi_k(t) + \sin \Delta\phi_1(t) \sin \Delta\phi_k(t) \rangle. \tag{11.63}$$

Random variables $\Delta\phi_1(t)$ and $\Delta\phi_k(t)$ are identically distributed and independent since channels corresponding to separate antennas are assumed to be uncorrelated. Therefore we obtain

$$\mathrm{Re}\left(\mathrm{E}\left\langle e^{j(\phi_1(t)-\phi_k(t)-\psi_k)}\right\rangle\right) = c_{N_{\mathrm{ph}}}\left(\mathrm{E}\left\langle\cos(\Delta\phi(t))\right\rangle^2 + \mathrm{E}\left\langle\sin(\Delta\phi(t))\right\rangle^2\right), \quad (11.64)$$

where the notation has been simplified by dropping unnecessary indices.

Let us now study the phase difference $\Delta\phi(t) = \phi(t) - \phi(t-\tau)$. If the velocity of the mobile is v it moves the distance $d = v\tau$ during the time delay τ and we see that the problem is equivalent to the one where spatially (distance d) separated antennas receive the same sinusoidal wave so that $\phi(t) = \phi_{t-\tau} + \frac{2\pi}{\lambda}d\cos\psi$, where ψ is uniformly distributed on $[-\pi, \pi)$ (we do not know the direction where the mobile moves). Thus, the problem is equivalent to the derivation of (2.7), and using the above formulas and (2.7) we notice that

$$\mathrm{E}\left\langle A_2\right\rangle = \frac{N_{\mathrm{t}}-1}{2N_{\mathrm{t}}}\pi c_{N_{\mathrm{ph}}}J_0\left(\frac{2\pi}{\lambda}v\tau\right)^2. \quad (11.65)$$

It remains to find $\mathrm{E}\left\langle A_3\right\rangle$. This case is studied in a similar manner as the case $\mathrm{E}\left\langle A_2\right\rangle$. Now

$$\begin{aligned}\arg\{w_m w_k^* h_m(t)h_k^*(t)\} &= \phi_m(t) - \phi_k(t) - \psi_m + \psi_k\\ &= \Delta\phi_m(t) - \Delta\phi_k(t) + \varphi_m - \varphi_k,\end{aligned} \quad (11.66)$$

where

$$\begin{aligned}\varphi_m &= \phi_m(t-\tau) - \phi_1(t-\tau) - \psi_m,\\ \varphi_k &= \phi_k(t-\tau) - \phi_1(t-\tau) - \psi_k.\end{aligned} \quad (11.67)$$

Variables φ_m and φ_k are independent and uniformly distributed on $\left(-\frac{\pi}{2^{N_{\mathrm{ph}}}}, \frac{\pi}{2^{N_{\mathrm{ph}}}}\right)$. Hence we obtain

$$\mathrm{E}\left\langle\cos(\varphi_m - \varphi_k)\right\rangle = \mathrm{E}\left\langle\cos\varphi_m\right\rangle\mathrm{E}\left\langle\cos\varphi_k\right\rangle = c_{N_{\mathrm{ph}}}^2. \quad (11.68)$$

Finally, we get

$$\mathrm{E}\left\langle A_3\right\rangle = \frac{(N_{\mathrm{t}}-1)(N_{\mathrm{t}}-2)}{4N_{\mathrm{t}}}\pi c_{N_{\mathrm{ph}}}^2 J_0\left(\frac{2\pi v\tau}{\lambda}\right)^2, \quad (11.69)$$

and the SNR gain γ is given by [260]

$$\gamma = 1 + \frac{N_{\mathrm{t}}-1}{2N_{\mathrm{t}}}\pi c_{N_{\mathrm{ph}}}\left(1 + \frac{N_{\mathrm{t}}-2}{2}c_{N_{\mathrm{ph}}}\right)J_0\left(\frac{2\pi v\tau}{\lambda}\right)^2, \quad (11.70)$$

where λ refers to the the wavelength of the carrier.

In the 3GPP FDD WCDMA system the feedback rate with closed-loop transmit diversity is 1500 bits/s. Thus, there is a single feedback bit available in each 2/3 ms slot of uplink control channel. Moreover, in the control channel there are also two feedback bits (in each slot for fast power control). If these bits are also available for the co-phase algorithm the feedback rate becomes 4500 bits/s. When the capacity of the feedback channel is less than $N_{\mathrm{ph}}(N_{\mathrm{t}}-1)$ bits/slot the transmitter may choose to update the transmit weight vector whenever a new bit is received from the feedback channel. In this case the elements of $\mathbf{w}$ experience different delays, and the SNR gain becomes [260]

$$\gamma = 1 + \frac{\pi c_{N_{\mathrm{ph}}}}{2N_{\mathrm{t}}}\left(\sum_{k=2}^{N_{\mathrm{t}}}J_0\left(\frac{2\pi v\tau_k}{\lambda}\right)^2 + \sum_{m=3}^{N_{\mathrm{t}}}\sum_{k=2}^{m-1}J_0\left(\frac{2\pi v\tau_m}{\lambda}\right)J_0\left(\frac{2\pi v\tau_k}{\lambda}\right)\right) \quad (11.71)$$

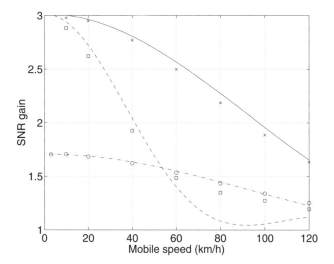

Fig. 11.16: Analytical SNR gains (linear scale) of Mode 1 (- · ·-), and co-phase algorithm with four transmit antennas and 1500 (- -) and 4500 (—) feedback bits per second, and simulated SNR gains of Mode 1 (○), $N_t = 4$, 1500 bps (□), $N_t = 4$, 4500 bps (×).

Figure 11.16 compares the SNR gains of the WCDMA closed-loop transmit diversity Mode 1 and co-phase algorithm with feedback rates 1500 and 4500 bits/s, one slot (2/3 ms) signalling delay, and $N_t = 4$, $N_{ph} = 2$ assuming single-path Rayleigh fading channels. Furthermore, $N_{ph} = 2$ in the co-phase algorithm is achieved by interpolating two one-bit feedback messages in base station as with Mode 1. We approximate the interpolation by the delay of a finite impulse response (FIR) filter with a symmetric impulse response so that the corresponding feedback delays become $1\frac{1}{2}$ slot with Mode 1 and 4500 bits/s feedback rate with $N_t = 4$. In the case of 1500 bits/s and $N_t = 4$ the delays in (11.71) are 2.5, 3.5 and 4.5 slots. It is seen that the analysis and simulations match perfectly in the case of Mode 1, but with $N_t = 4$ the approximation of the interpolation by the additional delay causes some bias error. The effect becomes more pronounced with 1500 bits/s feedback rate, when the phase of the channel changes rapidly with respect to the feedback rate. When in-phase and quadrature components of a transmit weight experience different delays, the feedback is able to track the channel better than in the case of approximating the effect by a delay caused by linear filtering.

In FDD WCDMA fast power control is typically applied to uplink control channel carrying the feedback information. Power control attempts to shorten the correlation time of the channel and maintain a fixed received power, and therefore we make the same simplifying assumption as in section 11.5 that feedback bit errors are uniformly distributed in time. A typical (uncoded) feedback error probability p is, e.g., 0.04 which roughly corresponds to the order of 10^{-3} coded BER in a voice system. In the following we study the effect of feedback errors to the SNR gain γ together with feedback latency.

The feedback errors affect the distribution of the variables

$$\varphi_m = \phi_1(t - \tau) - \phi_m(t - \tau) + \psi_m \tag{11.72}$$

More precisely, now the probability that φ_m is uniformly distributed on the correct interval $(-\frac{\pi}{2}, \frac{\pi}{2})$ is $(1 - p)$ while the probability that φ_m is uniformly distributed on the incorrect interval $(-\frac{\pi}{2}, \frac{\pi}{2})^c$ is p, where $(\cdot)^c$ means the complement of the set. Thus we find that

$$c_1 = \mathrm{E}\langle\cos\varphi_m\rangle = \frac{2}{\pi}(1 - 2p), \tag{11.73}$$

and when $N_{\mathrm{ph}} = 1$, the expected SNR gain γ for the co-phase algorithm is given by

$$\gamma = 1 + \frac{N_t - 1}{N_t}(1 - 2p)\left(1 + \frac{N_t - 2}{\pi}(1 - 2p)\right)J_0(\frac{2\pi v\tau}{\lambda})^2. \tag{11.74}$$

In the same vein, it is straightforward to show that when $N_{\mathrm{ph}} = 2$, substituting c_2 as $c_2 \longrightarrow (1 - 2p)c_2$ in (11.70) gives the expected SNR gain in feedback bit errors and the QPSK feedback constellation. The same principle applies to $N_{\mathrm{ph}} = 3$ and Gray-coded feedback messages as well.

Figure 11.17 shows the analytical and simulated performance of the three feedback schemes with 4% feedback bit error probability. Again, analysis and simulations match well when $N_t = 2$. Furthermore, it is noticed that Mode 1 is more robust to feedback errors than the four transmit antenna schemes, and with high mobile speeds the feedback delay is more critical to performance than the nominal 4% feedback bit error probability.

Figure 11.18 shows SNR gains for $N_{\mathrm{ph}} = 1, 2, 3$ with mobile speeds 10 and 40 km/h using the same parameters as above. At 10 km/h the gains increase together with the number of the feedback bits, but at 40 km/h the SNR gains start to decrease when the length of the feedback word increases If Othe number of feedback bits are considered, $N_{\mathrm{ph}} = 1$ provides the largest SNR gain.

The derivation of SNR gain as a function of mobile velocity assumed only the co-phase algorithm. While instantaneous gains and phases are independent when assuming Rayleigh fading, gains and phases are not independent as random processes, and therefore a similar study for Mode 2 would require multidimensional integration.

The effect of feedback latency to the BEP of closed-loop transmit diversity has been extensively studied in [243] assuming unquantized feedback signalling, perfect channel state information in the receiver, and error-free feedback. In case of two transmit antennas and a flat fading environment, STTD outperforms unquantized closed-loop transmit diversity when

$$f_D \geq \frac{1}{2\pi\tau}J_0^{-1}(\rho), \tag{11.75}$$

where f_D is the fading rate, and

$$\rho = \sqrt{\sqrt{\frac{(\gamma + 2)^3(\gamma + 6)^2}{(\gamma + 4)^3}} - (\gamma + 2)}. \tag{11.76}$$

When $\gamma \to 0$, $\rho \to \sqrt{3/\sqrt{2} - 2}$, and with $\tau = 2/3$ ms (one slot) and 2150 MHz downlink carrier frequency, $f_D \approx 455$ Hz, which corresponds to as much as 226 km/h mobile speed. However, link-level simulation results [232] taking into account system imperfections and quantization of the feedback message, indicate a much lower switching point, 70 km/h, from Mode 1 to STTD.

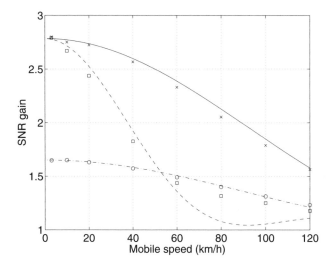

Fig. 11.17: Analytical SNR gains (linear scale) of Mode 1 (- · ·-), and the co-phase algorithm with four transmit antennas and 1500 (- -) and 4500 (—) feedback bits per second, and simulated SNR gains of Mode 1 (○), $N_t = 4$, 1500 bps (□), $N_t = 4$, 4500 bps (×). Probability of feedback errors $p = 0.04$.

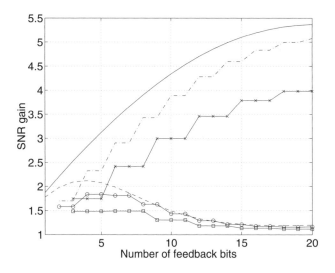

Fig. 11.18: SNR gains (linear scale) of the co-phase algorithm as a function of the number of feedback bits, $v = 10$ km/h and $N_{ph} = 1$ (—), $v = 40$ km/h and $N_{ph} = 1$ (- -), $v = 10$ km/h and $N_{ph} = 2$ (- · ·-), $v = 40$ km/h and $N_{ph} = 2$ (-○-), $v = 10$ km/h and $N_{ph} = 3$ (-x-), $v = 40$ km/h and $N_{ph} = 3$ (-□-).

11.7 BIT-ERROR PROBABILITY

In the conventional approach the bit-error probability $P(e)$ for a certain SNR distribution is computed by using the formula

$$P(e) = \int_{\mathbb{R}} p(\xi)g(\xi)d\xi, \tag{11.77}$$

where $g(\cdot)$ is the error rate of the modulation in terms of the received SNR and $p(\cdot)$ is the probability density function of the SNR. In the selection algorithm closed–form solution (11.50) is available, but the co-phase algorithm implies $2N_t - 1$ dimensional integration over N_t amplitudes and $N_t - 1$ phase differences due to the common reference antenna, which makes it difficult to obtain closed–form expressions.

Consider the BEP of BPSK modulation and closed-loop algorithms given by

$$f(\boldsymbol{\alpha}, \boldsymbol{\theta}) = \frac{1}{2}\text{erfc}\left(\sqrt{R(\boldsymbol{\alpha}, \boldsymbol{\theta})^2 \xi}\right), \tag{11.78}$$

where ξ is the received SNR, $\boldsymbol{\alpha} = (\alpha_1, \alpha_2, \dots, \alpha_{N_t})$, $\boldsymbol{\theta} = (\theta_1, \theta_2, \dots, \theta_{N_t})$ and $R(\cdot, \cdot)$ is defined by

$$R(\boldsymbol{\alpha}, \boldsymbol{\theta}) = \left| \sum_{m=1}^{N_t} w_{*,m} h_m \right| = \left| \sum_{m=1}^{N_t} \alpha_m u_m e^{j\theta_m} \right|, \tag{11.79}$$

where the weights $w_{*,m}$ have been selected by a closed-loop algorithm. Since $\boldsymbol{\alpha}$ and $\boldsymbol{\theta}$ are independent, the joint density function can be expressed as a product of probability density functions of $\boldsymbol{\alpha}$ and $\boldsymbol{\theta}$.

Let us denote by $q_\theta(\cdot)$ the $(N_t - 1)$-dimensional product density of $\boldsymbol{\theta}$, and by $\pi(\mathbf{r})$ the product of the components of a vector. It can be shown [258] that in case of large SNR $\xi_{N_t} = E_b/(N_t N_0) \gg 1$ and the co-phase algorithm with $(N_t - 1)N_{ph}$ feedback bits, the asymptotic bit error probability for BPSK signal is given by

$$P(e) = \binom{2N_t - 1}{N_t} \left(\frac{C_{N_t, N_{ph}}}{4\xi_{N_t}}\right)^{N_t}, \tag{11.80}$$

where the constant $C_{N_t, N_{ph}}$, referred to as an asymptotic BEP gain, attains the form

$$C_{N_t, N_{ph}} = \left\{ \int_{\mathbb{R}^{N_t-1}} q_\theta(\boldsymbol{\theta}) \left(\int_{\mathbb{R}_+^{N_t-1}} \frac{(N_t - 1)! 2^{N_t - 1} \pi(\mathbf{r})}{(N_t \cdot R(\mathbf{r}, \boldsymbol{\theta})^2)^{N_t}} d\mathbf{r} \right) d\boldsymbol{\theta} \right\}^{\frac{1}{N_t}}, \tag{11.81}$$

and $\mathbf{r} = (r_1, \dots, r_{N_t-1}, 1)$, $d\mathbf{r} = dr_1 \dots dr_{N_t-1}$.

It is seen that the asymptotic formula is similar to the known asymptotic BEP of receiver-maximal ratio combining (2.18). Open-loop schemes provide only pure diversity gain, and their asymptotic BEP gain is equal to one. In closed-loop transmit diversity schemes, the asymptotic BEP gain results from constructive channel adjustments. When $N_t = 2$, $C_{2, N_{ph}}$ is given by

$$C_{2, N_{ph}} = \left\{ \frac{1 - \frac{2^{N_{ph}-1}}{\pi} \sin\frac{\pi}{2^{N_{ph}-1}}}{2 \sin^2 \frac{\pi}{2^{N_{ph}}}} \right\}^{\frac{1}{2}}. \tag{11.82}$$

It is interesting to compare the BEP gain of the co-phase algorithm to the upper bound of BEP gain which is obtained when using ideal phase and amplitude adjustments with respect to both phases and amplitudes (complete channel state information in the transmitter). Moreover, the gain from ideal co-phasing is of some interest. Ideal co-phasing is simila to receiver-equal gain combining, for which the asymptotic bit error rate is well known. The result can be deduced in the same vein as before, and for ideal feedback $C_{\text{ideal}} = 1/N_t$ while the BEP gain for ideal co-phasing is of the form

$$C_{N_t,\infty} = \left\{ \frac{2^{N_t-1}(N_t-1)!}{(2N_t-1)!} \right\}^{\frac{1}{N_t}}. \tag{11.83}$$

Let us next study the order and co-phase algorithm when $N_t = 2$ and the amplitude weights are obtained from (11.35). The probability density function is of the form

$$p(\alpha_1, \alpha_2) = \begin{cases} 2\frac{\alpha_1\alpha_2}{\sigma^4} e^{-(\alpha_1^2+\alpha_2^2)/2\sigma^2}, & \alpha_1 > \alpha_2, \\ 0, & \text{otherwise.} \end{cases} \tag{11.84}$$

The BEP gain is now of the form

$$\tilde{C}_{2,N_{\text{ph}}}^2 = \frac{2^{N_{\text{ph}}+1}}{\pi} \int_0^{\frac{1}{2}} \int_0^{\frac{\pi}{2^{N_{\text{ph}}}}} \frac{r\,dr\,d\theta}{(u_1^2 + u_2^2 r^2 + 2u_1 u_2 r \cos\theta)^2}, \tag{11.85}$$

where $0 < u_2^2 < u_1^2 < 1$. Closed-form expression in terms of elementary functions for the integral (11.85) is also available, but the resulting expression is long and complicated and therefore is omitted here. Table 11.4 presents the BEP gains of co-phase and order and co-phase algorithms when $N_t = 2$ with respect to STTD, which offers the same diversity gain but no SNR gain. Note that the BEP gains are different from the SNR gains discussed in sections 11.2 and 11.3. Amplitude weights in the order and co-phase algorithm have been selected using (11.35). Weights agree with those obtained by numerically finding the minimum of (11.85) with respect to u_1 and u_2, so that when $N_t = 2$ the transmit weights obtained by maximizing the received SNR or minimizing the BEP are the same.

Table 11.4: BEP gains (linear scale) of co-phase algorithm and order and co-phase algorithm when $N_t = 2$

N_{ph}	1	2	3	∞
$C_{2,N_{\text{ph}}}$	1.41	1.66	1.71	1.73
$\tilde{C}_{2,N_{\text{ph}}}$	1.66	1.86	1.91	1.93

At low and medium SNR regions, the BEP of the two-transmit antenna closed-loop algorithms can be approximated by using the formula

$$P_{app}(e) = \frac{1}{4}(1-\mu)^2(2+\mu), \qquad \mu = \sqrt{\frac{\xi_2}{C+\xi_2}}. \tag{11.86}$$

For STTD $C = 1$ and for maximal ratio combining in the transmitter $C = 1/2$. For the co-phase and the order and co-phase algorithms $C = 1/C_{2,N_{\text{ph}}}$ or $C = 1/\tilde{C}_{2,N_{\text{ph}}}$. Hence

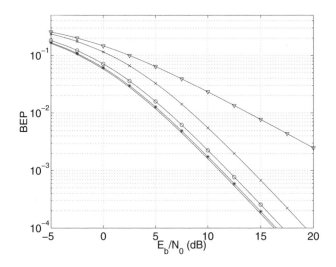

Fig. 11.19: BEP curves as a function of SNR for single antenna transmission ($-\nabla$-), STTD (-x-), Mode 1 (-o-), Mode 2 (-+-), and ideal feedback (—).

we approximate the BEP of the feedback algorithms by a suitable scaled BEP of MRC using the BEP gain in high SNR region for the whole SNR range. This approximation is well defined and we can write $P(e) = P_{app}(e) + E$, where the error term is asymptotically of order ξ_2^{-3} [258]. Hence, the error term vanishes when $\xi_2 \to \infty$.

Figure 11.19 displays the bit error curves as a function of SNR when $N_t = 2$ according to approximation (11.86). Comparing simulation results and Equation (11.86) has shown that the approximation is accurate in low SNR region as well. It is seen that the WCDMA closed-loop transmit diversity modes provide a performance which is very near to optimal. However, in practice the performance deteriorates due to channel estimation, antenna verification, feedback latency and feedback bit errors, as discussed in sections 11.5 and 11.6.

11.8   TRANSMIT WEIGHT GENERATION FOR THE ORDER AND CO-PHASE ALGORITHM

So far the performance of the order and co-phase algorithm has been studied in case of uncorrelated transmit antennas. However, amplitude weights can be determined in a similar manner when the channel correlation matrix is not diagonal. In this case we are not able to calculate SNR gain analytically, but the correlation matrix in order domain can be determined adaptively.

Assume that $\|\mathbf{h}_{(1)}\| \geq \cdots \geq \|\mathbf{h}_{(N_t)}\|$, where brackets in subscript refer to the order of norms of vectors. Then we have

$$\gamma = \mathrm{E}\left\langle \|\mathbf{H}_{(\cdot)}\mathbf{w}\|^2 \right\rangle = \mathbf{u}^{\mathrm{T}}\mathbf{C}\mathbf{u}, \tag{11.87}$$

where $\mathbf{H}_{(\cdot)} = (\mathbf{h}_{(1)}, \cdots, \mathbf{h}_{(N_t)})$, $\mathbf{u} = [u_1, \ldots, u_{N_t}]^T$, $u_m = |w_m|$, $\psi_m = \arg(w_m)$, and the elements of the correlation matrix $\mathbf{C} = (c_{m,k})_{m,k=1}^{N_t}$ are given by

$$c_{m,m} = \mathrm{E}\left\langle \|\mathbf{h}_{(m)}\|^2 \right\rangle$$

$$c_{m,k} = \mathrm{Re}\{\mathrm{E}\left\langle \mathbf{h}_{(m)}^\dagger \mathbf{h}_{(k)} e^{j(\psi_k - \psi_m)} \right\rangle\}. \tag{11.88}$$

Moreover, employing the notation $\phi_{m,k} = \arg(\mathbf{h}_{(m)}^\dagger \mathbf{h}_{(k)})$ we obtain

$$c_{m,k} = \mathrm{E}\left\langle |\mathbf{h}_{(m)}^\dagger \mathbf{h}_{(k)}| \cos(\phi_{m,k} - \psi_m + \psi_k) \right\rangle. \tag{11.89}$$

If the channels are uncorrelated then phases $\phi_{m,k}$ are uniformly distributed on $[-\pi, \pi)$. If we choose to adjust phases of $\{\mathbf{h}_{(m)}\}_{m=2}^{N_t}$ relative to $\mathbf{h}_{(1)}$ we can let $\psi_1 = 0$ without losing generality. In phase adjustments we choose ψ_m such that phases $\phi_{1,m} + \psi_m$ are uniformly distributed in the interval $(-\frac{\pi}{2^{N_{ph}}}, \frac{\pi}{2^{N_{ph}}})$, where N_{ph} is the number of feedback bits corresponding to each relative phase. There holds

$$c_{1,k} = c_{k,1} = c_{N_{ph}} \mathrm{E}\left\langle |\mathbf{h}_{(1)}^\dagger \mathbf{h}_{(k)}| \right\rangle, \quad k = 2, \cdots, N_t, \tag{11.90}$$

where $c_{N_{ph}} = \mathrm{E}\left\langle \cos(\phi_{1,k} + \psi_k) \right\rangle = \mathrm{sinc}(2^{-N_{ph}})$. Moreover, when $m, k \neq 1$ we write

$$\phi_{m,k} - \psi_m + \psi_k = \Phi_{m,k} - \Psi_m + \Psi_k, \tag{11.91}$$

where $\Psi_k = \phi_{1,k} + \psi_k$, $\Psi_m = \phi_{1,m} + \psi_m$ are uniformly distributed on $[-\frac{\pi}{2^{N_{ph}}}, \frac{\pi}{2^{N_{ph}}})$ and $\Phi_{m,k} = \phi_{m,k} + \phi_{1,k} - \phi_{1,m}$. In single-path channels $\Phi_{m,k} = 0$ and

$$c_{m,k} = c_{k,m} = c_{N_{ph}}^2 \mathrm{E}\left\langle |h_{(m)}^* h_{(k)}| \right\rangle, \quad k, m \neq 1. \tag{11.92}$$

However, in multipath channels $\Phi_{m,k} \neq 0$ and we obtain

$$c_{m,k} = c_{N_{ph}}^2 \mathrm{E}\left\langle \cos(\Phi_{m,k}) \right\rangle \mathrm{E}\left\langle |\mathbf{h}_{(m)}^\dagger \mathbf{h}_{(k)}| \right\rangle, \quad k, m \neq 1. \tag{11.93}$$

Since $\Phi_{m,k} = -\Phi_{k,m}$, $\mathbf{C}$ is also real and symmetric in multipath fading. Furthermore, all vectors $\mathbf{h}_m$ are identically distributed and therefore

$$\mathrm{E}\left\langle \cos(\Phi_{m_1,k_1}) \right\rangle = \mathrm{E}\left\langle \cos(\Phi_{m_2,k_2}) \right\rangle =: b \tag{11.94}$$

whenever $m_1, m_2 \neq 1$ and $k_1, k_2 \neq 1$. The values of b and $\mathrm{E}\left\langle |\mathbf{h}_{(m)}^\dagger \mathbf{h}_{(k)}| \right\rangle$ can be obtained from Monte Carlo simulations. A similar procedure can also be applied when the channels are correlated.

Feedback latency can also be taken into account when determining the transmit weights. In this case the elements of $\mathbf{C}$ become

$$c_{m,m}(\tau) = \mathrm{E}\left\langle \|\mathbf{h}_{(m)}(t - \tau)\|^2 \right\rangle$$

$$c_{m,k}(\tau) = \mathrm{Re}\{\mathrm{E}\left\langle \mathbf{h}_{(m)}(t - \tau)^\dagger \mathbf{h}_{(k)}(t - \tau) e^{j(\psi_k(t-\tau) - \psi_m(t-\tau))} \right\rangle\}, \tag{11.95}$$

where the notation $\|\mathbf{h}_{(\cdot)}(t - \tau)\|^2$ indicates that the columns of $\mathbf{H}$ have been ordered at time $t - \tau$, but the ordering is applied to the channel vectors at time t. Furthermore, if the system has an estimate of feedback error probability and of the corresponding probability density function, the effect of errors can be simulated. This makes the weight generation algorithm more robust to feedback errors and feedback latency by driving all transmit amplitude weights toward the same value when the quality of the feedback information becomes poor.

Transmit weight generation can be combined with some additional transformation $\mathbf{E}$ of the communication channel, so that $\gamma = \|\mathbf{G}_{()}\mathbf{w}\|^2$, $\mathbf{G} = \mathbf{HE}$, and $\mathbf{G}_{()}$ contains the elements of $\mathbf{G}$ ordered according to the vector norms. The transform matrix $\mathbf{E}$ can consist of the eigenvectors of the correlation matrix $\mathbf{R}$ as discussed in section 10.2.2. This is useful when order and co-phase feedback is combined with a long-term feedback scheme. Apart from eigenvalue decomposition, it is possible to use, independent component analysis (ICA) either directly to the channel matrix or to the downlink correlation matrix. This typically results in a set of non-orthogonal long-term beams, which can have different relative powers.

Note that when transmit weights are determined adaptively in mobile station, also short-term feedback schemes require signalling of long-term information consisting of the quantization points $\mathbb{W}$. Alternatively, quantization points can be determined in base station based on uplink measurements, but then, in order to utilize common pilot channels for channel estimation in mobile station, $\mathbb{W}$ has to be signalled to mobile station.

11.9   SNR GAIN IN CORRELATED RAYLEIGH FADING CHANNELS AS A FUNCTION OF MOBILE SPEED

Figure 11.20 shows SNR gains of different feedback schemes in flat Rayleigh fading environment (pico cell model) as a function of mobile speed using four-element transmit antenna array where the radiating elements are separated by $\lambda/2$ where λ denotes the wavelength of the carrier frequency. Angular spread is $360°$, feedback capacity is 1500 bps (one bit/slot) as defined in [261], and feedback bit-error probability is 4%. Choose 2 and choose 3 refer to the feedback schemes in [237], which select two or three from four transmit antennas based on the feedback and quantize the phase differences by two bits/antenna. The order and co-phase algorithm using two-bit gain signalling No. 2 in Table 11.1 and one-bit phase/antenna and filtering in BS, similar to extended Mode 1 in [235], gives good results with low mobile–speeds while extended Mode 1 for four antennas is better with higher speeds due to its shorter feedback word length. Amplitude weights for the order and co-phase algorithm have been determined adaptively, as explained in section 11.8. Choosing two or three antennas is inferior to both schemes while choosing the strongest transmit antenna is not a good idea with low mobile speeds. Different order signalling in Table 11.1 results in different performance but since the differences are small all simulations employed gain signalling No. 2. The interpretation of the feedback message remains the same and there is no need to worry about frame boundaries with the five-bit feedback word, because the WCDMA uplink frame consists of 15 slots, where each slot carries one feedback bit. Due to the wide angular spread, eigenvector transform would not bring any improvement to the performance. Figure 11.21 shows SNR gains in four uncorrelated transmit antennas and with flat Rayleigh fading. The SNR gains are slightly inferior to the ones in Figure 11.20 with high mobile speeds.

Figure 11.22 shows the results in macro channel model [239] with $10°$ angular spread. Now extended Mode 1 and the adaptive order and co-phase algorithm are almost the same as in Figure 11.22, because the signals correlate strongly and the transmit weights adapt to the same point. Short-term feedback works better than in Figure 11.20 with high mobile speeds due to the strong correlation between transmit antennas. The increase in SNR gain after 3 km/h mobile speed is explained by a short simulation period, which fails to capture enough channel modes at low mobile speeds.

Figure 11.23 repeats the previous experiment but now applies eigenvector transform to the channel in the same way as described in section 10.2.2. The order and co-phase algorithm works in a similar manner with low mobile speeds, but beam selection provides good results while phase adjustment is the worst alternative. Long-term beam-forming improves the performance of the short-term feedback schemes within high mobile speeds, but on the other hand with $10°$ angular spread transmit beam-forming alone is able to provide good performance without short-term feedback. However, simulations assumed error-free signalling of unquantized eigenbeams while the feedback error probability was set to 4%. Furthermore, long-term feedback information did not reduce the capacity of the short-term feedback, so that the short-term feedback rate is 1500 bps irrespective of long-term feedback. In practice, long-term feedback would occupy the same feedback channel, thereby reducing the rate of the short-term feedback. Although long-term and short-term feedback show good performance at high mobile speeds, the assumption that the correlation matrix does not change during the simulation may not be realistic.

Figures 11.24 and 11.25 repeat the previous experiments, but now with micro channel model [239] with $45°$ angular spread without and with eigenbeam-forming front end, respectively. Due to the larger angular spread and feedback latency, the SNR gains are not as large as in Figures 11.22 and 11.23. However, diversity gain is proportional to the angular spread, and comparing SNR gains between different spatial channel models does not provide enough information on overall link level performance.

Figure 11.26 shows the four amplitude weights as a function of mobile speed with different gain-signalling schemes and angular spreads, where the elements of the ordered correlation matrix $\mathbf{C}$ have been determined according to (11.95). The amplitude weights of the full gain signalling in the top left figure converge toward the same value when the mobile speed is increasing, indicating that the channel coherence time is becoming too short for gain signalling requiring five slots. Reduced gain signalling No. 2 is more robust with respect to speed. Amplitude weights of signalling types 2 and 4 look different, although their performance is similar, indicating that the performance of the adaptive order and co-phase algorithm is robust with respect to amplitude weights. Finally, with $10°$ angular spread the transmit amplitude weights effectively select the strongest beam. This is not surprising because after the eigenbeam transform, the channel correlation matrix contains one dominating eigenvalue.

11.9.1 Quantization of Long-term Feedback

So far we have assumed 4% nominal feedback error probability for short-term feedback, but long-term feedback has been assumed to be ideal. 3GPP contribution [238] quantizes gain and phase in long-term feedback with 3 and 5 bits, respectively, so that the transmission of one eigenbeam requires $4 \times 3 + 3 \times 5 = 27$ bits. Assuming uniform phase quantization, the order and co-phase algorithm requires only signalling of the amplitude weights so that the corresponding long-term feedback needs considerably fewer bits than eigenbeam-forming, because phase information need not be signalled. Furthermore, the structure of amplitude weights facilitates several error-checking features utilizing, e.g., $\sum w_i = 1$, $w_1 \geq w_2 \geq \cdots w_{N_t}$, or hierarchical coding where w_1 is coded using some error correction coding, and

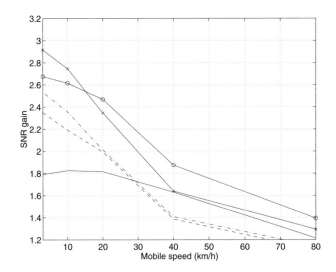

Fig. 11.20: SNR gains (linear scale) as a function of the mobile speed in flat Rayleigh fading channel and four Tx antennas with $\lambda/2$ spacing. (—) choose strongest, (- -) choose 2, (- · -) choose 3, (–o–) extended Mode 1, (–x–) adaptive order and co-phase algorithm with order signalling No. 2 in Table 11.1.

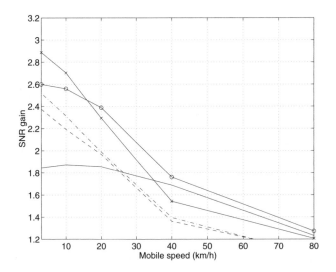

Fig. 11.21: SNR gains (linear scale) as a function of the mobile speed in flat Rayleigh fading channel and four uncorrelated Tx antennas. (—) choose strongest, (- -) choose 2, (- · -) choose 3, (–o–) extended Mode 1, (–x–) adaptive order and co-phase algorithm with order signalling No. 2 in Table 11.1.

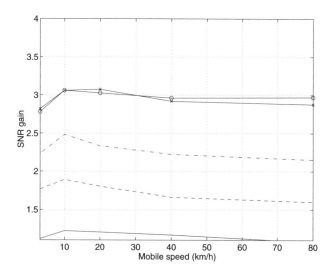

Fig. 11.22: SNR gains (linear scale) as a function of the mobile speed in flat Rayleigh fading channel with $10°$ angular spread macro channel model and four Tx antennas with $\lambda/2$ spacing. (—) choose strongest, (- -) choose 2, (- · -) choose 3, (–o–) extended Mode 1, (–x–) adaptive order and co-phase algorithm with order signalling No. 2 in Table 11.1.

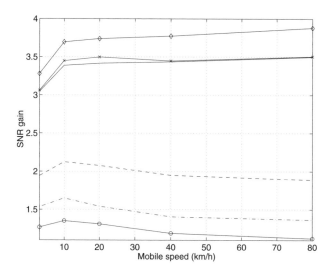

Fig. 11.23: SNR gains (linear scale) as a function of the mobile speed in flat Rayleigh fading channel with $10°$ angular spread macro channel model and four Tx antennas with $\lambda/2$ spacing applying transmit eigenbeam-forming. (—) choose strongest, (- -) choose 2, (- · -) choose 3, (–o–) extended Mode 1, (–x–) order signalling No. 2 in Table 11.1, (–◇–) transmit eigenbeam-forming only.

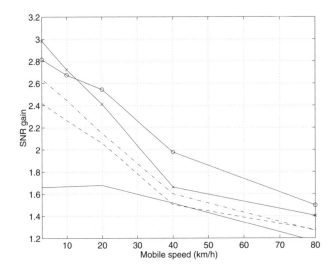

Fig. 11.24: SNR gains (linear scale) as a function of the mobile speed in flat Rayleigh fading channel with $45°$ angular spread micro channel model and four Tx antennas with $\lambda/2$ spacing. (—) choose strongest, (- -) choose 2, (- · -) choose 3, (–o–) extended Mode 1, (–x–) adaptive order and co-phase algorithm with order signalling No. 2 in Table 11.1.

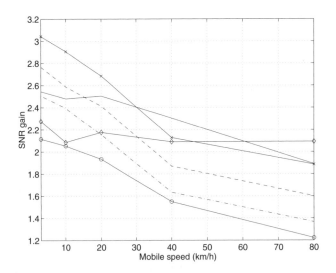

Fig. 11.25: SNR gains (linear scale) as a function of the mobile speed in flat Rayleigh fading channel with $45°$ angular spread micro channel model and four transmit antennas with $\lambda/2$ spacing applying transmit eigenbeam-forming. (—) choose strongest, (- -) choose 2, (- · -)choose 3, (–o–) extended Mode 1, (–x–) adaptive order and co-phase algorithm with order signalling No. 2 in Table 11.1, (–◇–) transmit eigenbeam-forming only.

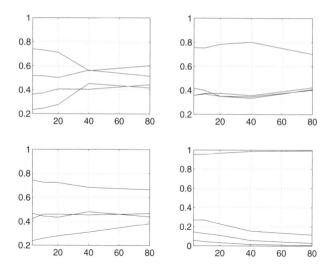

Fig. 11.26: Amplitude weights as a function of mobile speed. Top left: full order signalling, 360° angular spread. Top right: reduced signalling No. 2 in Table 11.1, 360° angular spread. Bottom left: reduced signalling No. 4 in Table 11.1, 360° angular spread. Bottom right: reduced signalling No. 2 in Table 11.1, 10° angular spread.

$w_2, \cdots, w_{N_t}$ are communicated to BS using the difference $w_i - w_1$. A suitable quantization can be determined using, e.g., Lloyd algorithm [244]. In [252] a subspace tracking algorithm is described where the long-term feedback consists of $\frac{N_t(N_t+1)}{2}$ angles defining Givens rotors. The paper provides an example where the angles are quantized with three bits. The number of bits of the long-term feedback is much less than in [238] but still more than in the order and co-phase algorithm.

Instead of using the same number of quantization bits for all eigenvectors in long-term feedback, a better strategy would be to allocate more feedback bits to the eigenvectors corresponding to larger eigenvalues. Thus, the resolution of the quantizer would in general depend on the magnitude of the eigenvalues.

Figure 11.27 shows the effect of three-bit uniform quantization in range [0,1] of the transmit amplitude weights when full-order signalling and reduced signalling schemes 2 and 4 are used in the same setting as in Figure 11.20. It is noticed that the simple quantization affects the SNR gains within slow mobile speeds. The amount of signalling of the amplitude weights only is small when compared to the signalling of the eigenbeams. Using 27 bits to signal three three-bit amplitude weights (it is not actually necessary to signal, say, the weakest one) with 4% nominal feedback error probability the transmission of the weights is practically error free. (Nominal error probability drops to 0.12% with simple repetition coding.)

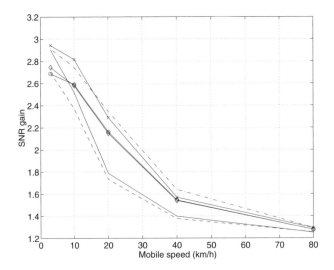

Fig. 11.27: SNR gains (linear scale) as a function of the mobile speed in flat Rayleigh fading channel and four transmit antennas with $\lambda/2$ spacing. (—) full-order signalling without quantization, (- -) full order signalling and three-bit quantization, (- · -) order signalling No. 2 without quantization, (–o–) order signalling No. 2 and three-bit quantization, (–x–) order signalling No. 4 without quantization, (–◇–) order signalling No. 4 and three-bit quantization.

11.10 SUMMARY

This chapter examined general short-term feedback transmit diversity concepts. It was shown analytically that the order and co-phase algorithm provides the expected received SNR gains, which are comparable to the case where the transmitter is equipped with complete channel state information. Also, asymptotic bit-error probabilities for the co-phase algorithm and the order and co-phase algorithm were presented.

In cases of feedback latency and when the feedback channel is subject to errors, the performance deteriorates, and asymptotic diversity gain of the closed-loop algorithms is equal to unity. This sounds like a serious drawback of closed-loop schemes, but even in the presence of feedback errors, closed-loop schemes provide SNR gain, and the applicability of closed-loop transmit diversity depends on the operation point of the system.

In cases of correlated transmit antennas or multipath channels when $N_t > 2$, analytical results for SNR gain are not available. However, using the general design principles it is possible to adapt the transmit weights to different propagation environments. When the channel coherence time does not exceed the feedback latency, adaptive short-term feedback results in good performance in terms of SNR gain. In case of correlated channels, combining short-term and long-term feedback (eigenbeam-forming) improves the performance at high mobile speeds when compared with pure short-term feedback concept. Optimal joint short-term and long-term feedback remains an open problem, though, and the development of such a scheme would require more detailed channel modelling, where the geometric properties of the spatial channel change as a function of time.

12

Hybrid Closed-loop and Open-loop Methods

It was shown in Chapters 10 and 11 that even when the feedback message is heavily quantized, closed-loop transmit diversity techniques typically outperform open-loop ones when the quality of the feedback information is adequate. However, with imperfect feedback signaling, the transmission schemes can be further improved by balancing SNR gains (due to channel-state feedback) and diversity gains (due to open-loop modulation concepts). A simple example of such a solution was provided in section 10.1.3.2 (see also [25]).

In this chapter, we review the principles behind hybrid transmission schemes, and develop various hybrid transceivers that combine open-loop schemes developed in chapters 3-9 and simple channel state feedback (or partial channel state information). Partial channel state information is modelled by quantized short-term channel information or by assuming that the transmitter knows (or estimates) the stochastic (long-term) channel model, but does not know the instantaneous channel realization. In addition, we introduce new adaptive transmission techniques for use with ARQ applications and adaptive modulation arrangements.

12.1 A COMPARISON OF OPEN-LOOP AND CLOSED-LOOP SYSTEMS

The performance of hybrid open-loop and closed-loop MISO systems with partial channel state information has been addressed from a capacity point of view in [90, 92, 93, 246], from a pairwise codeword error probability point of view in [27], and from a bit error probability point of view in [28, 262]. The results in these studies suggest that hybrid closed-loop and open-loop transceiver concepts are able to maintain a high diversity order, even in the presence of imperfect channel state information at the transmitter. Related power allocation solutions for MIMO systems have been addressed in [263, 264].

Partial channel state information may be modelled with a stochastic channel characterization, or stochastic channel state information. Two partial feedback strategies, known as mean feedback and covariance feedback, were considered in [92]. With mean feedback the transmitter assumes that the channel coefficients are distributed according to multivariate complex Gaussian distribution $N(\mathbf{w}, \delta^2 \mathbf{I})$, where $\mathbf{w}$ and δ^2 can be interpreted as the estimate of the channel based on the feedback and the variance of the estimate, respectively. In covariance feedback, the channel (known at the transmitter) is distributed as $N(\mathbf{0}, \mathbf{R}_{Tx})$. This can be interpreted to model a case where the channel is changing very rapidly and the mean feedback is unable to track or model the instantaneous channel dynamics. However, the geometrical properties of the channel change slowly, and it is in many cases realistic to assume that the transmitter knows the channel covariance matrix $\mathbf{R}_{Tx}$.

Capacity Expressions: In a MISO system, the capacity expression (2.21) becomes

$$C = \max_{\mathrm{Tr}(\mathbf{Q})=P} \mathrm{E} \left\langle \log(1 + \frac{\mathbf{h}^\dagger \mathbf{Q} \mathbf{h}}{\sigma^2}) \right\rangle, \tag{12.1}$$

where $\mathbf{Q}$ is the covariance of the transmitted signal and P is the total transmit power to be allocated to different channels. In mean feedback the optimum solution is to transmit a single data stream in the direction of $\mathbf{w}$ or, depending on SNR and the quality of feedback information, allocate transmit power between $\mathbf{w}$ and the remaining $N_t - 1$ orthogonal directions. The $N_t - 1$ directions are allocated equal power of the remaining power budget. In covariance feedback, only long-term information of the channel structure is available. The capacity-optimal solution sets $\mathbf{Q} = \mathbf{R}_{Tx}$ and transmits independent Gaussian streams along the eigenvectors of $\mathbf{R}_{Tx}$. Transmit power allocation of the channels is then solved numerically such that, in general, eigenvectors corresponding to larger eigenvalues receive more power.

Necessary and sufficient conditions for beam-forming based on mean or covariance feedback to provide capacity-optimizing transmission strategies were derived in [93]. This means that instead of vector coding, scalar coding can be used to transmit to the direction of the principal eigenvector in covariance feedback and to the direction of the feedback vector in mean feedback. The results for covariance feedback apply directly to long-term feedback in section 10.2.2.

Short–term feedback schemes discussed in Chapters 10 and 11 do not fall into the mean feedback category, because in order to shorten the length of the feedback word the short-term feedback schemes utilize a common reference antenna for phase and gain adjustment. This makes the elements of $\mathbf{w}$ dependent. A modification that yields independent w_m given independent h_m (see (11.6)) is not desirable, since this would decrease the performance of closed-loop algorithms at high mobile speeds. Furthermore, with low bit-rate feedback w_m and h_m are not jointly Gaussian distributed. While mean feedback and short-term feedback are different, we can use the same principle as in [93] when studying the WCDMA closed-loop schemes.

Following (2.28), let us decompose the covariance matrix of the transmitted signal as $\mathbf{Q} = \mathbf{W}^\dagger \mathbf{P} \mathbf{W}$, $\mathbf{W} = (\mathbf{w}_1^T, \cdots, \mathbf{w}_{N_t}^T)^T$, where $\mathbf{w}_1$ is the transmit weight vector corresponding to a feedback algorithm. Using this, the capacity can be written as

$$\begin{aligned}
C &= \max_{\mathrm{Tr}(\mathbf{P})=P} \mathrm{E} \left\langle \log(1 + \frac{\tilde{\mathbf{h}}^\dagger \mathbf{P} \tilde{\mathbf{h}}}{\sigma^2}) \right\rangle \\
&= \max_{\sum_{m=1}^{N_t} P_m = P} \mathrm{E} \left\langle \log(1 + \frac{\sum_{m=1}^{N_t} P_m \alpha_m}{\sigma^2}) \right\rangle,
\end{aligned}$$

where $\tilde{\mathbf{h}} = \mathbf{W}\mathbf{h}$, $\alpha_m^2 = |\tilde{h}_m|^2$, and P_m gives the transmit power allocation of channel m. Suppose that power $P_1 = P - p$ is allocated for $\mathbf{w}_1$ and p is evenly distributed to $\mathbf{w}_2, \cdots, \mathbf{w}_{N_t}$. Because of the limited feedback information, only $\mathbf{w}_1$ is known at the transmitter, and $\mathbf{w}_2, \cdots, \mathbf{w}_{N_t}$ are orthonormal channels spanning the null space of $\mathbf{w}_1^\dagger \mathbf{w}_1$. Otherwise $\mathbf{w}_2, \cdots, \mathbf{w}_{N_t}$ are arbitrary. The mutual information becomes

$$C(p) = \mathrm{E}\left\langle \log\left(1 + \frac{(P-p)\alpha_1^2}{\sigma^2} + \frac{p}{\sigma^2(N_t-1)}\sum_{m=2}^{N_t}\alpha_m^2\right)\right\rangle \tag{12.2}$$

Following the reasoning in [265], beam-forming is optimal (rank$(\mathbf{Q}) = 1$) when $\frac{dC(p)}{dp}\big|_{p=0} \leq 0$ (necessary condition) and $\frac{d^2C(p)}{d^2p} \leq 0$, $p \in (0, P)$ (sufficient condition), which results in

$$\mathrm{E}\left\langle \frac{1}{1 + \frac{P}{\sigma^2}\alpha_1^2} \right\rangle \leq \frac{1}{1 + \frac{P}{\sigma^2(N_t-1)}\sum_{m=2}^{N_t}\mathrm{E}\left\langle\alpha_m^2\right\rangle}. \tag{12.3}$$

Now $\gamma = \mathrm{E}\left\langle\alpha_1^2\right\rangle$ is equal to the SNR gain of the feedback algorithms which can be calculated analytically in a multitude of cases (see Chapter 11). However, the evaluation of the left-hand side of (12.3) in closed form seems to be rather challenging, and therefore we calculate it numerically. Furthermore, $\sum_{m=2}^{N_t}\mathrm{E}\left\langle\alpha_m^2\right\rangle = N_t - \mathrm{E}\left\langle\alpha_1^2\right\rangle$, so that the right-hand side of (12.3) can be calculated analytically for different closed-loop schemes. Applying Jensen's inequality to the left-hand side of Equation (12.3) gives the condition $\mathrm{E}\left\langle\alpha_1^2\right\rangle \geq \frac{N_t}{2}$ for beam-forming to be the optimal transmit strategy, which appears to be a rather loose bound.

When beam-forming is not optimal with long-term feedback, the transmitter allocates power $P - p$, $p > 0$ to the principal eigenvector of $\mathbf{Q} = \mathbf{R}_{\mathrm{Tx}}$ and power p to the channel corresponding to the second largest eigenvalue. Beam-forming remains optimal ($p = 0$) when [93]

$$\mathrm{E}\left\langle \frac{1}{1 + \frac{P}{\sigma^2}\alpha_1^2} \right\rangle \leq \frac{1}{1 + \frac{P}{\sigma^2}\lambda_2}, \tag{12.4}$$

where α_1^2 is exponentially distributed, showing that the optimality of beam-forming depends on the total transmit power P and the magnitudes of the two largest eigenvalues λ_1 and λ_2 of the channel correlation matrix. The expectation on the left-hand side can be further evaluated to yield [93]

$$\frac{\sigma^2}{P\lambda_1}e^{\frac{\sigma^2}{P\lambda_1}}\Gamma(0, \frac{\sigma^2}{P\lambda_1}) \leq \frac{1}{1 + \frac{P}{\sigma^2}\lambda_2}, \tag{12.5}$$

where $\Gamma(\cdot, \cdot)$ refers to the incomplete Gamma function. The optimality conditions have been generalized in [246] to show when it is optimal to send m parallel data streams for a given channel and operation point. There, the condition in Equation (12.5) arises as a special case when $m = 1$. According to the numerical results, when SNR increases, the number of parallel channels increase as well.

Short-term feedback– Mode 1 vs. Mode 2: The SNR gain expression $\gamma = \mathrm{E}\left\langle\alpha_1^2\right\rangle$ is calculated analytically in Chapter 11 for different feedback algorithms with various imperfections. These can be combined with the capacity expressions to gain insight into the sensitivity of different closed-loop concepts. Indeed, Figure 12.1 demonstrates SNR regions

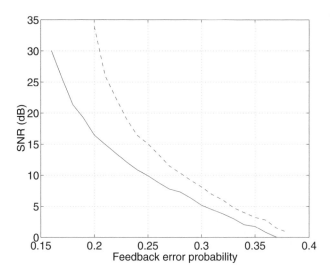

Fig. 12.1: SNR regions (below the curves) where Mode 1 (—) and Mode 2 (- -) are optimal information transfer strategies as a function of feedback error probability.

where Mode 1 and Mode 2 are optimal transmit strategies in a flat Rayleigh fading environment as a function of the probability of feedback bit errors. When the expected received SNR is below a curve, the corresponding closed-loop mode is better than a vector coding scheme. Without feedback errors, closed-loop modes provide full diversity and SNR gain, and therefore a vector coding scheme, providing only diversity, cannot be better. When the quality of the feedback information decreases, feedback modes loose diversity (see section 11.5) and $E\langle\alpha_1^2\rangle$ decreases. When the error probability is small, the closed-loop modes are still optimal while with high error probabilities the closed-loop modes are never optimal. In between, the closed-loop modes are optimal below the SNR values shown in Figure 12.1. Above the curves a vector coding scheme utilizing the quantized channel state information of the closed-loop modes provides a better mutual information. Mode 2 provides a larger SNR gain than Mode 1, and even when the feedback word in Mode 2 is longer, Mode 2 has a larger optimal SNR region in case of feedback bit errors. Both modes are better than vector coding in a capacity sense when considering the nominal 4% feedback error probability utilized in WCDMA link–level simulations.

The order between the modes changes when capacity region is depicted as a function of mobile speed (Figure 12.2) with 1500 bps feedback capacity and one slot (2/3 ms) feedback latency. Now Mode 1 outperforms Mode 2 because of its shorter feedback word. Note that the optimality region for capacity grossly overestimates the performance of closed-loop schemes, because in link-level simulations the switching point between STTD and closed-loop is already at 70 km/h [232].

In MIMO systems and with covariance feedback, the power allocation problem becomes

$$C(p) = \max_{\mathrm{Tr}(\mathbf{P})=P} E\left\langle \log\det(\mathbf{I} + (P-p)\frac{\tilde{\mathbf{h}}_1 P_1 \tilde{\mathbf{h}}_1^\dagger}{\sigma^2} + p\frac{\tilde{\mathbf{h}}_2 P_2 \tilde{\mathbf{h}}_2^\dagger}{\sigma^2})\right\rangle \qquad (12.6)$$

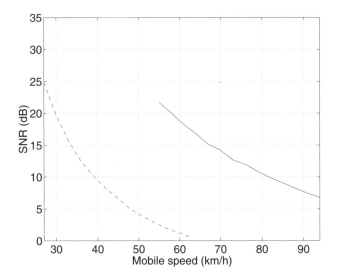

Fig. 12.2: SNR regions (below the curves) where closed-loop Mode 1 (—) and Mode 2 Mode 2 (- -) are optimal information transfer strategies as a function of mobile speed.

Now the evaluation of the derivative of $C(p)$ is more complicated than with one receive antenna. In this case [263] developed separate necessary and sufficient conditions for beam-forming to be the optimal transmission strategy. The conditions were later refined in [264].

12.2 TRANSCEIVER CONCEPTS

In this section we consider different transceiver concepts that combine orthogonal and non-orthogonal matrix modulation and channel information in different formats. Channel information is either obtained in real time from the terminal via short-term feedback, or terminal signals only channel information pertaining to the statistical long-term channel character-istics. As stated above, the statistical or geometrical characteristics vary slowly in time, and require infrequent updating. Therefore, also the required feedback capacity is signifi-cantly lower than that needed for fast feedback. In Chapter 10 we touched on a similar topic using single-stream transmission and feedback-based beam-forming. Here, we assume that multiple beams are used due to imperfect channel state information.

12.2.1 Correlated Channels, Precoding and Matrix Modulation

The correlation properties are characterized in analogy with the statistical channel models adopted in 3GPP work, as described in Chapter 2. The channel models considered in Chapters 2 and 10 explicitly model the statistical correlation between each transmit–receive antenna pair. These models account for certain practical deployment aspects. For example, if the inter-element spacing in the transmitting or receiving antennas is narrow, channel coefficients

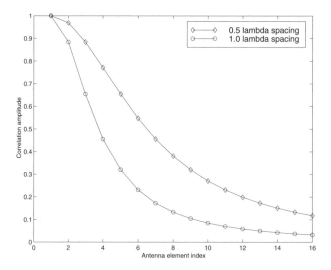

Fig. 12.3: Correlation magnitudes between antenna element 1 and the other antenna elements in a spatially correlated channel with inter-element spacing λ and $\lambda/2$. The correlation matrix corresponds to a case where the angle of arrival is 20 degrees and the angle spread is 10 degrees.

can exhibit significant correlation. In addition, the number of scatterers may be small and dominant multipath components may only arrive from certain directions. This limits the angular spread, and again correlation increases. It is of interest to see whether high-rate matrix modulation schemes are applicable in channels where significant channel correlations prevail.

To obtain transmit diversity, the different antenna elements cannot be fully correlated. The long-term or average correlation matrix

$$\mathbf{R}_{\text{Tx}} = \mathrm{E} \left\langle \mathbf{H}\mathbf{H}^{\dagger} \right\rangle \tag{12.7}$$

depends on the geometric properties of the channel and the geometry of the antenna array, as described above. While the channel itself is beyond the control of an average communication engineer, the array structure is not. For example, the inter-element distance can be varied. Different inter-element distances lead to different channel correlation matrices. This is shown in Figure 12.3, which depicts the characteristic magnitudes of the correlation coefficients between antenna element 1 and the remaining 15 ULA elements in one example. The physical channel model hypothesized in this example assumes a direction-of-arrival of 20 degrees and angle spread of 10 degrees. It is seen that when antenna separation increases, the correlation between the channels between different elements generally decreases.

In MIMO transmissions a more relevant characterization is related to the eigenvalue spread of the channel correlation matrix. Figure 12.4 demonstrates the cumulative sum of eigenvalues for inter-element spacings $\lambda/2$ and λ. It is seen that when inter-element spacing is increased from $\lambda/2$ to λ, the number of dominant eigenvalues increases. When the antenna spacing is $\lambda/2$ the channel is strongly correlated and has only a few dominant eigenvalues. In this case MIMO transmission may not be efficient, and conventional beam-

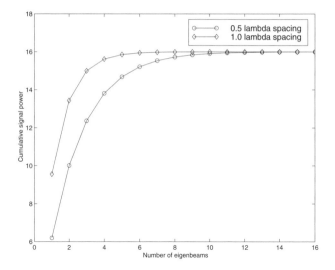

Fig. 12.4: Cumulative signal power when a different number of eigenbeams are used in a spatially correlated channel with different inter-element spacing.

forming, or closed-loop beam-forming (in the spirit of Chapters 10 and 11) is more amenable for information transfer.

The long-term channel correlation matrix of the terminal can be constructed analogously. In the MIMO example later in this chapter we assume that the two antennas are correlated according to Equation (2.11) in Chapter 2, where the correlation coefficient between the two elements is -0.304.

The capacity results in the previous section suggest that (with optimal coding) eigenvalues and eigenvectors of the channel correlation matrix $\mathbf{R}_{\mathrm{Tx}}$, or its sample estimate, play a significant role when maximizing capacity. The rows of matrix $\mathbf{E}$ in the eigenvalue decomposition (EVD)

$$\mathbf{R}_{\mathrm{Tx}} = \mathbf{E}^{\dagger} \Lambda \mathbf{E} \tag{12.8}$$

reveal the dominant eigenvectors or eigenbeams of the channel correlation matrix. The dominant beam is defined as the eigenvector corresponding to the largest eigenvalue in matrix $\Lambda = \mathrm{diag}(\lambda_1, ..., \lambda_{N_t})$. The average power amplification of eigenbeam k is λ_k. We choose to use eigenbeams as an inherent part of the precoding matrix $\mathbf{W}$, as described below. To recapitulate, the received signal is

$$\mathbf{Y} = \mathbf{XWH} + \text{noise}, \tag{12.9}$$

where matrix $\mathbf{X}$ is the space–time modulation matrix, $\mathbf{W}$ is a precoding matrix at the transmitter, and matrix $\mathbf{H}$ contains the MIMO channel coefficients, as described in Equation (1.1) in Chapter 1. The eigenbeams enter into the factorization for the beam-forming matrix,

$$\mathbf{W} = \Pi \mathbf{AE}, \tag{12.10}$$

where the rows of $\mathbf{E}$ are eigenvectors of the channel correlation matrix, $\mathbf{A}$ is a diagonal matrix that specifies the transmit amplitudes for each eigenbeam, and Π is a permutation matrix that can be used to shuffle the columns of $\mathbf{X}$ to appropriate eigenbeams. Therefore, the average received signal power for the kth eigenbeam is $\lambda_k a_k^2$. For a given channel realization $\mathbf{H}$, the receiver experiences channel

$$\tilde{\mathbf{H}} = \mathbf{W}\mathbf{H}. \tag{12.11}$$

The transmitter is provided with information pertaining to $\mathbf{W}$. This information can specify $\mathbf{E}, \mathbf{\Pi}$, or $\mathbf{A}$ in the adopted factorization (12.10). The information can be obtained partly from blind uplink measurements and partly using feedback from the receiver. In a spatially white channel the natural eigenbeams are $\mathbf{E} = \mathbf{I}_{N_t}$. However, in a spatially correlated channel all N_t dimensions are not equally relevant.

From a performance point of view, $\mathbf{X}$ and $\mathbf{W}$ should be defined jointly. If we first fix the modulation matrix $\mathbf{X}$ we need to adjust the "free parameters" in $\mathbf{W}$ factorization in order to match them appropriately to $\mathbf{X}$. With long-term beam-forming, where the eigenvectors are not refreshed after each channel realization, the remaining free parameters enable power control or power allocation via matrix $\mathbf{A}$ and antenna or beam shuffling via matrix $\mathbf{\Pi}$.

12.2.2 Long-term Beam-forming with MISO Transmission

STTD-PHOP in Correlated Channels: In Chapter 8 we introduced the STTD-PHOP transmitter structure [24] and demonstrated its performance in an i.i.d. Rayleigh fading channel. Recall from Chapter 8 that the modulation matrix adopted for STTD-PHOP is a product of

$$\mathbf{X} = \frac{1}{\sqrt{2}} \begin{bmatrix} \mathbf{X}_A & \mathbf{X}_A \end{bmatrix}, \tag{12.12}$$

and the right action matrix is

$$\mathbf{W}_1 = \begin{bmatrix} \mathbf{I}_2 & \mathbf{0}_2 \\ \mathbf{0}_2 & \Theta \end{bmatrix}, \tag{12.13}$$

where the phasor matrix is

$$\Theta = \begin{bmatrix} e^{j\theta_{t,3}} & 0 \\ 0 & e^{j\theta_{t,4}} \end{bmatrix}. \tag{12.14}$$

Applying $\mathbf{W}_1$ on the modulation matrix (12.12) rotates the symbols in the third and fourth columns of the modulation matrix. With long-term beam-forming $\mathbf{W}_1$ can be considered as a factor in the overall beam-forming matrix $\mathbf{W}$,

$$\mathbf{W} = \mathbf{W}_1 \mathbf{\Pi}\mathbf{A}\mathbf{E}. \tag{12.15}$$

Using this, the received signal is

$$\mathbf{y} = \mathbf{X}\mathbf{W}\mathbf{h} + \text{noise}, \tag{12.16}$$

where we assume, for simplicity, that only one receive antenna is used. Here, the four STTD-PHOP branches are transmitted via four dominant eigenbeams. Therefore, we call this beam-space STTD-PHOP in the following. The receiver is interested in the channel correlations, which are dictated by vector $\tilde{\mathbf{h}} = \mathbf{\Pi}\mathbf{A}\mathbf{E}\mathbf{h}$. Using this and matrix $\mathbf{W}_1$ in (12.13), the equivalent channel correlation matrix for beam-space STTD-PHOP is

$$\mathcal{D}_{\tilde{\mathbf{h}}} = \begin{bmatrix} |\tilde{h}_{i_1} + \tilde{h}_{i_3}e^{j\theta_{t,i_3}}|^2 & \mathbf{0}_2 \\ \mathbf{0}_2 & |\tilde{h}_{i_2} + \tilde{h}_{i_4}e^{j\theta_{t,i_4}}|^2 \end{bmatrix} \tag{12.17}$$

$$\mathcal{S}_{\tilde{\mathbf{h}}} = 0, \tag{12.18}$$

wherein the indices $i_1, ..., i_4$ are determined by the permutation matrix $\mathbf{\Pi}$. The permutation matrix plays a role in optimizing performance when the received channels have different powers or statistics.

Permutation Matrix Without Fast Feedback: It is easy to convince oneself that the average received power and the transmit diversity order can be controlled not only by power allocation, but also by the permutation matrix $\mathbf{\Pi}$. For example, if the eigenvalues corresponding to two of the four eigenbeams are zero, different permutation matrices retard the concept either to beam-space STTD or PSTD, described in Chapter 3. Naturally, with these two options, STTD is preferred due to its higher transmit diversity order. A permutation matrix that achieves this goal is one that shuffles the two branches of the STTD to the two strongest eigenbeams. These are then paired with phase-hopped copies transmitted via the third and fourth eigenbeams. Clearly, the transmit amplitudes, contained in matrix $\mathbf{A}$, should also be optimized, e.g. according to water-filling or some other criteria [88]. However, initial studies suggest that the performance improvement is not large when compared to equal power transmission or to some crude power allocation policy that uses a discrete set of possible transmit powers per beam. The most notable effect of matrix $\mathbf{A}$ is that it allows conventional eigenbeam-forming (transmission to one beam only), beam-space STTD (transmission to two strongest beams), and beam-space STTD-PHOP (transmission to more than two beams) to take place in one single model.

Permutation Matrix and Fast Feedback: With sufficient channel state information at the transmitter, the co-phasing terms $e^{j\theta_{t,3}}$ and $e^{j\theta_{t,3}}$ can be optimized, such that the instantaneous received signal power is maximized. The solution can be obtained in analogy with WCDMA closed-loop transmit diversity concepts. With phase feedback, the feedback is determined from

$$\theta_{t,3} = \arg[\tilde{\mathbf{h}}_{i_1}[t]^\dagger \tilde{\mathbf{h}}_{i_3}[t]] \tag{12.19}$$

$$\theta_{t,4} = \arg[\mathbf{h}_{i_2}[t]^\dagger \mathbf{h}_{i_4}[t]]. \tag{12.20}$$

Here, however, the permutation matrix should be defined differently, when compared to cases without fast feedback. The beam-forming gain is maximized by co-phasing the eigenbeams with the two largest eigenvalues.

Reducing the Required Feedback: The concept in Equations (12.19) and (12.20) controls two phases simultaneously. This doubles the required feedback signalling overhead, if the control delay is not increased. To avoid this, the receiver can determine a single co-phasing coefficient that maximizes the received power when integrated over the whole space–time block code. The solution is obtained from

$$\theta_{t_3} = \theta_{t_4} = \arg[(\tilde{\mathbf{h}}_{i_1}[t] + \tilde{\mathbf{h}}_{i_3}[t])^\dagger (\tilde{\mathbf{h}}_{i_2}[t] + \tilde{\mathbf{h}}_{i_4}[t])] \tag{12.21}$$

In this case the shuffling (permutation) should be such that channel power is spread as evenly as possible to channels with odd and even indices.

12.2.3 Long-term Beam-forming with Non-orthogonal Matrix Modulation

The symbol rate two matrix modulation schemes considered below are DSTTD and DABBA. These can be described using modulation matrix

$$\mathbf{X}(x_1, ..., x_8) = \begin{bmatrix} \tilde{\mathbf{X}}_A & \tilde{\mathbf{X}}_C \\ \tilde{\mathbf{X}}_D & \tilde{\mathbf{X}}_B \end{bmatrix}. \tag{12.22}$$

where $\tilde{\mathbf{X}}_A$ and $\tilde{\mathbf{X}}_B$ are individually orthogonal space–time block codes with rotated symbol constellations, as described in Chapter 8. These two matrices together carry four symbols. To increase the symbol rate to two $\tilde{\mathbf{X}}_C$ modulates symbols x_5 and x_6, and $\tilde{\mathbf{X}}_D$ modulates symbols x_7 and x_8. These are transmitted in parallel. The scheme is equivalent to DSTTD when rotated constellations are not used.

Unlike Chapters 8 and 9, the four columns are transmitted via different beams. Naturally, the same principle can be applied to any high symbol rate scheme. The channel sensed by the receiver is again $\tilde{\mathbf{W}} = \mathbf{WH}$. The choice of $\mathbf{W}$ affects the choice for $\mathbf{X}$ since different channel profiles favour schemes where interference is mitigated when there are large power differences between the eigenbeams.

12.3 PERFORMANCE

Numerous power-efficient matrix modulation schemes with a high symbol rate were proposed and analysed in Chapter 9 in an i.i.d Rayleigh fading channel. However, a practical MIMO channel is rarely i.i.d. Rayleigh. Below, we construct specific examples where a symbol rate two matrix modulation is pushed through a correlated MIMO channel.

Beam-space STTD-PHOP: First, we study performance when STTD, STTD-PHOP and feedback STTD-PHOP are pushed through a correlated MISO channel with $N_t = 8$ or $N_t = 16$ array elements using either $\lambda/2$ or λ spacing, as in section 12.2.1. The number of receive antennas $N_r = 1$. The channel is quasi-static flat Rayleigh fading, and the BICM transmitter uses rate 3/4 turbo coding and a frame size of 378 bits, in analogy with Chapter 10. With QPSK modulation we achieve 1.5 bps/Hz spectral efficiency. In each case we adopt equal power transmission to two or four beams, with STTD and STTD-PHOP, respectively. The transmitter is assumed to know up to four eigenbeams, and the order of the corresponding eigenvalues. These are needed when selecting the permutation matrix and eigenvectors in the precoding matrix $\mathbf{W}$ factorization (12.10).

We notice from Figures 12.5 and 12.6 that with $N_t = 8$ STTD is superior to STTD-PHOP. This is expected, since equal power transmission over four beams is not efficient in this case, as one eigenvalue dominates (see Figure 12.4). However, when $N_t = 16$ the performance of STTD-PHOP is somewhat better than STTD, especially at the high signal-to-noise region. The real performance boost, in the order of 5 dB, comes from simple feedback using two phase feedback bits in this example.

Beam-space DABBA/DSTTD: In the following example, we concatenate a symbol rate two matrix modulation (DSTTD and DABBA) and eigenbeam-forming. The long-term beams are assumed to be known at the transmitter. They can be estimated at the terminal and signalled to the base station with sufficient accuracy. Different columns of the modulation

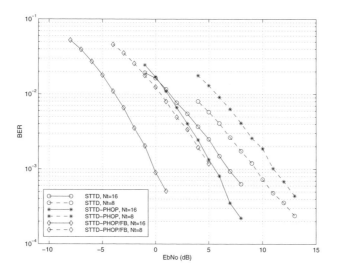

Fig. 12.5: Bit-error rates in correlated channels with $N_t = 8$ and inter-element spacing $\lambda/2$, and $N_t = 16$ with inter-element spacing λ. Eigenbeam-forming STTD, STTD-PHOP, and STTD-PHOP with phase feedback (STTD-PHOP/FB), each with 1.5 bps/Hz spectral efficiency using rate 3/4 turbo code.

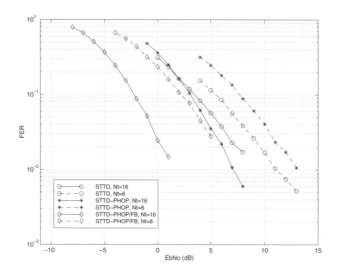

Fig. 12.6: Frame-error rates corresponding to different cases in Figure 12.5 using a frame size of 378 bits.

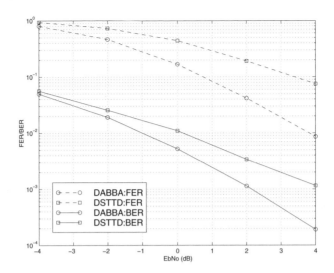

Fig. 12.7: Frame-error rates and bit-error rates in a spatially correlated channel for DSTTD and DABBA with 4 bps/Hz spectral efficiency. The transmitter has a 16-element ULA with λ spacing, and the receiver has two elements with $\lambda/2$ spacing. Transmission is to four dominant long-term beams with equal power.

matrix are mapped to different eigenbeams with even power allocation. The base station is assumed to deploy a uniform linear antenna array with 16 elements with λ spacing to create a sufficient number of dominant eigenbeams (see Figure 12.4). The receiver has two antenna elements. Recall that with a symbol rate two transmission, we obtain 4 bps/Hz spectral efficiency, whereas in the previous example the uncoded spectral efficiency was only 2 bps/Hz. In this example we do not utilize fast feedback, nor any channel code. The number of uncoded bits (frame size) used to calculate FER is 128.

Figure 12.7 reveals that 10% FER is achieved when $E_b/N_0 = 0.7$ dB and 3.3 dB, respectively for DABBA and DSTTD. The relative performance gain in favour of DABBA is due to full diversity, which is beneficial when the eigenbeam powers are different. With DSTTD, low-power substreams dominate the error events. BER of 10^{-3} is achieved at 2 and 4 dB with DABBA and DSTTD, respectively. These numbers can be compared to the performance results in Chapter 10, obtained in an i.i.d. quasi-static channel without eigenbeam-forming using four transmit antennas, where DABBA and DSTTD were shown to achieve bit-error-rate 10^{-3} at approximately $E_b/N_0 = 6.5$ dB and $E_b/N_0 = 8$ dB, respectively.

12.4   SPACE–TIME RETRANSMISSION

Significant effort has been directed in recent years toward developing Automatic Repeat Request (ARQ) protocols for 3G systems. Indeed, the Hybrid ARQ protocol developed for HSDPA [3], summarized in Chapter 1, forms an important part of the Release 5 WCDMA specification. One typical feature of these recent ARQ solutions, when compared to earlier releases, is a reduced control delay between each transmissions. In addition, increasingly efficient and complex signal combining strategies are advocated. An optimal signal combining strategy should collect information from all transmissions, without discarding erroneous packets.

To reduce control delay, the Hybrid ARQ process in Release 5 operates at the physical layer, and is controlled by the base station. Eventually, in some future system, the control delay may be short enough such that the retransmissions occur predominantly within the channel coherence time. This clearly reduces time diversity, but many other forms of diversity are still available.

The most notable diversity resource considered in this book is space diversity. We have learned that in the absence of alternative diversity resources, spatial diversity may be exploited by careful utilization of transmit and receive antennas, when combined with efficient (matrix) modulation and coding concepts. In addition, we have learned that multiple-input multiple-output (MIMO) channels enable very high symbol and data rates, provided that the channels impinging on these antennas are sufficiently uncorrelated.

Recall that the high-rate MIMO transmission methods, necessarily lead to non-orthogonal matrix modulation alphabets. From the receiver viewpoint, this is manifested by the off-diagonal correlation coefficients in the equivalent channel matrix, as shown in Chapter 8. Then, the success of signal reception depends essentially on both the SNR and the rank (or the condition number) of the equivalent channel matrix. In a properly designed space–time retransmission concept both the SNR and the condition number (or orthogonality) should increase after each retransmission. It is shown below that this can be guaranteed under quasi-static fading assumption, where the channel remains invariant until the packet is correctly received. Under this scenario, the space–time retransmission concept initially attempts some MIMO or MISO high-rate matrix modulation, but eventually with sufficient retransmissions, reduces to some full diversity orthogonal space–time block code. We assume the simplest possible feedback concept, in which only one bit (NACK/ACK) is sent to the transmitter, and where the transmitter operates without explicit channel state information. Combining space diversity solutions in this way with retransmission concepts leads to a *space–time adaptive retransmission* (STAR) paradigm [266], in which

- space–time modulation is used for the initial transmission,

- space–time modulation used for retransmission is different from that used in the initial transmission.

- when space–time modulation matrices are combined over the transmissions, the diagonal dominance (see section 4.6) of the resulting transmission scheme is increased.

This rather loose description is clarified below using examples.

12.4.1   Retransmission with Symbol Rate One Schemes

Recall the received signal model

$$\mathbf{y}[t] = \mathbf{X}_{tr}[t]\mathbf{h}[t] + \mathbf{n}[t], \tag{12.23}$$

used in Chapter 8. We assume, without loss of generality, that $\mathbf{X}_{tr}$ is the transformed STBC-OTD defined by Equations (8.35) and (8.36). The equivalent channel correlation matrix for one space–time block transmitted at time epoch t, say, is

$$\mathcal{R}[t] = \mathbf{U}(\mu[t], \nu[t])^\dagger \mathcal{H}^\dagger[t] \mathcal{H}[t] \mathbf{U}(\mu[t], \nu[t]) = \mathcal{D}_\mathbf{h}[t] + \mathcal{S}_\mathbf{h}[t], \tag{12.24}$$

where

$$\mathcal{D}_\mathbf{h}[t] = \begin{bmatrix} a_1[t] & 0 \\ 0 & a_2[t] \end{bmatrix} \otimes \mathbf{I}_{N_t/2}, \tag{12.25}$$

and

$$\mathcal{S}_\mathbf{h}[t] = \begin{bmatrix} 0 & b[t] \\ b[t]^* & 0 \end{bmatrix} \otimes \mathbf{I}_{N_t/2}, \tag{12.26}$$

with

$$a_1[t] = p_1[t]|\mu[t]|^2 + p_2[t]|\nu[t]|^2 \tag{12.27}$$

$$a_2[t] = p_2[t]|\mu[t]|^2 + p_1[t]|\nu[t]|^2 \tag{12.28}$$

$$b[t] = (p_2[t] - p_1[t])\mu[t]^*\nu[t] \tag{12.29}$$

$$p_1[t] = \sum_{j=1}^{N_r}\sum_{i=1}^{N_t/2} |h_{i,j}[t]|^2 \tag{12.30}$$

$$p_2[t] = \sum_{j=1}^{N_r}\sum_{i=N_t/2+1}^{N_t} |h_{i,j}[t]|^2. \tag{12.31}$$

We have appended explicitly the parameter t to model to capture the time dynamics inherent in the retransmission process.

Assume that the initial transmission occurs at epoch t_1. If the signal is detected correctly the throughput remains at one. Otherwise, a retransmission trigger is signalled to the base station as a negative acknowledgement (NACK). Consequently, retransmission is scheduled at the base station for time epoch t_2. Conventionally, assuming that the signals are combined at the terminal at symbol level, the equivalent channel correlation matrix after epoch t_2 is

$$\mathcal{R}[t_1, t_2] \equiv \mathcal{S}_\mathbf{h}[t_1] + \mathcal{S}_\mathbf{h}[t_2] + \mathcal{D}_\mathbf{h}[t_1] + \mathcal{D}_\mathbf{h}[t_2]. \tag{12.32}$$

If the channel is static the correlation properties of $\mathcal{R}[t_1, t_2]$ are the same as for $\mathcal{R}[t_1]$ or for $\mathcal{R}[t_2]$. Therefore, if signal detection or decoding failed at t_1 due to dominant self-interference matrix $\mathcal{D}$, it may also fail at t_2.

The target with space–time retransmission is to guarantee that the equivalent channel correlation after the retransmission, or more generally after N retransmissions, is more amenable for detection. Indeed, we can adapt either the beam-forming matrix $\mathbf{W}$ in (1.1), the modulation matrix, or the symbol rotation parameters (μ, ν) for each retransmission separately, such that the diagonal dominance of matrix

$$\mathcal{R}[t_1, ..., t_N] \equiv \sum_{l=1}^{N} \mathcal{R}[t_l] \tag{12.33}$$

is increased. The diagonal elements (12.33) grow larger simply due to the increased signal power in retransmission. However, the off-diagonal power can be altered for each transmission as follows: at t_1 and t_2 we apply the precoding matrix parameters

$$(\mu[t_1], \nu[t_1]) = (1/\sqrt{2}, e^{j\phi}/\sqrt{2}) \tag{12.34}$$

and

$$(\mu[t_2], \nu[t_2]) = (1/\sqrt{2}, e^{-j\phi}/\sqrt{2}), \tag{12.35}$$

respectively. Then, it is easy to see that correlation coefficient b associated with matrix (12.33) vanishes, assuming that the channel power does not change between t_1 and t_2. Alternatively, we could apply antenna indices $(1, 2, 3, 4)$ and $(3, 4, 1, 2)$ in epochs t_1 and t_2. This can be accomplished by appropriate definitions of (permutation matrices) $\mathbf{W}[t_1]$ and $\mathbf{W}[t_2]$. After one retransmission and signal combing, transformed STBC-OTD is converted to a symbol rate half orthogonal full diversity space–time block code, where the decoding delay is eight symbols.

This example is intended to highlight the fact that ARQ and high-rate space–time matrix modulation concepts should be designed together. The effective symbol rate or throughput then naturally adapts itself to the prevailing channel conditions.

12.4.2 Retransmission with Symbol Rate Two Schemes

In the previous example the initial transmission utilized a non-orthogonal modulation matrix with symbol rate one. However, retransmission protocols tend to benefit from a higher peak rate. Conventionally, this is achieved by adopting high-order modulation alphabets, as in HSDPA [3] where 16-QAM modulation is needed. Another option is to use high-rate matrix modulation during the first transmission to increase the peak throughput. Subsequent space–time transmissions use different modulation matrices. In a way, the initial transmission is a submatrix of a sub-optimal scheme. With enough retransmissions a concatenation of these submatrices eventually completes the scheme.

A potential inherent limitation in applying STAR is that the high-rate matrix modulation schemes developed in Chapter 9 in this book assume a static channel for the entire duration of the space–time transmission. If the retransmission delay is long, this requirement is not generally fulfilled. On the other hand, in practice the constructions tolerate some additional interference and loss of diversity. In the end, there is a tradeoff between ARQ delay and space–time orthogonality. In practice, it is sufficient that channel coefficients or powers are sufficiently correlated in order to benefit from space–time adaptive retransmission. Thus, each retransmission should commence within channel coherence time.

BLAST → Time-reversal STTD: One delay sub-optimal space–time block code (as discussed in section 3.6.2) may assume the form

$$\begin{bmatrix} x_1 & x_3 & \cdots & x_{2N-1} & \text{control delay} & -x_{2N}^* & \cdots & -x_4^* & -x_2^* \\ x_2 & x_4 & \cdots & x_{2N} & \text{control delay} & x_{2N-1}^* & \cdots & x_3^* & x_1^* \end{bmatrix}^T, \tag{12.36}$$

where "control delay" refers to ARQ and scheduling delay, initial transmission is vector signalling with symbol rate two, and if a retransmission is needed the transmission collapses to the time-reversal space–time block code of [152]. The first part of the transmission can potentially be detected with sufficient receive antennas, in analogy with BLAST [21]. If not, symbol rate is reduced to one.

DSTTD → STTD-OTD: With DSTTD the processing delay can be reduced from eight to four, at the expense of performance. In this case, the initial transmission applies

$$\mathbf{X}[t_1] = \begin{bmatrix} \mathbf{X}_A & \mathbf{X}_B \end{bmatrix}. \tag{12.37}$$

If a retransmission is requested, it is carried out with matrix

$$\mathbf{X}[t_2] = \begin{bmatrix} \mathbf{X}_A & -\mathbf{X}_B \end{bmatrix}. \tag{12.38}$$

If $\mathbf{X}_A$ and $\mathbf{X}_B$ are both STTD blocks with QPSK symbol constellations, the bit rate during the first transmission is 4 bps/Hz. If a retransmission is needed the throughput is reduced to 2 bps/Hz. If the retransmission occurs within channel coherence time DSTTD is converted to STTD-OTD, as described in Chapter 8. Yet another retransmission can be carried out during the next two time epochs, using permuted antenna indices, leading to modulation matrix

$$\mathbf{X} = \begin{bmatrix} \mathbf{X}_A & \mathbf{X}_B \\ \mathbf{X}_A & -\mathbf{X}_B \\ \mathbf{X}_B & \mathbf{X}_A \\ \mathbf{X}_B & -\mathbf{X}_A \end{bmatrix}, \tag{12.39}$$

with diversity order four. The orthogonality of this rate half transmission follows immediately when summing up the two equivalent channel correlation matrices with different permutation matrices, as shown in section 8.1.2.1. Therefore, we again utilize submatrices of a particular delay sub-optimal space–time block code, such that initially only a fraction of the code is transmitted, and the retransmissions complete the code.

DSTTD $\to$ ABBA: If the receiver is committed to detect DSTTD, it should also be able to handle ABBA. If the initial transmission transmits two first rows of

$$\mathbf{X} = \begin{bmatrix} \mathbf{X}_A & \mathbf{X}_B \\ \mathbf{X}_B & \mathbf{X}_A \\ \mathbf{X}_B & \mathbf{X}_A \\ \mathbf{X}_A & \mathbf{X}_B \end{bmatrix}, \tag{12.40}$$

and the first retransmission the next two rows, and finally the final four rows, we observe that DSTTD is first converted to ABBA, then to a fully orthogonal full-diversity delay sub-optimal rate 0.25 transmission. Finally, if the baseline transmission method is a symbol rate two space–time modulation using rotated constellations, the diagonal retransmissions can use the same scheme but with different rotation parameters in retransmissions, as described by (12.34) and (12.35). In this case, the first transmission uses a modulation matrix such as

$$\mathbf{X}[t_1] = \begin{bmatrix} \tilde{\mathbf{X}}_{A,\mu_1,\nu_1} & \tilde{\mathbf{X}}_{C,\mu_1,\nu_1} \\ \tilde{\mathbf{X}}_{D,\mu_1,\nu_1} & \tilde{\mathbf{X}}_{B,\mu_1,\nu_1} \end{bmatrix}. \tag{12.41}$$

A possible retransmission at epoch t_2 uses matrix

$$\mathbf{X}[t_2] = \begin{bmatrix} \tilde{\mathbf{X}}_{A,\mu_2,\nu_2} & \tilde{\mathbf{X}}_{C,\mu_2,\nu_2} \\ \tilde{\mathbf{X}}_{D,\mu_2,\nu_2} & \tilde{\mathbf{X}}_{B,\mu_2,\nu_2} \end{bmatrix}. \tag{12.42}$$

After these two transmissions, and eight symbol epochs, self-interference between blocks $\mathbf{X}_A$ and $\mathbf{X}_B$ vanishes, as it does for blocks $\mathbf{X}_C$ and $\mathbf{X}_D$. Additional transmissions can operate analogously.

12.4.3 Numerical Example

Consider a numerical example involving a space–time architecture with two or four transmit antennas and two receive antennas. We consider the following alternatives:

- DABBA using two parallel QPSK streams,

- DABBA using two parallel QPSK streams with STAR, and

- Single stream 16-QAM using STTD.

In all cases the information bits are encoded with a rate 4/5 turbo code using random bit interleaving and a frame size of 378 bits. The peak rate for all transmission concepts is 3.2 bps/Hz.

If a retransmission is needed with DABBA, we modify the modulation matrix used in each retransmission to increase diagonal dominance of the combined transmission using the STAR concept. This is implemented by changing the parameters in the precoding matrix (as given in (12.34) and (12.35) for the first two transmissions; For the third and fourth transmission we operate analogously, except that then the signals transmitted from antennas three and four are multiplied by -1. This leads to an orthogonal code after four transmissions. Without STAR we simply repeat the same DABBA modulation matrix in all transmissions. When STTD is used we employ 16-QAM modulation to obtain diversity order four using a 2 Tx-2 Rx architecture. The maximum number of transmissions is set to four, and the throughput is defined as

$$\text{Throughput} = R \sum_{l=1}^{4} \pi_l / l \quad [\text{bps/Hz}] \tag{12.43}$$

where R is the peak rate of the given concept, π_l the probability that the lth transmission is successful.

Figure 12.8 reveals that symbol rate two space–time modulation DABBA using STAR (legend "DABBA/STAR") provides superior performance in a quasi-static Rayleigh fading channel. When STAR is not used with DABBA (legend "DABBA") performance degrades especially when the E_b/N_0 is low. STTD transmission using 16-QAM (and Chase combining) is clearly inferior. For example, it requires about 2 dB more power than DABBA/STAR to reach 2 bps/Hz.

The throughput gains are apparent, and the gains are attained without the need to change the modulation order for retransmissions. A future system that supports both matrix modulation and ARQ can benefit from the proposed concept. This allows implicit rate adaptation in arbitrarily correlated channels. With sufficient retransmissions signal reception will become independent of the channel rank.

12.5   ADAPTIVE SPACE–TIME MODULATION ARRANGEMENTS

Channel Quality Indicators (CQI) with corresponding Adaptive Modulation and Coding (AMC) arrangements are another form of feedback. WCDMA Release 5 [3] makes extensive use of these concepts in providing high–speed downlink packet access.

The AMC arrangements in [3] are geared for single antenna transmission and reception. A straightforward extension of the idea of adaptive modulation to the space–time domain is Adaptive Space–Time Modulation Arrangement (ASTMA).

The most important additional channel information arising from a MIMO channel is the channel rank, or more precisely the practical rank (Prank) of the channel, see section 2.4. This number, conditioned on the family of space–time modulators available, indicates how many parallel streams the channel can support.

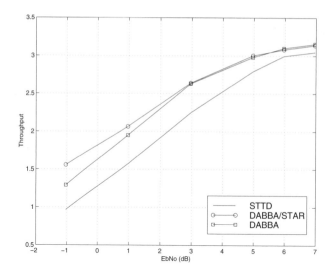

Fig. 12.8: Throughput for space–time retransmission with different space–time modulation schemes.

As the simplest example of ASTMA with CQI based on Prank, consider a $N_t = N_r = 2$ case with 4 bps/Hz. In section 9.2, a symbol rate $R_s = 2$ scheme (9.13), transmitting four QPSK symbols in one space–time block extending over $T = 2$ symbol periods was optimized. This scheme was found to outperform 16-QAM STTD by a margin of 0.6 dB. With the observation in section 2.4 that even in uncorrelated channels, a sizeable fraction of channel matrices have reduced Prank, this margin may be increased with an one-bit CQI feedback indicating which scheme to choose. In slow fading, the feedback capacity required for such a scheme would be low, CQI needs to be transmitted once per packet, or in HSDPA, once per transport time interval (TTI).

In this case, a Prank measurement has only two possible values (1 and 2), and it reduces to measuring a condition number. Three practical options may be considered for measuring Prank. These are:

- a relative condition number, based on the ratio of the two singular values of the MIMO channel matrix (see (2.22)).

- a Demmel condition number, based on the ratio of the smallest absolute value to the Frobenius norm (square root of total channel power) of the channel;

- a computationally simple CQI which does not require solving for singular values, obtained by taking the determinant of H and comparing it to the total channel power.

In [97], using the Demmel condition number to choose between a vector modulation and a transmit diversity scheme was considered. The motivation for this was that a perfect Tx diversity scheme, e.g. STTD, maximum-ratio combines all channels, so the performance is characterized by the total channel power. In the example considered here, however, the $R_s = 2$ scheme is a judiciously designed matrix modulation, which captures full diversity just as STTD does, see Figure 9.1. Thus the performance of both schemes in this example

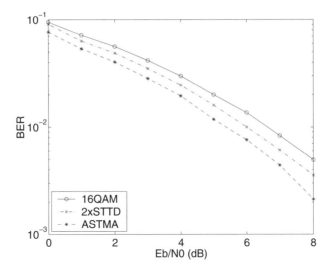

Fig. 12.9: Performance of a 4 bps/Hz $N_t = N_r = 2$ ASTMA with one-bit CQI. Compared to the constituent space–time modulation schemes.

of ASTMA is affected by the total channel power. This leads to the conjecture that a relative condition number would be better for CQI. Indeed, the performance of (9.13) depends on how well conditioned the equivalent channel matrix is. A straightforward calculation shows that the four singular values of the (complex symbol) equivalent channel matrix consist of two copies of each of the singular values of the channel matrix. This is a consequence of using a full Frobenius orthogonal basis of 2×2 matrices (see section 9.1) for the space–time modulation (9.13).

Numerical experiments reveal that all three options above provide comparable gains, with a slight penalty for the sub-optimal determinant CQI. In all cases, optimal Prank thresholds were such that for 30–40% of the channels, 16-QAM STTD was chosen, and for the rest, the two-stream scheme (9.13) was chosen. BER performance of the two constituent matrix modulation schemes, compared to ASTMA based on a relative condition number Prank with 40% threshold, can be found in Figure 12.9. The gain from this one-bit feedback is about 0.6 dB.

12.6 SUMMARY

In this chapter we have considered hybrid transmission methods that combine several transmission techniques introduced elsewhere in the book. In particular, we explored the interplay between $\mathbf{X}$ and $\mathbf{W}$ when the transmitter has partial channel knowledge. These two matrices jointly define the transmitted matrix-valued signal and need to be designed together. With perfect channel knowledge, $\mathbf{X}$ reduces in a MISO system to a complex scalar, and $\mathbf{W}$ to a vector, but with imperfect channel knowledge the dimensions of both matrices expand, holding pure open-loop concepts as a special case in the absence of channel state information. In a MIMO system, there is an even wider spectrum of possibilities. Imperfect channel state information reduces the beam-forming gain and motivates the use of hybrid open-loop closed-loop transmission systems that capture both diversity and SNR or array gain.

One particular application assumes that only a statistical model related to the transmission channel is known, not the channel realizations *per se*. Then, eigenbeam-forming is a well-founded technical solution. The eigenbeams may be determined at the terminal and signalled back to the base station, much in the same way as the instantaneous weights are signalled within the current WCDMA specification. However, when compared to instantaneous channel state feedback, the required feedback signalling rate is reduced. In some cases, the parameters of the statistical channel model can be acquired directly from the received signal statistics without explicit signalling bits.

The different branches of a space–time modulation matrix can be mapped to the eigenbeams in order to capture part of the array gain. Simultaneously, to capture the transmit diversity benefit the channels should be uncorrelated. Decorrelation can be intentionally increased simply by increasing the inter-element distances in the transmitting antenna array. It was shown that sufficient decorrelation through eigenbeam-forming enables the use of high symbol-rate space–time modulators in correlated channels. Such solutions are of significant interest when developing enhancements to 3G wireless systems or novel 4G air interfaces. Eigenbeams were adopted in this chapter mainly for ease of discussion. More practical concepts may be constructed around a fixed beam antenna architecture, where the elements of space–time modulation matrices are distributed to the strongest fixed beams.

Finally, we discussed two emerging topics. The first is a hybrid solution where repeat requests and space–time modulation matrices are effectively combined. It was shown that high-rate space–time modulation can be used to increase both throughput and diversity, in particular if the retransmissions are properly designed. The second considers a simple selection of a matrix modulation according to the instantaneous channel state. Bearing in mind the integral part that advanced ARQ and AMC concepts play in currently evolving systems, and the recent surge of efficient MIMO transmission schemes, such a combination may pave the way for increasingly efficient spectrum utilization in future wireless systems.

Appendix A
Symmetries, Invariants and Inequalities

A.1 SYMMETRIES AND TRANSFORMATIONS

In this book, the measures used when comparing different transmission schemes are functions of the parameters of the transmission, and the MIMO channel,

$$f = f(\mathbf{X}, \mathbf{W}, \mathbf{H}) . \tag{A.1}$$

Usually we are interested in ergodic measures, where we average over the channel using an appropriate pdf. Thus the bit error probability, for example, is averaged over all channel realizations, as well as the ergodic information measures used in this book.

With multi-antenna channels we are in the realm of multivariate analysis, and we shall borrow the terminology from the pertinent literature [214, 215].

Invariance Under a Transformation: A function $f(x), x \in \mathcal{X}$ is invariant under a set of transformations G if

$$f(g(x)) = f(x) \text{ for all } x \in \mathcal{X} \text{ and } g \in G \tag{A.2}$$

Group of Transformations: If G is closed, i.e. the transformation $g_2 \circ g_1$ obtained by first applying g_1, then g_2, is in the set G, and there exists an inverse transformation for all $g \in G$, then G is a group. All transformations considered in this book are of this kind.

Symmetry of a Quantity: A group of transformations leaving a quantity f invariant according to (A.2) is called a symmetry group of f, or symmetry of f, for short.

Example: Bit error Probability. The average bit error probability is an integral over $\mathbf{H}$ of the bit error probability conditioned on $\mathbf{H}$, with the appropriate channel pdf. For i.i.d. Rayleigh fading, the channel pdf is

$$p(\mathbf{H}) = \left(\frac{1}{2\pi\sigma^2} \right)^{N_t N_r} e^{- \operatorname{Tr} \mathbf{H}^\dagger \mathbf{H}/2\sigma^2}, \tag{A.3}$$

which directly shows that $p(\mathbf{WH}) = p(\mathbf{H})$ for $N_t \times N_t$ unitary $\mathbf{W}$. Thus any unitary transformation of the channel matrix is a symmetry of the pdf. Now consider a linear matrix modulator (4.4), with the $T \times N_t$ basis matrices $\mathbf{B}^{(k)}$ and the maximum likelihood detection metric $\Omega(\hat{\mathbf{x}})$ of (4.64). The dependence of Ω on the structure of the matrix modulation comes through the equivalent channel correlation matrix with elements (4.58), of the form

$$\bar{\mathcal{R}}_{jk} = \tfrac{1}{2} \operatorname{Tr} \left[\left(\mathbf{B}^{(j) \dagger} \mathbf{B}^{(k)} + \mathbf{B}^{(k) \dagger} \mathbf{B}^{(j)} \right) \mathbf{H} \, \mathbf{H}^\dagger \right]$$

First, this shows that the Ω, and thus any error probability (bit, symbol, block) related to Ω, is invariant under any left unitary transformation

$$\mathbf{B}^{(k)} \mapsto \mathbf{V} \, \mathbf{B}^{(k)} \quad \text{for all } k, \; \mathbf{V} \text{ unitary } T \times T \tag{A.4}$$

This *left unitary symmetry* is thus a symmetry of performance for any channel statistics and any channel realization. Furthermore, a right unitary transformation

$$\mathbf{B}^{(k)} \mapsto \mathbf{B}^{(k)} \mathbf{W} \quad \text{for all } k, \; \mathbf{W} \text{ unitary } N_t \times N_t \tag{A.5}$$

can be considered as acting on the channel $\mathbf{H}$. As observed above, this is a symmetry of the channel pdf for i.i.d. Rayleigh fading, and thus also a symmetry of average performance. That is, (A.5) changes performance for a given channel realization, but averaged over all realizations, it leaves performance invariant. This is the *right unitary symmetry of performance*, discussed in Chapter 7. It should be noted that this is a symmetry only for i.i.d. Rayleigh fading. For correlated fading it changes performance. This is discussed in Chapter 8.

Similarly, the right unitary transformation is a symmetry of ergodic mutual information (4.71) for i.i.d. Rayleigh fading, and the left unitary transformation is a symmetry of non-ergodic mutual information for any channel realization.

A.2   UNITARY INVARIANTS AND DETERMINANT INEQUALITIES

As noted above, for i.i.d. Rayleigh fading channels, unitary transformations of the channels leave the channel pdf invariant. Thus for such channels, the information and performance measures can be expressed in terms of unitarily invariant functions of the matrix modulation. As the matrices are finite dimensional, such functions can be expressed in terms of invariant polynomials.

Unitarily invariant polynomials of matrix elements, or unitary invariants for short, are functions of matrices that are invariant under unitary similarity transformations

$$\mathbf{M} \mapsto \mathbf{U}^\dagger \, \mathbf{M} \, \mathbf{U} . \tag{A.6}$$

The best known unitary invariants are the trace and the determinant of a matrix. These are trivially invariant under (A.6). Moreover, a Hermitian matrix may be diagonalized with a unitary transformation of the form (A.6),

$$\mathbf{M} = \mathbf{U}^\dagger \Lambda \mathbf{U} , \qquad (A.7)$$

where Λ is a diagonal matrix of eigenvalues λ_i. Eigenvalues of a Hermitian matrix are trivially unitary invariants. [1] Another set of unitary invariants are the trace invariants

$$t_n(\mathbf{M}) = \mathrm{Tr}\,[\mathbf{M}^n] = \sum_{i=1}^{N} \lambda_i^n , \; n = 1, \ldots, N . \qquad (A.8)$$

From cyclicity of trace it follows that they are unitary invariants of any $N \times N$ matrices $\mathbf{M}$. They can be considered as a complete set of unitary invariant variables.

As noted, the determinant is a unitary invariant, and can thus be expressed as a function of the trace invariants. More concretely, we are interested in the expansion of the determinant in powers of the trace. For this, denote

$$\tilde{t}_1 = \frac{1}{N}\,\mathrm{Tr}[\mathbf{M}] , \qquad (A.9)$$

and expand the determinant as

$$\det \mathbf{M} = \tilde{t}_1^N + \sum_{n=2}^{N} \frac{(-1)^{n-1}}{n}\, \tilde{t}_1^{N-n}\, \tilde{t}_n . \qquad (A.10)$$

The coefficients $\tilde{t}_n$, $n \geq 2$ of this expansion provide a very special set of independent unitary invariants. In mathematical terms, they can be defined in terms elementary symmetric polynomials (see [267]) of the *traceless part*

$$\widetilde{\mathbf{M}} = \mathbf{M} - \tilde{t}_1 \mathbf{I}_N . \qquad (A.11)$$

of $\mathbf{M}$. They can be expressed as

$$\tilde{t}_n = -n \sum_{2k_2 + 3k_3 + \ldots = n} \frac{(-1)^{\sum k_m}}{\prod_m{}' m^{k_m} k_m!} \prod_{m=2}^{N} \left(\mathrm{Tr}\, \widetilde{\mathbf{M}}^m \right)^{k_m} \qquad (A.12)$$

The sum is over all non-negative integers k_m, $m = 2, \ldots, N$, satisfying the constraint $\sum_{m=2}^{N} m\, k_m = n$. The primed product in the denominator is over all non-vanishing k_m. A few of the first symmetric polynomials are

$$
\begin{aligned}
\tilde{t}_2 &= \mathrm{Tr}\left[\widetilde{\mathbf{M}}^2\right] = \mathrm{Tr}\left[\mathbf{M}^2\right] - N\,\tilde{t}_1^2 \\
\tilde{t}_3 &= \mathrm{Tr}\left[\widetilde{\mathbf{M}}^3\right] \\
\tilde{t}_4 &= \mathrm{Tr}\left[\widetilde{\mathbf{M}}^4\right] - \tfrac{1}{2}\,\tilde{t}_2^2
\end{aligned}
$$

---

[1] Note that the rank of a matrix is invariant under unitary similarity transformations, but it is not a "unitary invariant" of the kind discussed here, i.e. an invariant polynomial. Rather, it is a feature of the set of invariant polynomials.

$$\tilde{t}_5 = \text{Tr}\left[\widetilde{\mathbf{M}^5}\right] - \tfrac{5}{6}\,\tilde{t}_2\,\tilde{t}_3 \tag{A.13}$$

Generically, the leading term in $\tilde{t}_n$ is $\text{Tr}\left[\widetilde{\mathbf{M}^n}\right]$, in which the leading term is t_n.

As there are only N independent unitary invariants for a $N \times N$ matrix, the spurious trace invariants t_n, $n > N$ may be expressed in terms of the independent invariants t_n, $n = 1, \ldots, N$. The Schur polynomials have the remarkable property that the spurious invariants vanish identically, $\tilde{t}_n = 0$, $\forall\, n > N$.

Hadamard Inequality: The usefulness of the expansion (A.10) stems from the fact that Hermitian matrices are "diagonal dominated". This diagonal dominance is neatly visible in the Hadamard inequality. For a positive semi-definite Hermitian matrix, the determinant is upper-bounded by the product of the diagonal elements:

$$\det \mathbf{M} \leq \prod_{i=1}^{N} m_{ii} \tag{A.14}$$

Schur Inequality: A corresponding lower bound is provided by an inequality due to Schur (see [268] p. 224). For a positive semi-definite Hermitian $N \times N$ matrix with $N > 2$, the determinant is lower bounded by

$$\det \mathbf{M} \geq \prod_{i=1}^{N} m_{ii} - \frac{\text{Tr}[\mathbf{M}]^{N-2}}{2(N-2)}\left(\text{Tr}\left[\mathbf{M}^2\right] - \sum_{i=1}^{N} m_{ii}^2\right). \tag{A.15}$$

Modified Hadamard Inequality: Positive semi-definite Hermitian matrices are not only "diagonal dominated", they are "trace dominated". Accordingly, corresponding inequalities may be written in terms of the symmetric polynomials $\tilde{t}_n$. Corresponding to (A.14) we have the modified Hadamard inequality. For a positive semi-definite Hermitian $N \times N$ matrix, the determinant is upper-bounded by

$$\det \mathbf{M} \leq u_{\text{det}} = \tilde{t}_1^N, \tag{A.16}$$

i.e. the first term in the expansion (A.10). Similar inequalities, alternatively lower- and upper-bounding the determinant, may be derived for the other terms in (A.10).

Appendix B
Matrix Representations of Clifford Algebras

Suppose you have K objects $\mathbf{G}^{(k)}$ that fulfil the defining relations (5.5). These are called generators of a Clifford algebra. They are fourth roots of unity (the identity matrix)

$$\left(\mathbf{G}^{(k)}\right)^2 = -\mathbf{I}, \tag{B.1}$$

and they anti-commute

$$\mathbf{G}^{(k)}\,\mathbf{G}^{(l)} = -\mathbf{G}^{(l)}\,\mathbf{G}^{(k)} \quad k \neq l. \tag{B.2}$$

We want to represent these objects as anti-Hermitian square matrices, and we are interested in the minimal matrix dimension where the algebraic rules (B.2) may be represented. For anti-Hermitian matrices, being second root of unity translates to unitarity. The enveloping algebra of these objects is the algebra Clifford$_K$. The basis of Clifford$_K$ has the algebraic dimension 2^K, and consists of the identity $\mathbf{I}$, the generators $\mathbf{G}^{(k)}$ and all their products. A clear distinction should be made between the dimension of the matrix representation and the algebraic dimension of a (matrix) algebra. The dimension of the matrix representation is the dimensionality of the matrices representing the algebraic relations. The algebraic dimension is the number of independent basis elements in the algebra. For a detailed description of properties of Clifford algebras, see [163]. Here we shall only concentrate on matrices that fulfil the rules (B.1) and (B.2).

In constructing matrix representations of Clifford algebras, we shall use the following anti-Hermitian matrices:

$$\mathbf{T}^{(2)} = \begin{bmatrix} j & 0 \\ 0 & -j \end{bmatrix}; \quad \mathbf{T}^{(3)} = \begin{bmatrix} 0 & 1 \\ -1 & 0 \end{bmatrix}; \quad \mathbf{T}^{(4)} = \begin{bmatrix} 0 & j \\ j & 0 \end{bmatrix}, \tag{B.3}$$

i.e. the three non-trivial matrices of the STTD basis (4.20).

B.1 DIMENSIONS OF MATRIX REPRESENTATIONS

First consider the case $K = 2$. The objects $\mathbf{G}^{(1)}$ and $\mathbf{G}^{(2)}$ anti-commute:

$$\mathbf{G}^{(1)}\mathbf{G}^{(2)} = -\mathbf{G}^{(2)}\mathbf{G}^{(1)}. \tag{B.4}$$

Clearly they cannot be represented as complex numbers. The minimal dimension to represent two anti-commuting objects is 2. It is easy to find two anti-commuting anti-Hermitian unitary 2×2 matrices; take for example

$$\mathbf{G}^{(1)} = \mathbf{T}^{(3)}; \quad \mathbf{G}^{(2)} = \mathbf{T}^{(4)}. \tag{B.5}$$

Now consider a Clifford algebra generated by three elements $\mathbf{G}^{(1)}, \mathbf{G}^{(2)}, \mathbf{G}^{(3)}$. From the anti-commutation relations it follows that the product $\mathbf{G}^{(1)}\mathbf{G}^{(2)}\mathbf{G}^{(3)}$ commutes with all generators of the algebra:

$$\left(\mathbf{G}^{(1)}\mathbf{G}^{(2)}\mathbf{G}^{(3)}\right) \mathbf{G}^{(k)} = \mathbf{G}^{(k)} \left(\mathbf{G}^{(1)}\mathbf{G}^{(2)}\mathbf{G}^{(3)}\right). \tag{B.6}$$

From Schur's lemma (see [267]) it follows that the matrix representation $\mathbf{G}^{(1)}\mathbf{G}^{(2)}\mathbf{G}^{(3)}$ is proportional to identity. In this case it means that

$$\mathbf{G}^{(1)}\mathbf{G}^{(2)}\mathbf{G}^{(3)} = -\mathbf{I}_2 \tag{B.7}$$

Thus $\mathbf{G}^{(3)}$ can be represented in the same matrix dimension as $\mathbf{G}^{(1)}$ and $\mathbf{G}^{(2)}$, and is simply proportional to their product. Corresponding to (B.5) we have e.g.

$$\mathbf{G}^{(3)} = \mathbf{G}^{(1)}\mathbf{G}^{(2)} = \mathbf{T}^{(2)}. \tag{B.8}$$

Taking the matrices $\mathbf{B}$ in Equation (4.4) to be the above representation of the three generators of Clifford$_3$, together with $\mathbf{I}_2$, we get exactly the 2×2 Alamouti code (STTD).

Next add a fourth generator to the Clifford algebra. Denoting $\tilde{\mathbf{G}}^{(3)} = -j\,\mathbf{G}^{(1)}\mathbf{G}^{(2)}\mathbf{G}^{(3)}$, $\tilde{\mathbf{G}}^{(4)} = -j\,\mathbf{G}^{(1)}\mathbf{G}^{(2)}\mathbf{G}^{(4)}$, we see from the anti-commutation relations that the sub-algebras generated by $\mathbf{G}^{(1)}$, $\mathbf{G}^{(2)}$ and $\tilde{\mathbf{G}}^{(3)}$, $\tilde{\mathbf{G}}^{(4)}$ commute. The two sub-algebras, on the other hand, are isomorphic to Clifford$_2$. Thus each of these sub-algebras can be represented using the matrices in Equation (B.3). The two commuting sub-algebras, however, have to be represented using the tensor product. Thus we get the following matrix representation for the generators of Clifford$_4$:

$$\mathbf{G}^{(1)} = \mathbf{I}_2 \otimes \mathbf{T}^{(3)}; \quad \mathbf{G}^{(2)} = \mathbf{I}_2 \otimes \mathbf{T}^{(4)} \tag{B.9}$$

$$\tilde{\mathbf{G}}^{(3)} = \mathbf{T}^{(3)} \otimes \mathbf{I}_2; \quad \tilde{\mathbf{G}}^{(4)} = \mathbf{T}^{(4)} \otimes \mathbf{I}_2. \tag{B.10}$$

Adding a fifth generator, we again note that $j\,\mathbf{G}^{(1)}\mathbf{G}^{(2)}\mathbf{G}^{(3)}\mathbf{G}^{(4)}\mathbf{G}^{(5)}$ commutes with all elements in the algebra, and its matrix representation can be taken to be proportional to the identity. Thus we can define e.g. $\mathbf{G}^{(5)} = -j\,\mathbf{G}^{(1)}\mathbf{G}^{(2)}\mathbf{G}^{(3)}\mathbf{G}^{(4)}$.

We have constructed the following representation of Clifford $_5$:

$$\mathbf{G}^{(1)} = \begin{bmatrix} 0 & 1 & 0 & 0 \\ -1 & 0 & 0 & 0 \\ 0 & 0 & 0 & 1 \\ 0 & 0 & -1 & 0 \end{bmatrix}, \quad \mathbf{G}^{(2)} = \begin{bmatrix} 0 & j & 0 & 0 \\ j & 0 & 0 & 0 \\ 0 & 0 & 0 & j \\ 0 & 0 & j & 0 \end{bmatrix}$$

$$\mathbf{G}^{(3)} = \begin{bmatrix} 0 & 0 & 1 & 0 \\ 0 & 0 & 0 & -1 \\ -1 & 0 & 0 & 0 \\ 0 & 1 & 0 & 0 \end{bmatrix}, \quad \mathbf{G}^{(4)} = \begin{bmatrix} 0 & 0 & j & 0 \\ 0 & 0 & 0 & -j \\ j & 0 & 0 & 0 \\ 0 & -j & 0 & 0 \end{bmatrix}$$

$$\mathbf{G}^{(5)} = \begin{bmatrix} j & 0 & 0 & 0 \\ 0 & -j & 0 & 0 \\ 0 & 0 & -j & 0 \\ 0 & 0 & 0 & j \end{bmatrix} \tag{B.11}$$

Using (4.4), we get the form (3.39) of the rate 3/4 orthogonal design (after a cyclic permutation of indices).

The same construction can be applied to find inductively the irreducible representations of a Clifford algebra generated by an arbitrary number of elements K. There are two different sets of induction steps. Matrix representations for even K are constructed from matrix representations of $K - 2$, whereas matrix representations for odd K are constructed from the representation for $K - 1$. For even K, the induction steps are the following:

1. Define $\tilde{\mathbf{G}}^{(K)} = (-j)^{K/2-1} \left(\prod_{i=1}^{K-2} \mathbf{G}^{(j)} \right) \mathbf{G}^{(K)}$,

   $\tilde{\mathbf{G}}^{(K-1)} = (-j)^{K/2-1} \left(\prod_{j=1}^{K-2} \mathbf{G}^{(j)} \right) \mathbf{G}^{(K-1)}$.

2. Note that $\{\mathbf{G}^{(j)}\}_{j=1}^{K-2}$ and $\{\tilde{\mathbf{G}}^{(K-1)}, \mathbf{G}^{(\tilde{K})}\}$ generate Clifford sub-algebras isomorphic to Clifford $_{K-2}$ and Clifford $_2$, respectively, and that these two sub-algebras commute.

3. Use a tensor product of existing lower-dimensional matrix representations $\mathcal{R}_{K-2}$ of Clifford $_{K-2}$ and $\mathcal{R}_2$ of Clifford $_2$ (the latter e.g. (B.5)) to represent these sub-algebras:

$$\begin{aligned} \mathcal{R}_K(\mathbf{G}^{(j)}) &= \mathbf{I}_2 \otimes \mathcal{R}_{K-2}(\mathbf{G}^{(j)}), \quad j = 1, \ldots K - 2 \\ \mathcal{R}_K(\tilde{\mathbf{G}}^{(K-1)}) &= \mathcal{R}_2(\mathbf{G}^{(1)}) \otimes \mathbf{I}_{\dim \mathcal{R}_{K-2}} \\ \mathcal{R}_K(\mathbf{G}^{(\tilde{K})}) &= \mathcal{R}_2(\mathbf{G}^{(2)}) \otimes \mathbf{I}_{\dim \mathcal{R}_{K-2}} \end{aligned} \tag{B.12}$$

4. Finally, invert the definitions of the $\tilde{\mathbf{G}}$s,

$$\mathcal{R}_K \left(\mathbf{G}^{(K-1,K)} \right) = j^{K/2-1} \, \mathcal{R}_K \left(\prod_{j=1}^{K-2} \mathbf{G}^{(j)} \, \tilde{\mathbf{G}}^{(K-1,K)} \right) \tag{B.13}$$

The induction steps from even $K - 1$ to odd K are the following

1. Define $\Lambda_K = (-j)^{(K+1)/2} \prod_{j=1}^{K} \mathbf{G}^{(j)}$

2. Note that Λ_K commutes with all $\{\mathbf{G}^{(j)}\}_{j=1}^{K}$. Represent it by the identity matrix in the dimension of the representation $\mathcal{R}_{K-1}$.

3. Take

$$
\begin{aligned}
\mathcal{R}_K(\mathbf{G}^{(j)}) &= \mathcal{R}_{K-1}(\mathbf{G}^{(j)}), \quad j = 1, \ldots K - 1 \\
\mathcal{R}_K(\mathbf{G}^{(K)}) &= j^{(K+1)/2} \, \mathcal{R}_{K-1}\left(\prod_{j=1}^{K-1} \mathbf{G}^{(j)} \right)
\end{aligned}
\tag{B.14}
$$

The induction steps above can be used to prove that the minimal matrix dimension to represent a Clifford algebra generated by K elements is $2^{\lfloor K/2 \rfloor}$. The minimal representation matrices are elements in the space of $\lfloor K/2 \rfloor$-fold tensor products of two-dimensional matrices.

$$
\mathcal{R}_K = \underbrace{\mathcal{R}_2 \otimes \mathcal{R}_2 \otimes \ldots \otimes \mathcal{R}_2}_{\lfloor K/2 \rfloor \text{ times}}
\tag{B.15}
$$

A 2^L dimensional matrix representation of Clifford$_{2L-1}$ constructed by applying the induction above is[1]

$$
\begin{aligned}
\mathbf{G}^{(2)} &= \underbrace{\mathbf{I}_2 \otimes \mathbf{I}_2 \otimes \mathbf{I}_2 \otimes \ldots \otimes \mathbf{I}_2}_{L-2 \text{ times}} \otimes \mathbf{T}^{(3)} \\
\mathbf{G}^{(3)} &= \underbrace{\mathbf{I}_2 \otimes \mathbf{I}_2 \otimes \mathbf{I}_2 \otimes \ldots \otimes \mathbf{I}_2}_{L-2 \text{ times}} \otimes \mathbf{T}^{(4)} \\
\mathbf{G}^{(4)} &= \underbrace{\mathbf{I}_2 \otimes \mathbf{I}_2 \otimes \ldots \otimes \mathbf{I}_2}_{L-3 \text{ times}} \otimes \mathbf{T}^{(3)} \otimes j\,\mathbf{T}^{(2)} \\
\mathbf{G}^{(5)} &= \underbrace{\mathbf{I}_2 \otimes \mathbf{I}_2 \otimes \ldots \otimes \mathbf{I}_2}_{L-3 \text{ times}} \otimes \mathbf{T}^{(4)} \otimes j\,\mathbf{T}^{(2)} \\
\mathbf{G}^{(6)} &= -\underbrace{\mathbf{I}_2 \otimes \ldots \otimes \mathbf{I}_2}_{L-4 \text{ times}} \otimes \mathbf{T}^{(3)} \otimes \mathbf{T}^{(2)} \otimes \mathbf{T}^{(2)} \\
\mathbf{G}^{(7)} &= -\underbrace{\mathbf{I}_2 \otimes \ldots \otimes \mathbf{I}_2}_{L-4 \text{ times}} \otimes \mathbf{T}^{(4)} \otimes \mathbf{T}^{(2)} \otimes \mathbf{T}^{(2)} \\
&\quad\vdots \\
\mathbf{G}^{(2k)} &= j^{k-1} \underbrace{\mathbf{I}_2 \otimes \ldots \otimes \mathbf{I}_2}_{L-1-k \text{ times}} \otimes \mathbf{T}^{(3)} \otimes \underbrace{\mathbf{T}^{(2)} \otimes \ldots \otimes \mathbf{T}^{(2)}}_{k-1 \text{ times}} \\
\mathbf{G}^{(2k+1)} &= j^{k-1} \underbrace{\mathbf{I}_2 \otimes \ldots \otimes \mathbf{I}_2}_{L-1-k \text{ times}} \otimes \mathbf{T}^{(4)} \otimes \underbrace{\mathbf{T}^{(2)} \otimes \ldots \otimes \mathbf{T}^{(2)}}_{k-1 \text{ times}} \\
&\quad\vdots \\
\mathbf{G}^{(2L-2)} &= j^{L-2} \mathbf{T}^{(3)} \otimes \underbrace{\mathbf{T}^{(2)} \otimes \ldots \otimes \mathbf{T}^{(2)}}_{L-2 \text{ times}}
\end{aligned}
\tag{B.16}
$$

---

[1] We have reordered the the numbering of $\mathbf{G}$ matrices to get orthogonal designs when using Equation(4.4).

$$\mathbf{G}^{(2L-1)} = \mathrm{j}^{L-2}\mathbf{T}^{(4)} \otimes \underbrace{\mathbf{T}^{(2)} \otimes \ldots \otimes \mathbf{T}^{(2)}}_{L-2 \text{ times}}$$

$$\mathbf{G}^{(1)} = \mathrm{j}^{L-1} \underbrace{\mathbf{T}^{(2)} \otimes \mathbf{T}^{(2)} \otimes \ldots \otimes \mathbf{T}^{(2)}}_{L-1 \text{ times}}$$

The anti-commutation relations (B.2) have the symmetry

$$\mathbf{G}^{(j)} \mapsto \mathbf{W}^{\dagger}\mathbf{G}^{(j)}\,\mathbf{W}, \; j = 1, \ldots, K \tag{B.17}$$

where $\mathbf{W}$ is an unitary $\dim\mathcal{R}_K \times \dim\mathcal{R}_K$ matrix. This symmetry is large enough to accommodate any choice of basis in the Clifford$_2$ sub-algebras. Together with the arbitrary choice of $\mathbf{B}^{(1)}$, this leads to the unitary left and right symmetries discussed in section 5.1.3.

B.2   CLIFFORD BASIS OF GENERIC MATRICES

The STTD basis of four 2×2 matrices

$$\mathcal{B}_2 = \{\mathbf{T}^{(1)}, \mathbf{T}^{(2)}, \mathbf{T}^{(3)}, \mathbf{T}^{(4)}\} \tag{B.18}$$

where $\mathbf{T}^{(1)} = \mathbf{I}_2$, and the others are spelled out in (B.3), is closed under multiplication. Moreover, these matrices are orthogonal with respect to the Frobenius norm (4.43). This means that $\mathcal{B}_2$ with complex coefficients forms an orthogonal basis for the space of complex 2×2 matrices, which has an algebraic dimensionality of four complex dimensions. Explicitly, this means that any complex 2×2 matrix may be expressed in the basis $\mathcal{B}_2$:

$$\begin{bmatrix} a & b \\ c & d \end{bmatrix} = \frac{a}{2}(\mathbf{I}_2 - \mathrm{j}\,\mathbf{T}^{(2)}) + \frac{b}{2}(\mathbf{T}^{(3)} - \mathrm{j}\,\mathbf{T}^{(4)}) + \frac{c}{2}(\mathbf{T}^{(3)} + \mathrm{j}\,\mathbf{T}^{(4)}) + \frac{d}{2}(\mathbf{I}_2 + \mathrm{j}\,\mathbf{T}^{(2)}).$$
$$\tag{B.19}$$

This may seem like a trivial exercise. The essence is that the basis $\mathcal{B}_2$ consists of unitary matrices, and can thus be used as a basis for constructing linear matrix modulation schemes which have maximal symbolwise diversity. To be used in the context of Equation (4.4), the basis should be considered with real coefficients. Thus the eight matrices $\mathcal{B}_2 \cup \mathrm{j}\mathcal{B}_2$ are a basis for complex 2×2 matrices considered with *real* coefficients, which has an algebraic dimensionality of eight real dimensions. With real coefficients, half of this basis, namely $\mathcal{B}_2$ alone, yields the Alamouti scheme (STTD).

This property may be extended to any matrix dimension 2^L where a Clifford algebra may be represented. As noted, Clifford$_K$ has algebraic dimension 2^K. The basis consists of the identity, the K generators and all their products. Taking the matrix representation of Clifford$_{2L}$ in terms of $2^L \times 2^L$ matrices, we thus have a set of $2^{2L} = (2^L)^2$ basis elements in the Clifford algebra. It is a trivial consequence of the algebraic relations (B.1) and (B.2) that all of these basis elements are unitary and orthogonal with respect to the Frobenius norm. Thus the basis of Clifford$_{2L}$ is an example of a complete unitary basis for complex $2^L \times 2^L$ matrices.

In particular, the basis for Clifford$_4$ gives a basis for 4×4 matrices. The algebra of complex 4×4 matrices has 16 complex dimensions, corresponding to the 16 complex matrix elements. This basis can be constructed from all products of the four first $\mathbf{G}^{(k)}$ in (B.11). There are 16 linearly independent matrices. Together with j times the same, we have 32 unitary matrices which can be used as a basis for complex 4×4 matrix modulation with real coefficients. These 32 matrices can be expressed as all 16 tensor products of the four elements in the 2×2 matrix basis $\mathcal{B}_2$:

$$\mathcal{B}_4 = \{\mathbf{T}^{(\mu)} \otimes \mathbf{T}^{(\nu)}\}_{\mu,\nu=1}^4 , \tag{B.20}$$

and j times the same. Explicitly, the 16 matrices in $\mathcal{B}_4$ read

$$\mathbf{B}^{(1\otimes1)} = \mathbf{I}_4; \qquad\qquad \mathbf{B}^{(1\otimes2)} = \begin{bmatrix} j & 0 & 0 & 0 \\ 0 & -j & 0 & 0 \\ 0 & 0 & j & 0 \\ 0 & 0 & 0 & -j \end{bmatrix},$$

$$\mathbf{B}^{(1\otimes3)} = \begin{bmatrix} 0 & 1 & 0 & 0 \\ -1 & 0 & 0 & 0 \\ 0 & 0 & 0 & 1 \\ 0 & 0 & -1 & 0 \end{bmatrix}, \qquad \mathbf{B}^{(1\otimes4)} = \begin{bmatrix} 0 & j & 0 & 0 \\ j & 0 & 0 & 0 \\ 0 & 0 & 0 & j \\ 0 & 0 & j & 0 \end{bmatrix},$$

$$\mathbf{B}^{(2\otimes1)} = \begin{bmatrix} j & 0 & 0 & 0 \\ 0 & j & 0 & 0 \\ 0 & 0 & -j & 0 \\ 0 & 0 & 0 & -j \end{bmatrix}, \qquad \mathbf{B}^{(2\otimes2)} = \begin{bmatrix} -1 & 0 & 0 & 0 \\ 0 & 1 & 0 & 0 \\ 0 & 0 & 1 & 0 \\ 0 & 0 & 0 & -1 \end{bmatrix},$$

$$\mathbf{B}^{(2\otimes3)} = \begin{bmatrix} 0 & j & 0 & 0 \\ -j & 0 & 0 & 0 \\ 0 & 0 & 0 & -j \\ 0 & 0 & j & 0 \end{bmatrix}, \qquad \mathbf{B}^{(2\otimes4)} = \begin{bmatrix} 0 & -1 & 0 & 0 \\ -1 & 0 & 0 & 0 \\ 0 & 0 & 0 & 1 \\ 0 & 0 & 1 & 0 \end{bmatrix},$$

$$\mathbf{B}^{(3\otimes1)} = \begin{bmatrix} 0 & 0 & 1 & 0 \\ 0 & 0 & 0 & 1 \\ -1 & 0 & 0 & 0 \\ 0 & -1 & 0 & 0 \end{bmatrix}, \qquad \mathbf{B}^{(3\otimes2)} = \begin{bmatrix} 0 & 0 & j & 0 \\ 0 & 0 & 0 & -j \\ -j & 0 & 0 & 0 \\ 0 & j & 0 & 0 \end{bmatrix},$$

$$\mathbf{B}^{(3\otimes3)} = \begin{bmatrix} 0 & 0 & 0 & 1 \\ 0 & 0 & -1 & 0 \\ 0 & -1 & 0 & 0 \\ 1 & 0 & 0 & 0 \end{bmatrix}, \qquad \mathbf{B}^{(3\otimes4)} = \begin{bmatrix} 0 & 0 & 0 & j \\ 0 & 0 & j & 0 \\ 0 & -j & 0 & 0 \\ -j & 0 & 0 & 0 \end{bmatrix},$$

$$\mathbf{B}^{(4\otimes1)} = \begin{bmatrix} 0 & 0 & j & 0 \\ 0 & 0 & 0 & j \\ j & 0 & 0 & 0 \\ 0 & j & 0 & 0 \end{bmatrix}, \qquad \mathbf{B}^{(4\otimes2)} = \begin{bmatrix} 0 & 0 & -1 & 0 \\ 0 & 0 & 0 & 1 \\ -1 & 0 & 0 & 0 \\ 0 & 1 & 0 & 0 \end{bmatrix},$$

$$\mathbf{B}^{(4\otimes3)} = \begin{bmatrix} 0 & 0 & 0 & j \\ 0 & 0 & -j & 0 \\ 0 & j & 0 & 0 \\ -j & 0 & 0 & 0 \end{bmatrix}, \qquad \mathbf{B}^{(4\otimes4)} = - \begin{bmatrix} 0 & 0 & 0 & 1 \\ 0 & 0 & 1 & 0 \\ 0 & 1 & 0 & 0 \\ 1 & 0 & 0 & 0 \end{bmatrix}. \tag{B.21}$$

Just as in (B.19) for complex 2×2 matrices, any complex 4×4 matrix can be expressed as a linear combination of the 16 matrices (B.21) with complex coefficients, or as a linear combination of the 32 matrices $\mathcal{B}_4 \cup j\mathcal{B}_4$ with real coefficients.

Two-symbol Quasi-orthogonal Subsets of the 4×4 Basis: For use in constructing quasi-orthogonal layered matrix modulation schemes with two complex symbols in each quasi-orthogonal layer, the full set of 4×4 basis matrices (B.21) and j times the same may be divided into subsets. With two complex symbols in each layer, one needs four basis matrices per layer. An especially transparent division of the full set of 32 basis matrices consist of the eight sets

$$\mathcal{B}_A, \mathcal{B}_B, \mathcal{B}_C, \mathcal{B}_D, \mathcal{B}_E, \mathcal{B}_F, \mathcal{B}_G, \mathcal{B}_H \ , \tag{B.22}$$

where

$$
\begin{aligned}
\mathcal{B}_A &= \{\mathbf{B}^{(1\otimes\mu)}\}_{\mu=1}^4 \\
\mathcal{B}_B &= \{-\mathrm{j}\,\mathbf{B}^{(4\otimes\mu)}\}_{\mu=1}^4 \\
\mathcal{B}_C &= \{-\mathrm{j}\,\mathbf{B}^{(2\otimes\mu)}\}_{\mu=1}^4 \\
\mathcal{B}_D &= \{\mathbf{B}^{(3\otimes\mu)}\}_{\mu=1}^4 \\
\mathcal{B}_E &= \mathrm{j}\,\mathcal{B}_A \\
\mathcal{B}_F &= \mathrm{j}\,\mathcal{B}_B \\
\mathcal{B}_G &= \mathrm{j}\,\mathcal{B}_C \\
\mathcal{B}_H &= \mathrm{j}\,\mathcal{B}_D \ .
\end{aligned}
\tag{B.23}
$$

When used as basis matrices for real symbols, these subsets yield the matrix modulations

$$
\begin{bmatrix} \mathbf{X}_A & \mathbf{0}_2 \\ \mathbf{0}_2 & \mathbf{X}_A \end{bmatrix}
\begin{bmatrix} \mathbf{0}_2 & \mathbf{X}_B \\ \mathbf{X}_B & \mathbf{0}_2 \end{bmatrix}
\begin{bmatrix} \mathbf{X}_C & \mathbf{0}_2 \\ \mathbf{0}_2 & -\mathbf{X}_C \end{bmatrix}
\begin{bmatrix} \mathbf{0}_2 & \mathbf{X}_D \\ -X_D & \mathbf{0}_2 \end{bmatrix}
$$

$$
\begin{bmatrix} \mathrm{j}\,\mathbf{X}_E & \mathbf{0}_2 \\ \mathbf{0}_2 & \mathrm{j}\,\mathbf{X}_E \end{bmatrix}
\begin{bmatrix} \mathbf{0}_2 & \mathrm{j}\,\mathbf{X}_F \\ \mathrm{j}\,\mathbf{X}_F & \mathbf{0}_2 \end{bmatrix}
\begin{bmatrix} \mathrm{j}\,\mathbf{X}_G & \mathbf{0}_2 \\ \mathbf{0}_2 & -\mathrm{j}\,\mathbf{X}_G \end{bmatrix}
\begin{bmatrix} \mathbf{0}_2 & \mathrm{j}\,\mathbf{X}_H \\ -\mathrm{j}\,X_H & \mathbf{0}_2 \end{bmatrix}
\tag{B.24}
$$

respectively. Each of the $\mathbf{X}_A, \mathbf{X}_B, \dots$ blocks is a 2×2 STTD block of the form (3.28), and $\mathbf{0}_2$ is a 2×2 matrix of zeros. Using (B.23), the set of 8 basis matrices (4.36) of ABBA (4.35) is $\mathcal{B}_A \cup \mathcal{B}_B$.

Three-symbol Quasi-orthogonal Subsets of 4×4 Basis: The reason to search for the anti-commuting anti-Hermitian generator matrices G_k was to use them together with the identity matrix to construct (complex) orthogonal designs. For 4×4 matrices, one example of the five generator matrices is (B.11). Together with the identity matrix, a set of six Radon–Hurwitz orthogonal basis matrices for a rate 3/4 orthogonal design is

$$
\begin{aligned}
\mathbf{B}^{(1)} &= \mathbf{I}_4 \\
\mathbf{B}^{(2)} &= \mathbf{G}^{(5)} = -\mathrm{j}\,\mathbf{B}^{(2\otimes2)} \\
\mathbf{B}^{(3)} &= \mathbf{G}^{(1)} = \mathbf{B}^{(1\otimes3)} \\
\mathbf{B}^{(4)} &= \mathbf{G}^{(2)} = \mathbf{B}^{(1\otimes4)} \\
\mathbf{B}^{(5)} &= \mathbf{G}^{(3)} = -\mathrm{j}\,\mathbf{B}^{(3\otimes2)} \\
\mathbf{B}^{(6)} &= \mathbf{G}^{(4)} = -\mathrm{j}\,\mathbf{B}^{(4\otimes2)} \ .
\end{aligned}
\tag{B.25}
$$

The full set of 32 Frobenius orthogonal basis matrices (B.21) and j times the same may be expressed as four subsets of six Radon–Hurwitz orthogonal matrices, plus eight extra matrices. One example of the splitting of $\mathcal{B}_{4\times4}$ to such subsets is

$$
\begin{aligned}
\mathcal{B}_M &= \{\mathbf{I}_4, \ \mathbf{B}^{(1\otimes3)}, \ -\mathrm{j}\,\mathbf{B}^{(3\otimes2)}, \ -\mathrm{j}\,\mathbf{B}^{(2\otimes2)}, \ \mathbf{B}^{(1\otimes4)}, \ -\mathrm{j}\,\mathbf{B}^{(4\otimes2)}\} \\
\mathcal{B}_N &= \mathbf{T}^{(2)} \otimes \mathbf{T}^{(2)}\, \mathcal{B}_M \\
\mathcal{B}_O &= \mathbf{T}^{(3)} \otimes \mathbf{T}^{(3)}\, \mathcal{B}_M \\
\mathcal{B}_P &= \mathbf{T}^{(4)} \otimes \mathbf{T}^{(4)}\, \mathcal{B}_M \\
\mathcal{B}_R &= \{\mathbf{B}^{(1\otimes2)}, \ \mathbf{B}^{(2\otimes1)}, \ \mathbf{B}^{(3\otimes4)}, \ \mathbf{B}^{(4\otimes3)}\} \cup \mathrm{j}\,\{\text{the same}\} \ .
\end{aligned}
\tag{B.26}
$$

Here the subset $\mathcal{B}_M$ consists of (B.25), and $\mathcal{B}_N, \mathcal{B}_O, \mathcal{B}_P$ are the other three subsets of six RH-orthogonal basis matrices. The remaining eight $\mathbf{B}^{(k)}$ are in $\mathcal{B}_R$. To be explicit, the orthogonal designs corresponding to these four subsets are

$$
\mathbf{X}_M = \begin{bmatrix} x_1 & x_2 & x_3 & 0 \\ -x_2^* & x_1^* & 0 & -x_3 \\ -x_3^* & 0 & x_1^* & x_2 \\ 0 & x_3^* & -x_2^* & x_1 \end{bmatrix}, \tag{B.27}
$$

$$
\mathbf{X}_N = \begin{bmatrix} -x_1 & -x_2 & -x_3 & 0 \\ -x_2^* & x_1^* & 0 & -x_3 \\ -x_3^* & 0 & x_1^* & x_2 \\ 0 & -x_3^* & x_2^* & -x_1 \end{bmatrix}, \tag{B.28}
$$

$$
\mathbf{X}_O = \begin{bmatrix} 0 & x_3^* & -x_2^* & x_1 \\ x_3^* & 0 & -x_1^* & -x_2 \\ x_2^* & -x_1^* & 0 & x_3 \\ x_1 & x_2 & x_3 & 0 \end{bmatrix}, \tag{B.29}
$$

$$
\mathbf{X}_P = \begin{bmatrix} 0 & -x_3^* & x_2^* & -x_1 \\ x_3^* & 0 & -x_1^* & -x_2 \\ x_2^* & -x_1^* & 0 & x_3 \\ -x_1 & -x_2 & -x_3 & 0 \end{bmatrix}, \tag{B.30}
$$

$$
\mathbf{X}_R = \begin{bmatrix} x_1 + x_2 & 0 & 0 & -x_3 - x_4 \\ 0 & x_1 - x_2 & x_3 - x_4 & 0 \\ 0 & -x_3 + x_4 & -x_1 + x_2 & 0 \\ x_3 + x_4 & 0 & 0 & -x_1 - x_2 \end{bmatrix}. \tag{B.31}
$$

The four first are rate $R_s = 3/4$ orthogonal designs, the last is a 1+1+1+1 layered rate one scheme. Together these five subsets with 32 free real parameters span the whole space of 4×4 complex matrices. It should be noted that four Frobenius orthogonal copies of a rate $R_s = 3/4$ orthogonal design were discussed in [119] as a method to expand constellations for set-partitioning and trellis coded modulation, i.e. as an extension to $N_t = 4$ of the coset space–time codes of [117] a.k.a. the super-orthogonal space-time codes of [118].

References

1. 3GPP, "Physical layer–general description," 3GPP technical specification, TS 25.202, March 2002, Available from www.3gpp.org.

2. 3GPP, "Physical channels and mapping of transport channels on to physical channels (FDD)," 3GPP technical specification, TS 25.211, March 2002, Available from www.3gpp.org.

3. 3GPP, "Physical layer aspects of UTRA high speed downlink packet access," 3GPP TSG-RAN technical report, TR 25.848, Ver. 4.0.0, 2001.

4. TIA/EIA/IS-95-B, "Mobile station–base station compatibility standard for wideband spread spectrum cellular systems," 3 February 1999.

5. TIA/EIA/IS-2000, "CDMA2000 standard for spread spectrum systems," 30 August 1999.

6. P. Bender, P. Black, M. Grob, R. Padovani, N. Sindhushayana and A. Viterbi, "CDMA/HDR: A bandwidth-efficient high-speed wireless data service for nomadic users," *IEEE Comm. Magazine*, pp. 70–77, July 2000.

7. C. Berrou, A. Glavieux and P. Thitimajshima, "Near Shannon limit error-correcting coding and decoding: Turbo codes," in *IEEE Int. Conf. on Comm.*, Geneva, Switzerland, 1993, pp. 1064–1070.

8. S. Benedetto and G. Montorsi, "Unveiling turbo codes: Some results on parallel concatenated coding schemes," *IEEE Trans. Inf. Th.*, Vol. 42, no. 2, pp. 409–429, March 1996.

9. C. E. Shannon, "A mathematical theory of communication," *Bell Syst. Tech. J.*, Vol. 27, pp. 379–423, 623–656, October 1948.

10. S. Alamouti, "A simple transmitter diversity scheme for wireless communications," *IEEE J. Sel. Areas Comm.*, Vol. 16, pp. 1451–1458, Oct. 1998.

11. 3GPP, "Physical layer–general description, v. 2.0.0," 3GPP technical specification, Ver. 2.0.0., April 1999, Available from www.3gpp.org.

12. Nokia, "Proposal to simplify the Tx diversity closed loop modes," Temporary document, 3GPP contribution, TSG-R1, Espoo, Finland, July 1999.

13. Y. L. Pezzennec, F. Boixadera, Y. Farmine and N. Whinnett, "A transmit adaptive antenna scheme with feedback mechanism for wireless communication systems," in *Proc. Allerton Conf. on Comm. Control Computing*, Illinois, September 1999.

14. Nokia, "Downlink transmit diversity," Temporary document, ETSI SMG1 contribution, Bocholt, Germany, 18-20 May 1998.

15. J. Winters, "On the capacity of radio communication systems with diversity in a Rayleigh fading environment," *IEEE J. Select. Areas Commun.*, Vol. 5, no. 5, pp. 871–878, June 1987.

16. E. Telatar, "Capacity of multi-antenna gaussian channels," *Eur. Trans. Telecomm.*, Vol. 10, no. 6, pp. 585–595, Nov/Dec 1999, based on AT& T Bell Laboratories, Internal Tech. Memo, June 1995.

17. G. Foschini and M. Gans, "On limits of wireless communication in a fading environment when using multiple antennas," *Wireless Pers. Comm.*, Vol. 6, no. 3, pp. 311–335, Mar. 1998.

18. V. Tarokh, H. Jafarkhani and A. Calderbank, "Space–time block codes from orthogonal designs," *IEEE Trans. Inf. Th.*, Vol. 45, no. 5, pp. 1456–1467, July 1999.

19. V. Tarokh, N. Seshadri and A. Calderbank, "Space–time codes for high data-rate wireless communication: Performance criterion and code construction," *IEEE Trans. Inf. Th.*, Vol. 44, no. 2, pp. 744–765, Mar. 1998.

20. G. Foschini, "Layered space–time architecture for wireless communication in a fading environment when using multi–element antennas," *Bell Labs Tech. J.*, pp. 41–59, 1996.

21. P. Wolniansky, G. Foschini, G. Golden and R. Valenzuela, "V-BLAST: An architecture for realizing very high data rates over the rich-scattering wireless channel," in *Proc. URSI Int. Symp. Sign., Syst. and Electr.*, Sept. 1998, pp. 295–300.

22. B. Hochwald and T. Marzetta, "Unitary space–time modulation for multiple-antenna communications in Rayleigh flat fading," *IEEE Trans. Inf. Th.*, Vol. 46, no. 2, pp. 543–564, Mar. 2000.

23. A. Hiroike, F. Adachi and N. Nakajima, "Combined effects of phase sweeping transmitter diversity and channel coding," *IEEE Trans. Veh. Tech.*, Vol. 41, no. 2, pp. 170–176, May 1992.

24. A. Hottinen, K. Kuchi and O. Tirkkonen, "A space–time coding concept for a multi-element transmitter," in *Proc. Canadian Workshop on Information Theory*, Vancouver, June 2001.

25. A. Hottinen, R. Wichman and D. Rajan, "Soft-weighted transmit diversity for WCDMA systems," in *Proc. Allerton Conf. on Comm. Control Computing*, Illinois, September 1999.

26. A. Hottinen and O. Tirkkonen, "A randomization technique for non-orthogonal space–time block codes," in *IEEE Veh. Tech. Conf.*, Spring, Rhodes, May 2001, pp. 1479–1482.

27. G. Jöngren, M. Skoglund and B. Ottersten, "Combining beam-forming and orthogonal space–time block coding," *IEEE Trans. Inf. Th.*, Vol. 48, no. 3, pp. 611–627, Mar. 2002.

28. S. Zhou and G. Giannakis, "Optimal transmitter eigen–beam-forming and space-time block coding based on channel correlations," in *IEEE Int. Conf. on Comm.*, 2002, pp. 553–557.

29. A. Hottinen, R. Wichman and O. Tirkkonen, "Closed-loop transmit diversity techniques for multi-element transceivers," in *Proc. IEEE Int. Conf. Veh. Tech.*, Boston, Mass., September 2000, pp. 70–73.

30. G. Caire, G. Taricco and E. Biglieri, "Bit interleaved coded modulation," *IEEE Trans. Inf. Th.*, Vol. 44, pp. 927–947, May 1998.

31. R. Knopp, Coding and Multiple-Access over Fading Channels, Ph.D. thesis, EPFL, Lausanne, Switzerland, 1997.

32. T. Heikkinen and A. Hottinen, "Space–time power allocation in CDMA networks," in *Princeton Conf. Inf. Sci. Sys.*, Princeton, NJ, March 1998, pp. 12–16.

33. P. Viswananath, D. Tse and R. Laroia, "Opportunistic beamforming using dumb antennas," *IEEE Trans. Inf. Th.*, Vol. 48, no. 6, pp. 1277–1294, June 2002.

34. I. Koutsopoulos and L. Tassiulas, "Adaptive resource allocation in SDMA-based wireless broadband networks with OFDM signaling," in *Proc. INFOCOM*, New York, New York, 2002, Vol. 3, pp. 1376–1385.

35. T. Heikkinen, "On congestion pricing in a wireless network," *Wireless Networks*, Vol. 8, no. 4, pp. 347–354, July 2002.

36. T. Heikkinen, "Price-based ad-hoc network formation," in *Princeton Conf. Inf. Sci. Sys.*, Princeton, NJ, March 2002.

37. S. Sandhu and A. Paulraj, "Space–time block codes: a capacity perspective," *IEEE Comm. Lett.*, Vol. 4, no. 12, Dec. 2000.

38. J. Hämäläinen and R. Wichman, "Closed-loop transmit diversity for FDD WCDMA systems," in *Asilomar Conference on Signals, Systems and Computers*, Oct. 2000.

39. Motorola, "Orthogonal transmit diversity for direct spread CDMA," Temporary document, ETSI contribution, SMG2 L1, Stockholm, Sweden, 15-17 September 1997.

40. Samsung, "Proposal for downlink time switched transmission diversity," Temporary document, ETSI contribution, SMG2 L1, Bocholt, Germany, 18-20 May 1998.

41. A. Dabak, S. Hosur and R. Negi, "Space–time block coded transmit antenna diversity scheme for WCDMA," in *IEEE Wireless Communications and Networking Conference*, New Orleans, 1999, pp. 1466–1469.

42. A. Hottinen and R. Wichman, "Transmit diversity by antenna selection in CDMA downlink," in *Proc. IEEE Int. Symp. Spr. Spect. Tech. Appl.*, Sun City, South Africa, September 1998, pp. 767–770.

43. 3GPP2, "Introduction to cdma2000 standards for spread spectrum systems," 3GPP2 technical specification, C.S0001-C, Version 1, May 2002, available www.3gpp2.org.

44. 3GPP2, "Physical layer standard for cdma2000 spread spectrum systems- release C," 3GPP2 technical specification, C.S0002-C, Version 1, May 2002, available from www.3gpp2.org.

45. B. Hochwald, T. Marzetta and C. Papadias, "A transmitter diversity scheme for wide-band CDMA systems based on space–time spreading," *IEEE J. Sel. Areas Comm.*, Vol. 19, no. 1, pp. 48–60, January 2001.

46. 3GPP2, "cdma 2000 high rate packet data air interface specification," 3GPP2 technical specification, C.S0024, Version 3, 5 December 2001, available from www.3gpp2.org.

47. E. Lindskog and A. Paulraj, "A transmit diversity scheme for channels with intersymbol interference," in *IEEE Int. Conf. on Comm.*, 2000, pp. 307–311.

48. E. Lindskog and D. Flore, "Time-reversal space–time block coding and transmit delay diversity-separate and combined," in *34th Asilomar Conference on Signals, Systems and Computers*, California, 2000, pp. 572–577.

49. L. Jalloul, K. Rohani, K. Kuchi and J. Chen, "Performance analysis of CDMA transmit diversity methods," in *IEEE Veh. Tech. Conf.*, Fall, 1999, Vol. 3, pp. 1326–1330.

50. A. Hottinen and O. Tirkkonen, "Space–time block code with symbol rate two," in *Princeton Conf. Inf. Sci. Sys.*, Princeton, NJ, March 2002.

51. O. Tirkkonen and A. Hottinen, "Improved MIMO performance with non-orthogonal space–time block codes," in *Proc. IEEE GLOBECOM*, Nov. 2001, Vol. 2, pp. 1122–1126.

52. M. Damen, K. Abed-Meraim and J.-C. Belfiore, "Diagonal algebraic space–time block code," *IEEE Trans. Inf. Th.*, Vol. 48, no. 3, pp. 628–636, Mar. 2002.

53. N. Sidiropoulous and R. Budampati, "A class of linear space–time codes with built-in blind channel identifibiality and rate-diversity flexibility," in *Princeton Conf. Inf. Sci. Sys.*, Princeton, NJ, March 2002.

54. N. Nakajima and Y. Yamao, "Development for 4th generation mobile communications," *Wireless Commun. Mob. Comp.*, Vol. 1, pp. 3–12, January-March 2001.

55. Y. Li, J. Chuang and N. Sollenberger, "Transmitter diversity for OFDM systems and its impact on high-rate data wireless networks," *IEEE J. Sel. Areas Comm.*, Vol. 17, no. 7, pp. 1233–1243, July 1999.

56. S. Kaiser, "Performance of multi-carrier CDM and COFDM in fading channels," in *Proc. IEEE GLOBECOM*, Amsterdam, the Netherlands, December 1999, pp. 847–851.

57. M. Nakagami, "The m–distribution—a general formula of intensity distribution of rapid fading," in *Stat. Meth. Radio Wave Prop.*, June 1958.

58. G. Tzeremes and C. Christodoulou, "Use of Weibull distribution for describing outdoor multipath fading," in *IEEE Antennas and Propagation Society International Symposium*, 2002, pp. 232–235.

59. R. Clarke, "A statistical theory of mobile–radio reception," *Bell System Technical Journal*, pp. 957–1000, 1968.

60. W. Jakes, Ed., Microwave Mobile Communications, Wiley, 1974.

61. 3GPP, "UE radio transmission and reception (FDD) (Release 5)," 3GPP TSG-RAN technical specification, TS 25.101, Ver. 5.3.0, 2002.

62. Siemens, "Simulation parameters for Tx diversity simulations using correlated antennas," Temporary document R1–00–1080, 3GPP TSG RAN WG1, Feb. 2001.

63. G. Golub and C. V. Loan, Matrix Computations, The John Hopkins University Press, Boston MA, 2nd edition, 1989.

64. P. Petrus, J. Reed and T. Rappaport, "Effects of directional antennas at the base station on the doppler spectrum," *IEEE Comm. Lett.*, Vol. 1, no. 2, pp. 40–42, Mar. 1997.

65. L. Schumacher, J. Kermoal, K. P. F. Frederiksen, A. Algans and P. Mogensen, "MIMO channel characterization," Tech. rep. IST–1999–11729 METRA, IST, Feb. 2002, Available at http://www.ist-imetra.org/.

66. J. Salz and J. Winters, "Effect of fading correlation on adaptive arrays in digital mobile radio," *IEEE Trans. Veh. Tech.*, Vol. 43, no. 4, pp. 1049–1057, Nov. 1994.

67. F. Adachi, M. Feeney, A. Williamson and J. Parsons, "Cross correlation between the envelopes of 900 mhz signals received at a mobile radio base station site," *IEE Proc. F*, Vol. 133, no. 6, pp. 506–512, Oct. 1986.

68. K. Pedersen, P. Mogensen and B. Fleury, "Spatial channel characteristics in outdoor environments and their impact on BS antenna system performance," in *Proc. IEEE Int. Conf. Veh. Tech.*, 1998, Vol. 2, pp. 719–723.

69. W. Lee, "Effects on correlation between two mobile radio base-station antennas," *IEEE Trans. Comm.*, Vol. 21, no. 11, pp. 1214–1224, Nov. 1973.

70. Lucent, Nokia, Siemens and Ericsson, "A standardized set of MIMO radio propagation channels," Temporary document R1–01–1179, 3GPP TSG RAN WG1, Nov. 2001.

71. M. Rapporteur, "MIMO discussion summary," Temporary document R1–02–0181, 3GPP TSG RAN WG1, 2002.

72. D.-S. Shiu, G. Foschini, M. Gans and J. Kahn, "Fading correlation and its effect on the capacity of multielement antenna systems," *IEEE Trans. Comm.*, Vol. 48, no. 3, pp. 502–513, Mar. 2000.

73. N. Chiurtu, B. Rimoldi and E. Telatar, "Dense multiple antenna systems," in *Proc. Information Theory Workshop 2001*, 2001, pp. 108–109.

74. J. Kermoal, L. Schumacher, K. Pedersen, P. Mogensen and F. Frederiksen, "A stochastic MIMO radio channel model with experimental validation," *IEEE J. Sel. Areas Comm.*, Vol. 20, no. 4, pp. 1211–1226, Aug. 2002.

75. K. Yu, M. Bengtsson, B. Ottersten, D. McNamara, P. Karlsson and M. Beach, "A wideband statistical model for NLOS indoor MIMO channels," in *IEEE Veh. Tech. Conf.*, Spring, May 2002, pp. 370–374.

76. G. Raleigh and J. Cioffi, "Spatio-temporal coding for wireless communications," *IEEE Trans. Comm.*, Vol. 46, no. 3, pp. 357–366, Mar. 1998.

77. H. Bölcskei, D. Gesbert and A. Paulraj, "On the capacity of OFDM-based spatial multiplexing systems," *IEEE Trans. Comm.*, Vol. 50, no. 2, pp. 225–234, Feb. 2002.

78. R. Müller, "A random matrix model of communication through antenna arrays," in *Proc. Allerton Conf. on Comm. Control Computing*, Oct. 2000.

79. R. Müller and H. Hofstetter, "Confirmation of random matrix model for the antenna array channel by indoor measurements," in *Int. Symp. Antennas and Propagation*, 2001, Vol. 1, pp. 472–475.

80. B. Lindmark and M. Nilsson, "On the available diversity gain from different dual–polarized antennas," *IEEE J. Sel. Areas Comm.*, Vol. 19, no. 2, pp. 287–294, Feb. 2001.

81. J. Proakis, *Digital Communications*, McGraw-Hill, 1995.

82. J. Saltz, "Digital transmission over cross-coupled linear channels," *Bell Labs Tech. J.*, July-August 1985.

83. A. Burr, "MIMO wireless systems: Potentials, problems and prospects," in *COST 273 workshop, Helsinki University of Technology, Helsinki, Finland*, May 2002.

84. T. Marzetta and B. Hochwald, "Capacity of mobile multiple-antenna communication link in Rayleigh flat fading," *IEEE Trans. Inf. Th.*, Vol. 45, no. 1, pp. 139–157, Jan. 1999.

85. L. Zheng and D. Tse, "Diversity and freedom: a fundamental tradeoff in multiple antenna channels," manuscript, 2002.

86. O. Tirkkonen and R. Kashaev, "Performance optimal and information maximal MIMO modulations," in *Proc. IEEE ISIT*, July 2002.

87. F. Farrokhi, G. Foschini, A. Lozano and R. A. Valenzuela, "Link-optimal BLAST processing with multiple-access interference," in *Proc. IEEE Int. Conf. Veh. Tech.*, Boston, Mass., September 2000.

88. A. Scaglione, P. Stoica, S. Barbarossa, G. Giannakis and H. Sampath, "Optimal designs for space–time linear precoders and decoders," *IEEE Trans. Sig. Proc.*, Vol. 50, no. 5, pp. 1051–1064, 2002.

89. D. Gerlach and A. Paulraj, "Base station transmitting antenna arrays for multipath environments," *Sig. Proc.*, Vol. 54, pp. 59–74, October 1996.

90. A. Narula, M. Lopez, M. Trott and G. Wornell, "Efficient use of side information in multiple-antenna data transmission over fading channels," *IEEE J. Sel. Areas Comm.*, Vol. 17, no. 8, pp. 1423–1436, Oct. 1998.

91. C. Brunner, Efficient Space–Time Processing Schemes for WCDMA, Ph.D. thesis, Technical University, Munich, Germany, May 2001.

92. E. Visotsky and U. Madhow, "Space–time transmit precoding with imperfect feedback," *IEEE Trans. Inf. Th.*, Vol. 4, no. 6, pp. 2632–2639, Sept. 2001.

93. S. Jafar and A. Goldsmith, "On optimality of beam-forming for multiple antenna systems with imperfect feedback," in *Proc. IEEE ISIT*, 2001.

94. A. Hottinen, R. Wichman and O. Tirkkonen, "Multi-antenna transmission with feedback for WCDMA systems," in *Proc. 3G Wireless*, San Francisco, May 2001.

95. M. Ivrlac, T. Kurpjuhn, C. Brunner and W. Utschick, "Efficient use of fading correlations in MIMO systems," in *Proc. IEEE Int. Conf. Veh. Tech.*, October 2001, pp. 2763–2767.

96. A. Narula, M. Trott and G. Wornell, "Performance limits of coded diversity methods for transmitter antenna arrays," *IEEE Trans. Inf. Th.*, Vol. 45, no. 7, pp. 2418–2433, Nov. 1999.

97. R. Heath, Jr. and A. Paulraj, "Characterization of MIMO channels for spatial multiplexing systems," in *IEEE Int. Conf. on Comm.*, June 2001, Vol. 2, pp. 591–595.

98. M.-S. Alouini and A. Goldsmith, "Capacity of rayleigh fading channels under different adaptive transmission and diversity–combining techniques," *IEEE Trans. Veh. Tech.*, Vol. 48, pp. 1165–1181, July 1999.

99. A. Wittneben, "Basestation modulation diversity for digital SIMULCAST," in *Proc. IEEE Int. Conf. Veh. Tech.*, 1991, pp. 848–853.

100. J. Winters, "The diversity gain of transmit diversity in wireless systems with Rayleigh fading," in *IEEE Int. Conf. on Comm.*, 1994, pp. 1121–1125.

101. W.-L. Kuo and M. Fitz, "Design and analysis of transmitter diversity using intentional frequency offset for wireless communications," *IEEE Trans. Veh. Tech.*, Vol. 46, no. 4, pp. 871–881, November 1997.

102. A. Narula, M. Trott and G. Wornell, "Information-theoretic analysis of multiple-antenna transmission diversity," in *Proc. Int. Symp. Inform. Th. Appl.*, Canada, September 1996.

103. A. Wittneben, "A new bandwidth efficient transmit antenna modulation diversity scheme for linear digital modulation," in *IEEE Int. Conf. on Comm.*, Geneva, Switzerland, June 1993, pp. 1630–1634.

104. J. Thompson, P. Grant and B. Mulgrew, "Downlink transmit diversity schemes for CDMA networks," *IEE Proc.-Commun.*, Vol. 147, no. 6, pp. 371–380, December 2000.

105. K. Ban, M. Katayama, W. Stark, T. Tamamoto and A. Ogawa, "Convolutionally coded DS/CDMA system using multi-antenna transmission," in *Proc. IEEE GLOBECOM*, Phoenix, Arizona, November 1997.

106. V. Weerackody, "Diversity for the direct-sequence spread spectrum system using multiple transmit antennas," in *IEEE Int. Conf. on Comm.*, Geneva, Switzerland, June 1993.

107. R. Knopp and P. Humblet, "On coding for block fading channels," *IEEE Trans. Inf. Th.*, Vol. 46, no. 1, pp. 189–205, January 2000.

108. K. Boulle, J. Belfiore, K. Abed-Meraim and A. Chkeif, "Modulation schemes designed for the Rayleigh fading channel," in *Princeton Conf. Inf. Sci. Sys.*, Princeton, NJ, March 1992.

109. D. Rainish, "Diversity transform for fading channels," *IEEE Trans. Comm.*, Vol. 44, no. 12, pp. 1653–1661, December 1996.

110. K. Kerpez, "Constellations for good diversity performance," *IEEE Trans. Comm.*, Vol. 41, pp. 1412–1421, September 1993.

111. J. Boutros and E. Viterbo, "Signal space diversity: a power and bandwidth efficient diversity technique for the Rayleigh fading channel," *IEEE Trans. Inf. Th.*, Vol. 44, pp. 1453–1467, July 1998.

112. X. Giraud, E. Boutillon and J.-C. Belfiore, "Algebraic tools to build modulation schemes for fading channels," *IEEE Trans. Inf. Th.*, Vol. 43, pp. 938–952, May 1997.

113. C. Schlegel, *Trellis Coding*, IEEE Press, 1997.

114. V. D. Silva and E. Sousa, "Fading-resistant modulation using several transmitter antennas," *IEEE Trans. Comm.*, Vol. 45, no. 10, pp. 1236–1244, October 1997.

115. A. Correia, A. Hottinen and R. Wichman, "Optimized constellations for transmitter diversity," in *IEEE Veh. Tech. Conf.*, Fall, Amsterdam, the Netherlands, 1999, Vol. 3, pp. 1785–1789.

116. A. Hammons and H. El Gamal, "On the theory of space–time codes for PSK modulation," *IEEE Trans. Inf. Th.*, Vol. 46, no. 2, pp. 524–542, 2000.

117. D. Ionescu, K. Mukkavili, Z. Yan and J. Lilleberg, "Improved 8- and 16- state space–time codes for 4PSK with two transmit antennas," *IEEE Comm. Lett.*, Vol. 5, no. 7, pp. 301–303, July 2001.

118. N. Seshadri and H. Jafarkhani, "Super-orthogonal space–time trellis codes," in *IEEE Int. Conf. on Comm.*, Apr. 2002, Vol. 3, pp. 1439–1443.

119. S. Siwamogsatham and M. Fitz, "Improved high-rate space–time codes via orthogonality and set partitioning," in *IEEE Wireless Communications and Networking Conference*, 2002, Vol. 1, pp. 264–270.

120. R. Srinivasan, M. Heikkilä and R. Pirhonen, "Performance evaluation of space–time coding for edge," in *IEEE Int. Conf. on Comm.*, 2001, pp. 3056–3060.

121. J. Terry and J. Heiskala, "Spherical space–time codes (sstc)," *IEEE Comm. Lett.*, Vol. 5, no. 3, 2001.

122. A. Stefanov and T. Duman, "Turbo–coded modulation for wireless communications with transmit diversity," in *IEEE Veh. Tech. Conf.*, Fall, Sept. 1999, Vol. 3, pp. 1565–1569.

123. Y. Liu, M. Fitz and O. Takeshita, "QPSK space–time turbo codes," in *IEEE Int. Conf. on Comm.*, June 2000, Vol. 1, pp. 292–296.

124. G. Bauch, "Concatenation of space–time block codes and "turbo–tcm"," in *IEEE Int. Conf. on Comm.*, 1999, Vol. 2, pp. 1202–1206.

125. K. Narayanan, "Turbo decoding of concatenated space–time codes," in *Proc. Allerton Conf. on Comm. Control Computing*, Sept. 1999.

126. D. Tujkovic, "Recursive space–time trellis code for turbo coded modulation," in *Proc. IEEE GLOBECOM*, Nov. 2000, Vol. 2, pp. 1010–1015.

127. D. Cui and A. Haimovich, "Design and performance of turbo space–time coded modulation," in *Proc. IEEE GLOBECOM*, Nov. 2000, Vol. 3, pp. 1627–1631.

128. B. Lu, X. Wang and K. Narayanan, "LDPC–based space–time coded OFDM systems over correlated fading channels: Performance analysis and receiver design," *IEEE Trans. Comm.*, Vol. 50, no. 1, Jan. 2002.

129. J.-C. Guey, M. Fitz, M. Bell and W.-Y. Kuo, "Signal design for transmitter diversity wireless communication systems over Rayleigh fading channels," in *IEEE Veh. Tech. Conf.*, Spring, 1996, pp. 136–140.

130. D. M. Ionescu, "New results on space–time code design criteria," in *IEEE Wireless Communications and Networking Conference*, Sept. 1999, pp. 684–687.

131. Z. Chen, J. Yuan and B. Vucetic, "Improved space–time trellis coded modulation scheme on slow Rayleigh fading channels," *Electronics Letters*, Vol. 37, no. 7, pp. 440–441, Mar. 2001.

132. O. Tirkkonen, A. Boariu and A. Hottinen, "Minimal non-orthogonality rate one space–time block code for 3+ Tx antennas," in *IEEE International Symposium on Spread Spectrum Techniques and Applications*, Sept. 2000, Vol. 2, pp. 429–432.

133. B. Hassibi and B. Hochwald, "High-rate codes that are linear in space and time," *IEEE Trans. Inf. Th.*, Vol. 48, no. 7, pp. 1804–1824, July 2002.

134. R. Heath, Jr., H. Bölcskei and A. Paulraj, "Space–time signaling and frame theory," in *IEEE International Conference on Acoustics, Speech, and Signal Processing*, 2001, Vol. 2, pp. 1194–1199.

135. O. Tirkkonen, "Maximal symbolwise diversity in non-orthogonal space–time block codes," in *Proc. IEEE ISIT*, June 2001.

136. Q. Yan and R. Blum, "Optimum space–time convolutional codes," in *IEEE Wireless Communications and Networking Conference*, 2000, Vol. 3, pp. 1351–1355.

137. S. Sandhu, R. Heath, Jr. and A. Paulraj, "Space–time block codes versus space–time trellis codes," in *IEEE Int. Conf. on Comm.*, June 2001, Vol. 4, pp. 1132–1136.

138. T. Liew and L. Hanzo, "Space–time codes and concatenated channel codes for wireless communications," *ieeeproc*, Vol. 90, no. 2, pp. 187–219, Feb. 2002, Simulation results of concatenated space-time and trellis codes.

139. G. Ganesan and P. Stoica, "Space–time diversity using orthogonal and amicable orthogonal designs," *Wireless Pers. Comm.*, Vol. 18, no. 2, pp. 165–178, August 2001.

140. O. Tirkkonen and A. Hottinen, "Complex modulation space–time block codes for four tx antennas," in *Proc. IEEE GLOBECOM*, Nov. 2000, Vol. 2, pp. 1005–1009.

141. O. Tirkkonen and A. Hottinen, "Square-matrix embeddable space–time block codes for complex signal constellations," *IEEE Trans. Inf. Th.*, Vol. 48, no. 2, pp. 384–395, Feb. 2002.

142. H. Jafarkhani, "A quasi-orthogonal space–time block code," *IEEE Trans. Comm.*, Vol. 49, no. 1, pp. 1–4, Jan. 2001.

143. U. Wachsmann, J. Thielecke and H. Schotten, "Exploiting the data rate potential of mimo channels: multi-stratum space–time coding," in *IEEE Veh. Tech. Conf.*, Spring, 2001, Vol. 1, pp. 199–203.

144. H. El Gamal and M. Damen, "An algebraic number theoretic framework for space–time coding," in *Proc. IEEE ISIT*, June 2002, p. 132.

145. J. Geng, U. Mitra and M. Fitz, "Optimal space–time block codes for CDMA systems," in *MILCOM 2000*, Apr. 2000, Vol. 1, pp. 387–391.

146. B. Hughes, "Differential space–time modulation," *IEEE Trans. Inf. Th.*, Vol. 46, no. 7, pp. 2567–2578, Nov. 2000.

147. V. Tarokh and H. Jafarkhani, "A differential detection scheme for transmit diversity," *IEEE J. Sel. Areas Comm.*, Vol. 18, no. 7, pp. 1169–1174, July 2000.

148. S. Sandhu, K. Pandit and A. Paulraj, "On non-linear space–time block codes," in *Proc. IEEE ISIT*, June 2002, p. 416.

149. B. Hassibi and B. Hochwald, "Cayley differential unitary space–time codes," *IEEE Trans. Inf. Th.*, Vol. 48, no. 6, pp. 1485–1503, June 2002.

150. B. Hassibi and Y. Jing, "Unitary space–time modulation via the Cayley transform," in *Proc. IEEE ISIT*, June 2002, p. 134.

151. C. Papadias and G. Foschini, "A space–time coding approach for systems employing four transmit antennas," in *IEEE International Conference on Acoustics, Speech, and Signal Processing*, 2001, Vol. 4, pp. 2481–2484.

152. E. Lindskog and A. Paulraj, "A transmit diversity concept for channels with intersymbol interference," in *IEEE Int. Conf. on Comm.*, 2000, pp. 307–311.

153. A. Naguib, "On the matched filter bound of transmit diversity techniques," in *IEEE Int. Conf. on Comm.*, June 2001, Vol. 2, pp. 596–603.

154. B. Raghothaman, A. Boariu, O. Tirkkonen and A. Hottinen, "Performance of simple space–time block code for more than two transmit antennas," in *Proc. Allerton Conf. on Comm. Control Computing*, Oct. 2000.

155. O. Tirkkonen and A. Hottinen, "Diversity/self-interference tradeoff in non-orthogonal space–time block codes," in *Proc. URSI Natl. Rad. Sci. Meeting 2001*, Jan. 2001, p. 72.

156. O. Tirkkonen, "Layered space–time block codes," manuscript, 2001.

157. R. Kashaev and O. Tirkkonen, "Recursive approach to MIMO capacity," in *Proc. Finnish Wireless Comm. Workshop'01*, Oct. 2001, pp. 63–65.

158. R. Kashaev and O. Tirkkonen, "On expansion of capacity in SNR," in *Proc. IEEE ISIT*, July 2002.

159. S. Verdú, "Spectral efficiency in the wideband regime," *IEEE Trans. Inf. Th.*, Vol. 48, no. 2, pp. 1319–1343, June 2002.

160. S. Schweber, *An Introduction to Relativistic Quantum Field Theory*, Evanston: Row, Peterson, 1961.

161. S. Benedetto and E. Biglieri, *Principles of Digital Transmission*, New York: Kluwer, 1999.

162. D. Tebbe and S. Dwyer III, "Uncertainty and probability of error," *IEEE Trans. Inf. Th.*, Vol. IT-14, pp. 516–518, May 1968.

163. P. Lounesto, *Clifford Algebras and Spinors*, Cambridge University Press, 2001.

164. O. Tirkkonen and A. Hottinen, "Tradeoffs between rate, puncturing and orthogonality in space–time block codes," in *IEEE Int. Conf. on Comm.*, June 2001, Vol. 4, pp. 1117–1121.

165. M. Damen, A. Tewfik and J.-C. Belfiore, "A construction of a space–time code based on number theory," *IEEE Trans. Inf. Th.*, Vol. 48, no. 3, pp. 753–760, Mar. 2002.

166. S. Verdú, Multiuser Detection, Cambridge University Press, 1998.

167. G. Mattellini, K. Kuchi and P.A.Ranta, "Space–time block codes for EDGE system," in *IEEE Veh. Tech. Conf.*, Fall, Atlantic City, NJ, October 2001, pp. 1235–1239.

168. N. Al-Dhahir, M. Uysal and C. Georghiades, "Three space–time block-coding schemes for frequency-selective fading channels with application to EDGE," in *IEEE Veh. Tech. Conf.*, Fall, Atlantic City, NJ, October 2001, pp. 1834–1838.

169. N. Nefedov and G.-P. Mattellini, "Evaluation of potential transmit diversity schemes with iterative receivers in EDGE," in *Proc. IEEE PIMRC*, Portugal, 2002.

170. M. Valenti and B. Woerner, "Refined channel estimation for coherent detection of turbo codes over flat fading channel," *Electronics Letters*, Vol. 34, pp. 1648–1649, August 1998.

171. H. Li, X. Lu and G. Giannakis, "Capon multiuser receiver for CDMA systems with space–time coding," *IEEE Trans. Sig. Proc.*, Vol. 50, no. 5, pp. 1193–1204, 2002.

172. V. Buchoux, O. Cappe, E. Moulines and A. Gorokhov, "On the performance of semi-blind subspace-based channel estimation," *IEEE Trans. Sign. Proc.*, Vol. 48, pp. 1750–1759, June 2000.

173. S. Zhou, B. Muquet and G. Giannakis, "Subspace-based (semi-) blind channel estimation for block precoded space-time OFDM," *IEEE Trans. Sign. Proc.*, Vol. 50, pp. 1215–1228, May 2002.

174. X. Wang and H. Poor, "Iterative(Turbo) soft interference cancellation and decoding for coded CDMA," *IEEE Trans. Comm.*, Vol. 47, no. 7, pp. 1046–1061, July 1999.

175. S. Verdú, "Minimum probability of error for asynchronous Gaussian multiple-access channels," *IEEE Trans. Inf. Th.*, Vol. 32, no. 1, pp. 85–9–6, January 1986.

176. R. Lupas and S. Verdú, "Linear multiuser detectors for synchronous code-division multiple-access channels," *IEEE Trans. Inf. Th.*, Vol. 35, no. 1, pp. 123–136, January 1989.

177. E. Fain and M. Varanasi, "Group metric decoding for synchronous frequency-selective Rayleigh fading multiple access channels," in *Proc. IEEE GLOBECOM*, 1997.

178. M. Pohst, "On the computation of lattice vectors of minimal length, successive minima and reduced basis with applications," *ACM SIGSAM Bull.*, Vol. 15, pp. 37–44, 1981.

179. U. Fincke and M. Pohst, "Improved methods for calculating vectors of short length in a lattice, including complexity analysis," *Math. Comput.*, pp. 463–471, April 1985.

180. E. Viterbo and E. Biglieri, "A universal lattice decoder," in 14eme *Colloq. GRETSI*, France, 1993, pp. 611–614.

181. L. Brunel and J. Boutros, "Euclidean space lattice decoding for joint detection in CDMA systems," in *Information Theory Workshop*, Kruger National Park, South Africa, June 1999, pp. 611–614.

182. O. Damen, A. Chkeif and J.-C. Belfiore, "Lattice code decoder for space–time codes," *IEEE Comm. Lett.*, Vol. 4, no. 5, pp. 161–163, May 2000.

183. J. Conway, The Sensual (Quadratic) Form, Carush Mathematical Monographs. The Mathematical Association of America, Washington, DC, 1997.

184. H. Poor and S. Verdú, "Probability of error in MMSE multiuser detection," *IEEE Trans. Inf. Th.*, Vol. 43, pp. 858–871, May 1997.

185. M. Honig, U. Madhow and S. Verdú, "Blind adaptive multiuser detection," *IEEE Trans. Inf. Th.*, Vol. 41, no. 3, pp. 944–960, 1995.

186. U. Madhow, "Blind adaptive interference suppression for the near-far resistant acquisition and demodulation of direct-sequence CDMA," *IEEE Trans. Sign. Proc.*, Vol. 45, no. 1, pp. 124–136, 1997.

187. X. Wang and H. Poor, "Blind equalization and multiuser detection for CDMA communications in dispersive channels," *IEEE Trans. Comm.*, Vol. 46, pp. 91–103, January 1998.

188. M. Varanasi and B. Aazhang, "Multistage detection in asynchronous code-division multiple-access communications," *IEEE Trans. Comm.*, Vol. 38, no. 4, pp. 509–519, April 1990.

189. A. Dempster, N. Laird and D. Rubin, "Maximum likelihood from incomplete data via the EM algorithm (with discussion)," *Journal of the Royal Statistical Society, ser. B*, Vol. 39, pp. 1–38, 1977.

190. L. Nelson and H. Poor, "Iterative multiuser receivers for CDMA channels: an EM-based approach," *IEEE Trans. Comm.*, Vol. 44, pp. 1700–1710, December 1996.

191. J. Hu and R. Blum, "A new algorithm for multiuser detection based on a gradient guided search," in *Princeton Conf. Inf. Sci. Sys.*, Princeton, NJ, March 2000.

192. X.-D. Li, H. Huang, G. Foschini and R. Valenzuela, "Effects of iterative detection and decoding on the performance of BLAST," in *Proc. IEEE GLOBECOM*, 2000, Vol. 2, pp. 1061–1066.

193. X.-D. Li, H. Huang, A. Lozano and G. Foschini, "Reduced-complexity dectection algorithms for systems using multi-element arrays," in *Proc. IEEE GLOBECOM*, 2000, Vol. 2, pp. 1072–1076.

194. G. Ginis and J. Cioffi, "On the relation between V-BLAST and the GFDE," *IEEE Comm. Lett.*, Vol. 5, no. 9, September 2001.

195. A. Viterbi, "Very low rate convolutional codes for maximum theoretical performance of spread-spectrum multiple-access channels," *IEEE J. Sel. Areas Comm.*, Vol. 8, pp. 641–649, September 1990.

196. T. Giallorenzi and S. Wilson, "Multiuser ML sequence estimator for convolutionally coded asynchronous DS-CDMA systems," *IEEE Trans. Comm.*, Vol. 44, no. 8, pp. 997–1008, 1996.

197. M. Moher, "An iterative multiuser decoder for near-capacity communications," *IEEE Trans. Comm.*, Vol. 46, no. 7, pp. 870–880, July 1998.

198. M. Reed, P. Alexander, J. Asenstorfer and C. Schlegel, "Near single user performance using iterative multi-user detection for CDMA with turbo decoders," in *Proc. IEEE PIMRC*, 1997, pp. 740–744.

199. N. Ibrahim and G. Kaleh, "Iterative decoding and soft interference cancellation for the Gaussian multiple access channel," in *URSI International Symp. on Signals, Systems, and Electronics*, Pisa, Italy, September-October 1998, pp. 156–160.

200. J. Thomas and E. Geraniotis, "Iterative MMSE multiuser interference cancellation for trellis coded CDMA systems in multipath fading environments," in *Johns Hopkins Conf. Inf. Sci. Sys.*, Baltimore, MD, March 1999.

201. A. Hottinen, "Iterative multiuser decoding in WCDMA systems," in *Johns Hopkins Conf. Inf. Sci. Sys.*, Baltimore, MD, March 1999.

202. S. Das, E. Erkip, J. Cavallaro and B. Aazhang, "Iterative multiuser detection and decoding," in *Proc. IEEE GLOBECOM*, Sydney, Australia, November 1998, pp. 1006–1111.

203. M. Valenti and B. Woerner, "Iterative multiuser detection for convolutionally coded asynchronous DS-CDMA," in *Proc. IEEE PIMRC*, September 1998, pp. 213–217.

204. A. Hafeez and W. Stark, "Combined decision-feedback multiuser/soft-decision decoding for CDMA channels," in *Proc. IEEE Int. Conf. Veh. Tech.*, 1996, pp. 382–386.

205. Y. Sanada and Q. Wang, "A co-channel interference cancellation technique using orthogonal convolutional codes on multipath Rayleigh fading channel," *IEEE Trans. Veh. Tech.*, Vol. 46, pp. 114–128, 1997.

206. C. Douillard, A. Picart, P. Didier, M. Jezequel and C. Berrou, "Iterative correction of intersymbol interference: Turbo-equalization," *Eur. Trans. Telecomm.*, Vol. 6, pp. 507–511, September 1995.

207. L. Bahl, J. Cocke, F. Jelinek and J. Raviv, "Optimal decoding of linear codes for minimizing symbol error rate," *IEEE Trans. Inf. Th.*, pp. 284–287, March 1974.

208. J. Hagenauer and P. Hoher, "A Viterbi algorithm with soft-decision outputs and its applications," in *Proc. IEEE GLOBECOM*, 1989, pp. 47.1.1–47.1.7.

209. Texas Instruments, "Double-STTD scheme for HSDPA systems with four transmit antennas: link level simulation results," Temporary document 21(01)-0701, Release 5 Ad hoc, 3GPP TSG RAN WG1, June 2001.

210. E. Onggosanusi, A. Dabak and T. Schmidl, "High rate space–time block coded scheme: performance and improvements in correlated fading channels," in *IEEE Wireless Communications and Networking Conference*, Florida, March 2002, pp. 194–199.

211. O. Tirkkonen, "Optimizing space–time block codes by constellation rotations," in *Proc. Finnish Wireless Comm. Workshop'01*, Oct. 2001, pp. 59–60.

212. Y. Xin, Z. Wang and G. B. Giannakis, "Space–time constellation-rotating codes maximizing diversity and coding gains," in *Proc. IEEE GLOBECOM*, Nov. 2001.

213. M. Damen and N. Beaulieu, "A study of some space–time codes with rates beyond one symbol per channel use," in *Proc. IEEE GLOBECOM*, Nov. 2001, Vol. 1, pp. 445–449.

214. A. James, "Distributions of matrix variates and latent roots derived from normal samples," *Ann. Math. Stat.*, Vol. 35, pp. 475–501, 1964.

215. R. Muirhead, *Aspects of Multivariate Statistical Theory*, New York: John Wiley and Sons, 1982.

216. O. Tirkkonen, "Optimizing space–time block codes within symmetries of information," manuscript, Jan. 2002.

217. Nokia, "Demonstration of a 4-Tx-STTD OL diversity scheme," Temporary document, 3GPP contribution, TSG-R1, Torino, Italy, August 2001.

218. V. Tarokh, A. Naguib, N. Seshadri and A. Calderbank, "Combined array processing and space–time coding," *IEEE Trans. Inf. Th.*, Vol. 45, no. 4, pp. 1121–1128, May 1999.

219. M. Uysal and C. Georghiades, "New space–time block codes for high throughput efficiency," in *Proc. IEEE GLOBECOM*, Nov. 2001, Vol. 2, pp. 1103–1107.

220. H. El Gamal and A. Hammons, "A new approach to layered space–time coding and signal processing," *IEEE Trans. Inf. Th.*, Vol. 47, no. 6, pp. 2321–2324, Sept. 2001.

221. S. Bäro, G. Bauch, A. Pavlic and A. Semmler, "Improving BLAST performance using space–time block codes and turbo decoding," in *Proc. IEEE GLOBECOM*, Nov. 2000, Vol. 2, pp. 1067–1071.

222. A. Naguib and A. Calderbank, "Space-time coding and signal processing for high data rate wireless communications," *Wireless Commun. Mob. Comp.*, Vol. 1, no. 1, pp. 13–53, January-March 2001.

223. Texas Instruments, "Improved double-STTD schemes using asymmetric modulation and antenna shuffling," Temporary document, 3GPP contribution TSG-R1, 21-24 May 2001.

224. R. Johannesson and K. Zigangirov, *Fundamentals of Convolutional Coding*, IEEE Press, 1999.

225. A. Hottinen, J. Vesma and O. Tirkkonen, "High bit rates for HSDPA using MIMO channels," *WSEAS Trans. Commun.*, Vol. 1, pp. 31–36, July 2002.

226. A. Hottinen, J. Vesma, O. Tirkkonen and N. Nefedov, "High bit rates for 3G and beyond using MIMO channels," in *Proc. IEEE PIMRC*, Portugal, 2002.

227. J. Winters, "Switched diversity with feedback for DPSK mobile radio," *IEEE Trans. Veh. Tech.*, Vol. 32, pp. 134–150, February 1983.

228. D. Gerlach and A. Paulraj, "Adaptive transmitting antenna arrays with feedback," *IEEE Sig. Proc. Letters*, Vol. 1, pp. 150–152, October 1994.

229. S. Fukumoto, K. Higuchi, M. Sawahashi and F.Adachi, "Field experiments on closed-loop mode transmit diversity in W-CDMA forward link," in *IEEE International Symposium on Spread Spectrum Techniques and Applications*, Parsippany,NJ, September 2000, pp. 433–438.

230. A. Hottinen and R. Wichman, "Enhanced filtering for feedback mode transmit diversity," in *Princeton Conf. Inf. Sci. Sys.*, Princeton, NJ, March 2000.

231. R. Wichman and A. Hottinen, "IMT–2000 transmit diversity concepts," in *Proc. IEEE PIMRC*, Sept. 1999.

232. S. Parkvall, M. Karlsson, M. Samuelsson, L. Hedlund and B. Goransson, "Transmit diversity in WCDMA: Link and system level results," in *IEEE Veh. Tech. Conf.*, Spring, 2000.

233. Motorola, "Verification algorithm for closed-loop transmit diversity mode 2," Temporary document R1–00–1087, 3GPP TSG RAN WG1, Aug. 2000.

234. A. Hottinen and R. Wichman, "Transmit diversity using filtered feedback weights in the FDD/WCDMA system," in *Proc. Int. Zurich Sem. Comm.*, Zurich, Switzerland, February 2000.

235. Nokia, "An extension of closed-loop Tx diversity mode 1 for multiple Tx antennas," Temporary document R1–00–0712, 3GPP TSG RAN WG1, 2000.

236. Motorola, "Closed-loop transmit diversity mode 2 with reduced states for 4 elements," Temporary document R1–00–1132, 3GPP TSG RAN WG1, Aug. 2000.

237. Samsung, "Performance results of basis selection transmit diversity for four antennas," Temporary document R1–00–1073, 3GPP TSG RAN WG1, Aug. 2000.

238. Siemens, "Description of the eigenbeam-former concept (update) and performance evaluation," Temporary document R1–01–0203, 3GPP TSG RAN WG1, Feb. 2001.

239. Siemens, "Simulation parameters for Tx diversity simulations using correlated antennas," Temporary document R1–00–1080, 3GPP TSG RAN WG1, Feb. 2001.

240. Fujitsu, "Simulation results of the Tx diversity scheme with beam-forming feature," Temporary document TSGR1#18(01)–0103, 3GPP TSG RAN WG1, Jan. 2001.

241. W. Utschick and C. Brunner, "Efficient tracking and feedback of DL-eigenbeams in WCDMA," in *Proc. 4th European Personal Mobile Communications Conference*, Vienna, Austria, February 2000.

242. M. Sandell, "Analytical analysis of transmit diversity in WCDMA on fading multipath channels," in *Proc. IEEE PIMRC*, 1999.

243. E. Onggosanusi, A. Gatherer, A. Dabak and S. Hosur, "Performance analysis of closed-loop transmit diversity in the presence of feedback delay," *IEEE Trans. Comm.*, Vol. 49, no. 9, pp. 1618–1630, Sept. 2001.

244. R. Gray and D. Neuhoff, "Quantization," *IEEE Trans. Inf. Th.*, Vol. 44, no. 6, pp. 2325–2383, 1998.

245. J. Choi, "Performance analysis for transmit antenna diversity with/without channel information," *IEEE Trans. Veh. Tech.*, Vol. 51, no. 1, pp. 101–113, Jan. 2002.

246. E. Jorswieck and H. Boche, "On transmit diversity with imperfect channel state information," in *IEEE International Conference on Acoustics, Speech, and Signal Processing*, 2002, pp. 2181–2184.

247. J. Bucklew and C. G. Jr., "Quantization schemes for bivariate Gaussian random variables," *IEEE Trans. Inf. Th.*, Vol. 25, no. 5, pp. 537–543, Sept. 1979.

248. C. Ling and Z. Chunning, "Low–complexity antenna diversity receivers for mobile wireless applications," *Wireless Pers. Comm.*, Vol. 14, no. 1, pp. 65–81, July 2000.

249. J. Hämäläinen and R. Wichman, "On the performance of FDD WCDMA closed-loop transmit diversity modes in Nakagami and Ricean fading channels," in *IEEE International Symposium on Spread Spectrum Techniques and Applications*, Sept. 2002.

250. M. Abramowitz and I. Stegun, Eds., Handbook of Mathematical Functions, National Bureau of Standards, Washington DC, 1972.

251. R. Heath, Jr. and A. Paulraj, "A simple scheme for transmit diversity using partial channel feedback," in *Asilomar Conference on Signals, Systems and Computers*, 1998, pp. 1073–1078.

252. C. Brunner, J. Hammerschmidt, A. Seeger and J. Nossek, "Space–time eigenrake and downlink eigenbeam-former: Exploiting long-term and short-term channel properties in WCDMA," in *Proc. IEEE GLOBECOM*, 2000.

253. J. Lieblein, "On moments of order statistics from the Weibull distribution," in *Ann. Math. Stat.*, 1955, Vol. 26, pp. 330–333.

254. S. Gupta, "Order statistics from the Gamma distribution," in *Technometrics*, 1960, Vol. 2, pp. 243–262.

255. S. Wilson, "Magnitude/phase quantization of independent Gaussian variates," *IEEE Trans. Comm.*, Vol. 28, pp. 1924–1929, Nov. 1980.

256. J. Hämäläinen and R. Wichman, "Feedback schemes for FDD WCDMA systems in multipath environments," in *IEEE Veh. Tech. Conf.,* Spring, May 2001.

257. S. Chennakeshu and J. Anderson, "Error rates for Rayleigh fading multichannel reception of MPSK signals," *IEEE Trans. Comm.*, Vol. 43, no. 2/3/4, pp. 338–346, February/March/April 1995.

258. J. Hämäläinen and R. Wichman, "Asymptotic bit error probabilities of some closed-loop transmit diversity schemes," in *Proc. IEEE GLOBECOM*, Nov. 2002.

259. J. Hämäläinen and R. Wichman, "Performance analysis of closed-loop transmit diversity in the presence of feedback errors," in *Proc. IEEE PIMRC*, Sept. 2002.

260. J. Hämäläinen and R. Wichman, "The effect of feedback delay to the closed-loop transmit diversity in FDD WCDMA," in *Proc. IEEE PIMRC*, Sept. 2001.

261. 3GPP, "Physical layer–general description," 3GPP technical specification, TS 25.201, Ver.4.0.0., 1999, Available from www.3gpp.org.

262. S. Zhou and G. Giannakis, "Optimal transmitter eigen–beam-forming and space-time block coding based on channel mean," in *IEEE International Conference on Acoustics, Speech, and Signal Processing*, 2002.

263. S. Jafar, S. Vishwanath and A. Goldsmith, "Channel capacity and beam-forming for multiple transmit and receive antennas with covariance feedback," in *IEEE Int. Conf. on Comm.*, 2001, pp. 2266–2270.

264. E. Jorswieck and H. Boche, "On the optimality of beamforming for MIMO systems with covariance feedback," in *Princeton Conf. Inf. Sci. Sys.*, Princeton, NJ, March 2002.

265. S. Jafar and A. Goldsmith, "Beam-forming capacity and SNR maximization for multiple antenna systems," in *IEEE Veh. Tech. Conf.,* Spring, 2001.

266. A. Hottinen, "Space-time retransmission and matrix modulation," manuscript, 2001.

267. D. Littlewood, *The Theory of Group Characters and Matrix Representations of Groups*, Oxford: Clarendon Press, 1940.

268. A. Marshall and I. Olkin, *Inequalities: Theory of Majorization and its Applications*, New York: Academic Press, 1979.

Index